Control, Identification, and Input Optimization

MATHEMATICAL CONCEPTS AND METHODS IN SCIENCE AND ENGINEERING

Series Editor: Angelo Miele
Mechanical Engineering and Mathematical Sciences
Rice University

Recent volumes in the series:

11 **INTEGRAL TRANSFORMS IN SCIENCE AND ENGINEERING** • *Kurt Bernado Wolf*

12 **APPLIED MATHEMATICS: An Intellectual Orientation** • *Francis J. Murray*

14 **PRINCIPLES AND PROCEDURES OF NUMERICAL ANALYSIS** • *Ferenc Szidarovszky and Sidney Yakowitz*

16 **MATHEMATICAL PRINCIPLES IN MECHANICS AND ELECTROMAGNETISM, Part A: Analytical and Continuum Mechanics** • *C. C. Wang*

17 **MATHEMATICAL PRINCIPLES IN MECHANICS AND ELECTROMAGNETISM, Part B: Electromagnetism and Gravitation** • *C. C. Wang*

18 **SOLUTION METHODS FOR INTEGRAL EQUATIONS: Theory and Applications** • *Edited by Michael A. Golberg*

19 **DYNAMIC OPTIMIZATION AND MATHEMATICAL ECONOMICS** • *Edited by Pan-Tai Liu*

20 **DYNAMICAL SYSTEMS AND EVOLUTION EQUATIONS: Theory and Applications** • *J. A. Walker*

21 **ADVANCES IN GEOMETRIC PROGRAMMING** • *Edited by Mordecai Avriel*

22 **APPLICATIONS OF FUNCTIONAL ANALYSIS IN ENGINEERING** • *J. L. Nowinski*

23 **APPLIED PROBABILITY** • *Frank A. Haight*

24 **THE CALCULUS OF VARIATIONS AND OPTIMAL CONTROL: An Introduction** • *George Leitmann*

25 **CONTROL, IDENTIFICATION, AND INPUT OPTIMIZATION** • *Robert Kalaba and Karl Spingarn*

Control, Identification, and Input Optimization

Robert Kalaba

University of Southern California
Los Angeles, California

and

Karl Spingarn

Hughes Aircraft Company
Los Angeles, California

PLENUM PRESS • NEW YORK AND LONDON

Library of Congress Cataloging in Publication Data

Kalaba, Robert E.

Control, identification, and input optimization.

(Mathematical concepts and methods in science and engineering; 25)

Includes bibliographical references and index.

1. Control theory. 2. System identification. I. Spingarn, Karl. II. Title. III. Series.

QA402.3.K27 629.8'312 81-23404

ISBN 0-306-40847-3 AACR2

A Division of Plenum Publishing Corporation
233 Spring Street, New York, N.Y. 10013

Printed in the United States of America

Preface

This book is a self-contained text devoted to the numerical determination of optimal inputs for system identification. It presents the current state of optimal inputs with extensive background material on optimization and system identification. The field of optimal inputs has been an area of considerable research recently with important advances by R. Mehra, G. C. Goodwin, M. Aoki, and N. E. Nahi, to name just a few eminent investigators. The authors' interest in optimal inputs first developed when F. E. Yates, an eminent physiologist, expressed the need for optimal or preferred inputs to estimate physiological parameters.

The text assumes no previous knowledge of optimal control theory, numerical methods for solving two-point boundary-value problems, or system identification. As such it should be of interest to students as well as researchers in control engineering, computer science, biomedical engineering, operations research, and economics. In addition the sections on beam theory should be of special interest to mechanical and civil engineers and the sections on eigenvalues should be of interest to numerical analysts. The authors have tried to present a balanced viewpoint; however, primary emphasis is on those methods in which they have had first-hand experience. Their work has been influenced by many authors. Special acknowledgment should go to those listed above as well as R. Bellman, A. Miele, G. A. Bekey, and A. P. Sage.

The book can be used for a two-semester course in control theory, system identification, and optimal inputs. The first semester would cover optimal control theory and system identification, and the second semester would cover optimal inputs and applications. Alternatively, for those students who have already been introduced to optimal control theory, the book can be used for a one-semester course in system identification and optimal inputs. The text can also be used as an introduction to control theory or to system identification.

The desired purpose of the text is to provide upper-division undergraduate and graduate students, as well as engineers and scientists in industry, with the analytical and computational tools required to compute optimal inputs for system identification. System identification is concerned with the estimation of the parameters of dynamic system models. The accuracy of the parameter estimates is enhanced by the use of optimal inputs to increase the sensitivity of the observations to the parameters being estimated.

The determination of optimal inputs for system identification requires a knowledge of dynamic system optimization techniques, numerical methods of solution, and methods of system identification. All of these topics are covered in the text.

Part I of the text is an introduction to the subject. Part II is Optimal Control and Methods of Numerical Solutions. The chapters on optimal control include the Euler–Lagrange equations, dynamic programming, Pontryagin's maximum principle, and the Hamilton–Jacobi equations. The numerical methods for linear and nonlinear two-point boundary-value problems include the matrix Riccati equation, method of complementary functions, invariant imbedding, quasilinearization, and the Newton–Raphson methods.

Part III is System Identification. The chapters in this part include the Gauss–Newton and quasilinearization methods for system identification. Applications of these methods are presented. Part IV is Optimal Inputs for System Identification. The equations for the determination of optimal inputs are derived and applications are given. Part V lists computer programs.

Optimal inputs have applications in such diverse fields as biomedical modeling and aircraft stability and control parameter estimation. The text is intended to provide a means of computing optimal inputs by providing the analytical and computational building blocks for this purpose, for example, control theory, computational methods, system identification, and optimal inputs.

The authors wish to thank the National Science Foundation, the National Institutes of Health, and the Air Force Office of Scientific Research for continuing support of our research efforts.

Robert Kalaba
Karl Spingarn

Contents

PART I. INTRODUCTION

PART II. OPTIMAL CONTROL AND METHODS FOR NUMERICAL SOLUTIONS

PART III. SYSTEM IDENTIFICATION

PART IV. OPTIMAL INPUTS FOR SYSTEM IDENTIFICATION

PART V. COMPUTER PROGRAMS

PART I

Introduction

1

Introduction

This chapter gives a brief outline of the three major topics in this book: optimal control, dynamic system identification, and optimal inputs for system identification. Modern control theory, two-point boundary value problems, and methods of numerical solutions are introduced in Section 1.1. Methods of system identification for the estimation of the unknown parameters of a dynamic system model are introduced in Section 1.2. Optimal inputs for increasing the accuracy of system identification are discussed in Section 1.3. The integration of differential equations, required to obtain the solutions to all of the above, is discussed in Section 1.4.

1.1. Optimal Control

Optimal control is concerned with finding the optimal input control which changes the state of a system in order to achieve a desired objective. The objective is often a quadratic performance criterion. For example, a typical optimal control problem consists of finding the control, $u(t)$, that minimizes the performance index or cost functional

$$J = \frac{1}{2} \int_0^T (u^2 + x^2)\, dt,$$

where the system state, $x(t)$, satisfies the dynamical equation

$$\dot{x} = ax + u, \qquad x(0) = c.$$

This is a problem in the calculus of variations. In this book the equations to be solved to obtain the solution are formulated using the Euler–

Lagrange equations, Pontryagin's maximum principle, dynamic programming, or Kalaba's initial value method. All of these methods involve the solution of two-point boundary-value problems. Numerical solutions are obtained using the method of complementary functions, the matrix Riccati equation, invariant imbedding, or analytical methods for linear systems and quasilinearization, or the Newton–Raphson method for nonlinear systems. The derivations of the equations are discussed in Chapter 2 and numerical solutions are discussed in Chapters 3 and 4. Numerical solutions for other two-point boundary-value problems, such as for buckling loads, are also discussed in Chapter 4.

1.2. System Identification

System identification is concerned with the determination of the parameters of a system. Consider for example the second-order system of linear differential equations

$$\dot{x}_1(t) = -ax_1(t) + bx_2(t), \tag{1.1}$$

$$\dot{x}_2(t) = -cx_1(t) - dx_2(t) + u(t), \tag{1.2}$$

where $x_1(t)$ and $x_2(t)$ are the state variables, $u(t)$ is the input, and a, b, c, and d are constant coefficients. The response of the system depends upon the parameters a, b, c, and d. The magnitudes of these parameters are often unknown. The estimation of the unknown parameters is called *parameter estimation* or *system identification.*

In the early days of electrical control engineering, identification was concerned with the determination of the transfer function of a system. The transfer function could be found by applying a known input signal, usually a sinusoidal signal, to a "black box" and measuring the response at the output. While this approach is useful in some cases, it cannot be easily extended to nonlinear systems or systems with measurement and process noise. Furthermore, the identification of the transfer function often does not provide insight into the basic underlying dynamic processes.

In time domain analyses, the parameters of the differential equations are desired, as in equations (1.1) and (1.2). In fields such as biomedical engineering, the parameters provide a unique insight into the physiological processes. Several books have been published recently on modern methods of system identification (see the references for Chapter 5).

In this book, we shall discuss two important methods of dynamic system identification: the *Gauss–Newton method* and *quasilinearization*. These methods are discussed in Chapters 5 and 6. Applications of system identification are discussed in Chapter 7.

We shall be concerned mostly with first-order and second-order linear dynamic systems of the form given by equations (1.1) and (1.2). The methods of system identification apply equally well, however, for higher-order systems and nonlinear systems.

1.3. Optimal Inputs

It is intuitively clear that the accuracy of the parameter estimates will depend upon the signal input. Thus, instead of arbitrarily choosing an input, such as a sinusoidal signal, an optimal input could be utilized. The performance criterion for the optimal input is related to the sensitivity of the system output to the unknown parameters. The optimal input which maximizes the sensitivity will maximize the parameter estimation accuracy.

Consider for example the linear dynamic system given by

$$\dot{x}(t) = -ax(t) + u(t), \tag{1.3}$$

with initial condition

$$x(0) = c, \tag{1.4}$$

where $x(t)$ is a scalar state variable, $u(t)$ is the input, and a is an unknown system parameter. The unknown parameter a is to be estimated from measurements of the state variable, $x(t)$. The measurements, $z(t)$, are assumed to be given by

$$z(t) = x(t) + v(t), \tag{1.5}$$

where $v(t)$ is the measurement noise. The optimal input is obtained by maximizing a functional of the sensitivity of the state variable, $x(t)$, to the unknown parameter a. Measurements of $x(t)$ utilized to estimate the parameter a will then yield maximum estimation accuracy.

Using a quadratic performance criterion, the optimal input which maximizes the integral of the square of the sensitivity is to be determined. Thus it is desired to maximize

$$J = \max_{u(t)} \int_0^T x_a^2(t)\, dt, \tag{1.6}$$

where

$$x_a(t) = \frac{\partial x(t)}{\partial a} \tag{1.7}$$

and T is the terminal time such that $0 \leq t \leq T$. The larger the magnitude of the input, the greater will be the sensitivity. Thus the input must be constrained; otherwise the maximization of equation (1.6) will yield the trivial solution that the optimal input is infinite. The input is constrained such that it is less than or equal to a constant E:

$$\int_0^T u^2(t)\, dt \leq E. \tag{1.8}$$

Maximizing equation (1.6), subject to constraint (1.8), is equivalent to the maximization of the performance index:

$$J = \max_{u(t)} \frac{1}{2} \int_0^T [x_a^2(t) - qu^2(t)]\, dt. \tag{1.9}$$

The constant q is the Lagrange multiplier whose magnitude must be chosen to satisfy constraint (1.8). In the above formulation, the measurement noise is not considered. The effect of the measurement noise will be considered in later developments.

In order to obtain $x_a(t)$, equations (1.3) and (1.4) are differentiated with respect to a. The derivative of $u(t)$ with respect to a is assumed to be zero. Thus we have

$$\dot{x}_a(t) = -x(t) - ax_a(t), \qquad x_a(0) = 0. \tag{1.10}$$

The performance index is maximized using Pontryagin's maximum principle. The Hamiltonian function is

$$H = \tfrac{1}{2}[-x_a^2 + qu^2] + p_1(-ax + u) + p_2(-x - ax_a). \tag{1.11}$$

The costate variables, $p_1(t)$ and $p_2(t)$, are the solutions of the differential equations

$$\dot{p}_1(t) = -\frac{\partial H}{\partial x} = ap_1(t) + p_2(t), \tag{1.12}$$

$$\dot{p}_2(t) = -\frac{\partial H}{\partial x_a} = x_a(t) + ap_2(t), \tag{1.13}$$

with boundary conditions

$$p_1(T) = 0, \qquad p_2(T) = 0. \tag{1.14}$$

The input $u(t)$ that maximizes H is

$$\frac{\partial H}{\partial u} = qu(t) + p_1(t) = 0, \qquad u(t) = -\frac{1}{q}\, p_1(t). \tag{1.15}$$

Upon substituting equation (1.15) into equation (1.3), equations (1.3), (1.10), (1.12), and (1.13) yield the two-point boundary-value equations with associated boundary conditions

$$\dot{x}(t) = -ax(t) - \frac{1}{q}\, p_1(t), \qquad x(0) = c, \tag{1.16}$$

$$\dot{x}_a(t) = -x(t) - ax_a(t), \qquad x_a(0) = 0, \tag{1.17}$$

$$\dot{p}_1(t) = ap_1(t) + p_2(t), \qquad p_1(T) = 0, \tag{1.18}$$

$$\dot{p}_2(t) = x_a(t) + ap_2(t), \qquad p_2(T) = 0. \tag{1.19}$$

Thus the derivation of the optimal input for the first-order linear differential equation requires the solution of a two-point boundary-value problem for four linear differential equations. In general an nth-order differential equation requires the solution of a two-point boundary-value problem for $4n$ differential equations.

Optimal inputs are discussed in Chapters 8 and 9. New results on eigenvalue problems are also discussed in Chapter 9. Applications of optimal inputs are discussed in Chapter 10. The solution to the above equations is obtained using the same numerical methods discussed for two-point boundary problems in Chapters 3 and 4. Computer programs are given in Chapter 11.

1.4. Computational Preliminaries

It is assumed that the reader is familiar with the numerical solution of initial-value ordinary differential equations, the Newton–Raphson method for finding the root of an equation, and the numerical solution of linear algebraic equations. These concepts are reviewed briefly here. For more information the reader is urged to consult References 1–5.

The simplest method for the numerical solution of ordinary differential equations is Euler's method. Given the initial-value first-order ordinary differential equation

$$\frac{dx}{dt} = f(x, t), \qquad x(t_0) = x_0, \tag{1.20}$$

the Euler equation is

$$x_{n+1} = x_n + hf(x_n, t_n), \tag{1.21}$$

where $n = 0, 1, 2, \ldots$ and the uniform step size h is equal to a small increment in t:

$$h = \Delta t. \tag{1.22}$$

A method which is more commonly used is the fourth-order Runge–Kutta method. The equations are

$$k_1 = f(x_n, t_n), \tag{1.23}$$

$$k_2 = f(x_n + \tfrac{1}{2}hk_1, t_n + \tfrac{1}{2}h), \tag{1.24}$$

$$k_3 = f(x_n + \tfrac{1}{2}hk_2, t_n + \tfrac{1}{2}h), \tag{1.25}$$

$$k_4 = f(x_n + hk_3, t_n + h), \tag{1.26}$$

$$x_{n+1} = x_n + \tfrac{1}{6}h(k_1 + 2k_2 + 2k_3 + k_4), \tag{1.27}$$

for $n = 0, 1, 2, \ldots$. This is the method used in the program listings in Chapter 11.

In the above equations, the solution is obtained by simply integrating forward starting with the initial condition. For two-point boundary-value problems, however, the boundary conditions are given both at the beginning and at the terminal time. Methods for converting these problems to initial-value problems must then be used as discussed in the following chapters.

To find the root of an equation, assume x_0 is the initial approximation to the root of the scalar equation

$$f(x) = 0. \tag{1.28}$$

Using the Newton–Raphson method, approximate $f(x)$ by the equation

$$f(x) \doteq f(x_0) + (x - x_0)f'(x_0). \tag{1.29}$$

Solving for x, the next approximation is

$$x_1 = x_0 - \frac{f(x_0)}{f'(x_0)}, \tag{1.30}$$

and in general the recurrence relation is

$$x_{n+1} = x_n - \frac{f(x_n)}{f'(x_n)}, \tag{1.31}$$

for $n = 0, 1, 2, \ldots$. The concept above is similar to that used in the Gauss–Newton and quasilinearization methods for system identification in Chapters 5 and 6.

The numerical solution of simultaneous linear algebraic equations can be obtained by methods such as the Gauss elimination method. However, in the programs given in Chapter 11, the solution is obtained simply by calling a subroutine for the inverse of a matrix.

Exercises

1. Use Newton's method to find the nonzero root of the equation

$$x = e^{-x}.$$

Assume the first approximation is $x = 0.65$.

2. Use Euler's method for the numerical integration of the ordinary differential equation

$$\frac{dx}{dt} = -x, \qquad x(0) = 0,$$

to obtain $x(t)$ from $t = 0$ to $t = 1$. Compare the results obtained using a step size, $\Delta t = 0.2$, with the results obtained from the analytical solution.

3. Given the two-point boundary-value problem

$$\frac{d^2x}{dt^2} = -x, \qquad x(0) = 1, \qquad x(1) = 0.$$

(a) What additional information is needed in order to use Euler's method to numerically integrate forward from $t = 0$ to $t = 1$?
(b) Analytically find the missing initial condition $\dot{x}(0)$.

PART II

Optimal Control and Methods for Numerical Solutions

2

Optimal Control

The derivations of the equations used in optimal control theory are given in this chapter. Numerical solutions are considered in Chapters 3 and 4. The simplest problem in the calculus of variations is discussed in Section 2.1. The Euler–Lagrange equations are derived using the classical calculus of variations (References 1–5) approach. The equations are then rederived using dynamic programming (References 6–8).

Pontryagin's maximum principle (Reference 9) for solving optimal control problems is introduced. A suspended cable (Reference 10) example is given, and new approaches to optimal control and filtering (Reference 11) are considered. A summary of the equations commonly used in optimal control is given at the end of the chapter.

2.1. Simplest Problem in the Calculus of Variations

The simplest problem in the calculus of variations involves the extremization of an integral, called a functional, in one independent variable and an unknown dependent function. To obtain the extremum, the unknown function must satisfy the Euler–Lagrange equations. These equations are derived, and dynamic programming and the Hamilton–Jacobi equations are introduced.

2.1.1. Euler–Lagrange Equations

Consider the problem of determining a function $y = y(t)$, $a \leq t \leq b$, which satisfies the boundary conditions

$$y(a) = y_0, \qquad y(b) = y_1 \tag{2.1}$$

and which minimizes the integral I,

$$I = \int_a^b F(t, y, \dot{y})\, dt, \tag{2.2}$$

where F is a given function of its three arguments, and

$$\dot{y} = dy/dt. \tag{2.3}$$

This problem is called the simplest problem in the calculus of variations. Let us see how it arises in a control theory setting.

Suppose that a system has the dynamical equation

$$\dot{y} = -ay + w(t), \qquad 0 \leq t \leq T. \tag{2.4}$$

The initial state of the system is

$$y(0) = c_1 \tag{2.5}$$

and the terminal state is

$$y(T) = c_2. \tag{2.6}$$

The cost of the control process is

$$c = \int_0^T (y^2 + \lambda w^2)\, dt, \qquad \lambda > 0. \tag{2.7}$$

We wish to determine the control function $w = w(t)$, $0 \leq t \leq T$, so that the process is carried out at least cost. By substituting from equation (2.4) into equation (2.7) we see that our problem becomes that of choosing the function $y = y(t)$ so that we minimize the cost

$$c = \int_0^T [y^2 + \lambda(\dot{y} + ay)^2]\, dt \tag{2.8}$$

subject to the boundary constraints

$$y(0) = c_1, \qquad y(T) = c_2. \tag{2.9}$$

In the geometric sphere we may consider the problem of finding the shortest path from one point to another. We wish to minimize the integral

$$\int_{x_0}^{x_1} [1 + (y')^2]^{1/2}\, dx, \tag{2.10}$$

where $y' = dy/dx$, through choice of the function $y = y(x)$, $x_0 \leq x \leq x_1$, with

$$y(x_0) = y_0, \qquad y(x_1) = y_1. \tag{2.11}$$

Let us now return to the simplest problem in the calculus of variations sketched above. Let us assume that the function which minimizes I is

$$y(t) = x(t), \qquad a \leq t \leq b. \tag{2.12}$$

Let us now consider a nearby function

$$y(t) = x(t) + \varepsilon\eta(t), \tag{2.13}$$

where ε is a small parameter, positive or negative, and $\eta = \eta(t)$, $a \leq t \leq b$, is a variation function which is arbitrary, except that

$$\eta(a) = \eta(b) = 0, \tag{2.14}$$

since only curves that fulfill the boundary conditions are admissible. Use of this nonoptimal curve will not lower the value of the integral I, since $x = x(t)$ is the minimizing function, so that

$$\int_a^b F(t, x, \dot{x})\,dt \leq \int_a^b F(t, x + \varepsilon\eta, \dot{x} + \varepsilon\dot{\eta})\,dt. \tag{2.15}$$

Expansion of the right-hand side of the above inequality using Taylor's theorem then yields

$$\int_a^b F(t, x, \dot{x})\,dt \leq \int_a^b F(t, x, \dot{x})\,dt + \varepsilon \int_a^b [\eta(t)F_x(t, x, \dot{x}) + \dot{\eta}(t)F_{\dot{x}}(t, x, \dot{x})]\,dt + o(\varepsilon), \tag{2.16}$$

where the term $o(\varepsilon)$ refers to terms involving ε to the second and higher powers. Through subtraction we see that

$$0 \leq \varepsilon \int_a^b [\eta(t)F_x(t, x, \dot{x}) + \dot{\eta}F_{\dot{x}}(t, x, \dot{x})]\,dt + o(\varepsilon). \tag{2.17}$$

Now for $|\varepsilon|$ sufficiently small, the sign of the right-hand side of the above relation is determined by the first term, so that we must have

$$0 \leq \varepsilon \int_a^b [\eta(t)F_x(t, x, \dot{x}) + \dot{\eta}F_{\dot{x}}(t, x, \dot{x})]\,dt. \tag{2.18}$$

But the only way that this relation can hold for all sufficiently small values of ε, both positive and negative, is that

$$0 = \int_a^b [\eta(t)F_x(t, x, \dot{x}) + \dot{\eta}(t)F_{\dot{x}}(t, x, \dot{x})]\, dt. \tag{2.19}$$

We now wish to make use of the arbitrariness of the function $\eta = \eta(t)$. Through integration by parts we find that

$$0 = \int_a^b \eta(t)\left[F_x(t, x, \dot{x}) - \frac{d}{dt} F_{\dot{x}}(t, x, \dot{x})\right] dt + \eta(t)F_{\dot{x}}(t, x, \dot{x})\Big|_a^b. \tag{2.20}$$

But we have $\eta(a) = \eta(b) = 0$, so that the boundary terms disappear, and we are left with the equation

$$0 = \int_a^b \eta(t)\left[F_x(t, x, \dot{x}) - \frac{d}{dt} F_{\dot{x}}(t, x, \dot{x})\right] dt. \tag{2.21}$$

Since this relation must hold for arbitrary choice of the function $\eta = \eta(t)$, we must have

$$F_x(t, x, \dot{x}) - \frac{d}{dt} F_{\dot{x}}(t, x, \dot{x}) = 0, \qquad a \leq t \leq b, \tag{2.22}$$

which is called the Euler–Lagrange equation for the simplest problem in the calculus of variations.

Written at length, using the chain rule of differentiation, this equation becomes

$$F_x - F_{\dot{x}t} - F_{\dot{x}x}\dot{x} - F_{\dot{x}\dot{x}}\ddot{x} = 0, \qquad a \leq t \leq b, \tag{2.23}$$

which is a second-order ordinary differential equation which the optimizing function $x = x(t)$ must satisfy in the interval $a \leq t \leq b$. In addition, of course, we have the two-point boundary conditions

$$x(a) = y_0, \qquad x(b) = y_1. \tag{2.24}$$

Except for certain special cases, this problem has to be solved via numerical means, a central theme of this book.

The Euler–Lagrange equation was derived above for fixed initial and terminal boundary conditions using the calculus of variations approach. In optimal control, the initial condition, $x(0)$, is often fixed with the terminal condition, $x(T)$, unspecified. The functional

$$J = \int_0^T F(t, x, \dot{x})\, dt \tag{2.25}$$

with boundary conditions

$$x(0) = c, \qquad x(T) = \text{unspecified}, \tag{2.26}$$

is then minimized by the solution of the Euler–Lagrange equation

$$F_x(t, x, \dot{x}) - \frac{d}{dt} F_{\dot{x}}(t, x, \dot{x}) = 0 \tag{2.27}$$

with associated transversality condition

$$x(0) = c, \qquad F_{\dot{x}}(t, x, \dot{x})|_{t=T} = 0. \tag{2.28}$$

Exercises

1. Work out the Euler–Lagrange equations in the two cases mentioned above.
2. Derive the equations for the case in which $x(a) = y_0$, and x is free at $t = b$.

2.1.2. Dynamic Programming

Now let us turn to an alternative approach in which the optimizing curve is viewed as the envelope of its tangents. This is the view of dynamic programming which makes essential use of Bellman's principle of optimality, which characterizes optimal processes. We view the determination of the curve $x = x(t)$ as a multistage decision process in which we start at $t = a$ and $x = y_0$. A decision is made as to the value of $\dot{x}$, call it w, a cost is accumulated, and the system moves to a new state, from which the process is to be continued in optimal fashion. In mathematical terms we define the optimal cost function

$f(c, \tau) =$ the cost of a process beginning with the system in state c at time τ and using an optimal arc which terminates at the fixed time b, $-\infty < c < +\infty$, $\tau \leq b$. (2.29)

First we shall obtain a partial differential equation for this function f, together with an initial condition at $\tau = b$. Then we shall rederive the Euler–Lagrange equation obtained above.

We write

$$f(c, \tau) = \min_{w} [F(\tau, c, w)h + f(c + wh, \tau + h) + o(h)]. \tag{2.30}$$

The first term in the brackets on the right-hand side is the immediate cost during the interval $(\tau, \tau + h)$ using $\dot{x} = w$, and the second is the minimum cost for the remainder of the process starting at $\tau + h$ with the system in state $c + wh + o(h)$. As usual, $o(h)$ refers to terms in h that are of higher order than the first. The value of w chosen is that which minimizes the sum of the immediate cost and the minimum future cost of the process starting from the new state at the new time. To continue the analysis, we expand using Taylor series to get

$$f(c, \tau) = \min_{w} [F(\tau, c, w)h + f(c, \tau) + whf_c + hf_\tau + o(h)] \tag{2.31}$$

or

$$0 = \min_{w} [F(\tau, c, w)h + whf_c + hf_\tau + o(h)]. \tag{2.32}$$

Dividing by h and letting $h \to 0$, we find

$$-f_\tau(c, \tau) = \operatorname*{limit}_{h\to 0} \min_{w} [F(\tau, c, w) + wf_c(c, \tau) + o(h)].$$

Assuming that we can interchange the limiting and minimization operations, we get the desired partial differential equation for the minimum cost function f

$$-f_\tau(c, \tau) = \min_{w} [F(\tau, c, w) + wf_c(c, \tau)]. \tag{2.33}$$

We assume that the minimization is accomplished by setting the partial derivative of the term in square brackets equal to zero, which yields

$$F_w(\tau, c, w) + f_c(c, \tau) = 0, \tag{2.34}$$

which we may call the optimal control law, for it gives, implicitly, the correct choice of w, the slope, for each value of c and of τ, provided that the function f is known. When w is chosen according to the above relation, then we also have

$$-f_\tau(c, \tau) = F(\tau, c, w) + wf_c(c, \tau), \qquad \tau \leq b. \tag{2.35}$$

Thus we have two partial differential equations for the two unknown functions f and w. By solving the optimal control law equation for w in terms of τ, c, and f_c and substituting into the cost equation, we can obtain a single partial differential equation for the cost function f, a point to which we shall return later. The terminal condition at $\tau = b$ is

$$f(c, b) = 0, \qquad -\infty < c < +\infty. \tag{2.36}$$

We now wish to show that we can derive the Euler–Lagrange equation, obtained earlier, from the optimal control law and the cost equation.

Keeping in mind the form of the Euler–Lagrange equation, equation (2.22), we begin by taking the total derivative with respect to τ of both sides of equation (2.34), which results in

$$\frac{d}{d\tau} F_w(\tau, c, w) + f_{cc} \frac{dc}{d\tau} + f_{c\tau} = 0. \tag{2.37}$$

Then we differentiate both sides of equation (2.35) partially with respect to c, which shows that

$$-f_{\tau c} = F_c + F_w w_c + w_c f_c + w f_{cc}. \tag{2.38}$$

But the terms in w_c disappear, as we see from the optimal control law, so that

$$-f_{\tau c} = F_c + w f_{cc}. \tag{2.39}$$

Thus

$$-F_c = f_{\tau c} + w f_{cc}. \tag{2.40}$$

But keeping in mind that

$$w = dc/d\tau, \tag{2.41}$$

we have

$$-F_c = f_{\tau c} + \frac{dc}{d\tau} f_{cc}. \tag{2.42}$$

Consequently substituting equation (2.42) into equation (2.37) we have

$$\frac{d}{d\tau} F_w(\tau, c, w) - F_c(\tau, c, w) = 0, \tag{2.43}$$

which is the Euler–Lagrange equation with $t = \tau$, $x = c$, and $\dot{x} = w$.

2.1.3. Hamilton–Jacobi Equations

Let us go on to show that we can replace the Euler–Lagrange equation, a single second-order ordinary differential equation which the optimal arc must satisfy, with a system of two first-order ordinary differential equations. We introduce a new variable p by means of the definition

$$p = F_w(\tau, c, w), \tag{2.44}$$

and consider that this equation determines w, implicitly, as a function of τ, c, and p,

$$w = w(\tau, c, p). \tag{2.45}$$

Then we introduce the function H,

$$H = H(\tau, c, p), \tag{2.46}$$

by means of the definition

$$H = pw(\tau, c, p) - F[\tau, c, w(\tau, c, p)]. \tag{2.47}$$

Taking the derivative of H with respect to p we obtain

$$H_p = w + pw_p - F_w w_p. \tag{2.48}$$

But the last two terms are zero, so

$$w = H_p, \tag{2.49}$$

which is one of our desired differential equations. Differentiating H with respect to c we find

$$H_c = pw_c - F_c - F_w w_c. \tag{2.50}$$

The terms involving w_c drop out, as we see from equation (2.44), so

$$H_c = -F_c. \tag{2.51}$$

From the Euler–Lagrange equation (2.43) we have

$$\frac{d}{d\tau} p = F_c, \tag{2.52}$$

so we obtain the second canonical equation

$$-\frac{d}{d\tau} p = H_c. \tag{2.53}$$

In terms of the function

$$H = H(\tau, c, p), \tag{2.54}$$

the two first-order ordinary differential equations that the functions c and

p, as functions of τ must satisfy along an optimal arc are

$$\frac{dc}{d\tau} = H_p(\tau, c, p) \tag{2.55}$$

and

$$-\frac{dp}{d\tau} = H_c(\tau, c, p), \tag{2.56}$$

which are called Hamilton's canonical equations. They have the advantage over the Euler–Lagrange equation that the derivatives $dc/d\tau$ and $dp/d\tau$ appear only on the left with τ, c, and p appearing on the right.

We now show Jacobi's method for integrating these canonical equations. First we observe that we can write the single partial differential equation for the function f in a convenient form, now that we have introduced the function H. It will be recalled that the functions f and w satisfy the system of partial differential equations

$$F_w(\tau, c, w) + f_c(c, \tau) = 0 \tag{2.57}$$

and

$$-f_\tau(c, \tau) = F(\tau, c, w) + wf_c(c, \tau). \tag{2.58}$$

Also we have introduced

$$p = F_w(\tau, c, w), \tag{2.59}$$

which implicitly determines

$$w = w(\tau, c, p), \tag{2.60}$$

and H is given by the relation

$$H(\tau, c, p) = pw(\tau, c, p) - F\big(\tau, c, w(\tau, c, p)\big). \tag{2.61}$$

It follows from equations (2.57), (2.58), and (2.59) that

$$\begin{aligned} f_\tau &= -wf_c - F \\ &= wp - F \end{aligned} \tag{2.62}$$

or

$$f_\tau(c, \tau) = H\big(\tau, c, -f_c(c, \tau)\big), \tag{2.63}$$

which is a first-order partial differential equation for the function f, a form of the Hamilton–Jacobi partial differential equation.

We shall now show that if we can find a solution of this partial differential equation which depends upon an arbitrary (nonadditive) constant α,

$$f = f(c, \tau, \alpha), \tag{2.64}$$

then a solution of the canonical equations, equations (2.55) and (2.56), is given in the form

$$f_\alpha(c, \tau, \alpha) = \beta, \tag{2.65}$$

where β is another arbitrary constant. In effect, this last equation yields c as a function of τ and two arbitrary constants α and β and determines the optimal trajectories. The constants, of course, have to be adjusted to fulfill the initial and terminal conditions on $c = c(\tau)$.

Differentiating equation (2.65) with respect to τ yields

$$f_{\alpha c}\frac{dc}{d\tau} + f_{\alpha\tau} = 0. \tag{2.66}$$

On the other hand since $f = f(\tau, c, \alpha)$ satisfied the Hamilton–Jacobi partial differential equation (2.63) we have

$$f_{\tau\alpha} = H_p(-f_{c\alpha}), \tag{2.67}$$

so that

$$\frac{dc}{d\tau} = H_p, \tag{2.68}$$

assuming that $f_{\alpha c} \neq 0$. Thus we have obtained one of the canonical equations, equation (2.55). Now we must demonstrate that the other one holds. From equations (2.57) and (2.59)

$$p = -f_c. \tag{2.69}$$

Then differentiating with respect to τ

$$\frac{dp}{d\tau} = -f_{cc}\frac{dc}{d\tau} - f_{c\tau}. \tag{2.70}$$

On the other hand partial differentiation of the Hamilton–Jacobi equation with respect to c yields

$$f_{\tau c} = H_c + H_p[-f_{cc}] = H_c + \frac{dc}{d\tau}[-f_{cc}], \tag{2.71}$$

Solving the above equation for H_c and substituting into equation (2.70) we have

$$\frac{dp}{d\tau} = -H_c, \tag{2.72}$$

as claimed. This completes the verification of Jacobi's method for integrating the canonical equations through the finding of a solution of the Hamilton–Jacobi partial differential equation depending upon an arbitrary parameter. As we shall see later, this can frequently be accomplished through use of the method of separation of variables.

2.2. Several Unknown Functions

Oftentimes we face the problems of minimizing an integral where the integrand involves more than one function to be determined. Suppose that we wish to determine functions $x = x(t)$ and $y = y(t)$ which minimize the integral

$$I = \int_{t_0}^{t_1} F(t, x, y, \dot{x}, \dot{y})\, dt, \tag{2.73}$$

where the boundary conditions are

$$x(t_0) = x_0, \qquad y(t_0) = y_0, \tag{2.74}$$

$$x(t_1) = x_1, \qquad y(t_1) = y_1. \tag{2.75}$$

By varying first x and then y we see that the two Euler equations

$$F_x - \frac{d}{dt} F_{\dot{x}} = 0 \tag{2.76}$$

and

$$F_y - \frac{d}{dt} F_{\dot{y}} = 0 \tag{2.77}$$

must be satisfied, together with the boundary conditions.

2.3. Isoperimetric Problems

In our later work we shall encounter problems in which we wish to minimize the integral

$$\int_{x_0}^{x_1} F(x, y, y')\, dx \tag{2.78}$$

subject to the constraint that $y = y(x)$ must be chosen so that

$$\int_{x_0}^{x_1} G(x, y, y')\, dx = c. \tag{2.79}$$

The problem of finding an arc $y = y(x)$ of given length

$$\int_{x_0}^{x_1} (1 + y'^2)^{1/2}\, dx = l, \tag{2.80}$$

with $y(x_0) = y_0$ and $y(x_1) = y_1$ and which maximizes the area

$$\int_{x_0}^{x_1} y\, dx \tag{2.81}$$

is of this type and gives this class of problems its name. To derive the basic equations we shall use the dynamic programming approach. We imbed the original problem within a class of problems in which the terminal point is $x = \xi$, $y = \eta$ and the constraint integral has the value c, all considered to be variables. Then we define the minimum cost function φ to be

$$\varphi(\xi, \eta, c) = \min \int_{x_0}^{\xi} F(x, y, y')\, dx, \tag{2.82}$$

where the function $y = y(x)$ is subject to the constraint

$$\int_{x_0}^{\xi} G(x, y, y')\, dx = c. \tag{2.83}$$

Then the principle of optimality leads directly to the functional equation

$$\varphi(\xi, \eta, c) = \min_{w} [F(\xi, \eta, w)h + \varphi(\xi - h, \eta - wh, c - hG(\xi, \eta, w)) + o(h)] \tag{2.84}$$

or

$$\varphi_\xi(\xi, \eta, c) = \min_{w} [F(\xi, \eta, w) - w\varphi_\eta - G(\xi, \eta, w)\varphi_c], \tag{2.85}$$

where $w = \dot{\eta}$ and $h = \Delta\xi$. This last equation results in the optimal control law for determining $w = w(\xi, \eta, c)$

$$0 = F_w - \varphi_\eta - G_w\varphi_c \tag{2.86}$$

and the cost equation

$$\varphi_\xi = F(\xi, \eta, w) - w\varphi_\eta - G\varphi_c. \tag{2.87}$$

We first show that along an optimal arc $\eta = \eta(\xi)$, with $c = c(\xi)$ we have

$$\frac{d}{d\xi}\,\varphi_c(\xi, \eta, c) = 0, \tag{2.88}$$

so that φ_c is constant along such an arc. The chain rule shows that

$$\frac{d}{d\xi}\,\varphi_c = \varphi_{c\xi} + \varphi_{c\eta}\eta' + \varphi_{cc}c', \tag{2.89}$$

where

$$\eta' = d\eta/d\xi \quad \text{and} \quad c' = dc/d\xi. \tag{2.90}$$

Then by partial differentiation of cost equation (2.87) with respect to c we obtain

$$\varphi_{\xi c} = F_w w_c - w_c\varphi_\eta - w\varphi_{\eta c} - G_w(\xi, \eta, w)w_c\varphi_c - G\varphi_{cc}. \tag{2.91}$$

The terms involving w_c on the right-hand side cancel, as the optimal control law, equation (2.86), shows and we have

$$0 = \varphi_{\xi c} + y'\varphi_{\eta c} + c'\varphi_{cc}, \tag{2.92}$$

since

$$w = \eta' \tag{2.93}$$

and

$$c'(\xi) = G(\xi, \eta, w), \tag{2.94}$$

which establishes the assertion.

Total differentiation of the optimal control law, equation (2.86), with respect to ξ shows that

$$\frac{d}{d\xi}\,\varphi_\eta = \frac{d}{d\xi}\,(F_w - \varphi_c G_w). \tag{2.95}$$

Partial differentiation of cost equation (2.87) with respect to η yields

$$\varphi_{\xi\eta} = F_\eta + F_w w_\eta - w_\eta\varphi_\eta - w\varphi_{\eta\eta} - (G_\eta + G_w w_\eta)\varphi_c - G\varphi_{c\eta}, \tag{2.96}$$

and employment again of the optimal control law leads to the cancellation of the terms involving w_η,

$$\varphi_{\xi\eta} = F_\eta - w\varphi_{\eta\eta} - \varphi_c G_\eta - G\varphi_{c\eta}, \tag{2.97}$$

or substituting equations (2.93) and (2.94) in equation (2.97) and rearranging

$$\varphi_{\xi\eta} + \eta'\varphi_{\eta\eta} + c'\varphi_{c\eta} = F_\eta - \varphi_c G_\eta. \tag{2.98}$$

Since the left side of equation (2.98) is equal to the derivative of ϕ_η with respect to ξ, equation (2.98) becomes

$$\frac{d}{d\xi}\varphi_\eta = F_\eta - \varphi_c G_\eta. \tag{2.99}$$

Thus along an optimal arc we have, substituting ϕ_η from equation (2.86) into equation (2.99),

$$F_\eta - \varphi_c G_\eta = \frac{d}{d\xi}(F_w - \varphi_c G_w), \tag{2 100}$$

where

$$\varphi_c = \text{const}, \tag{2.101}$$

which is the desired Euler–Lagrange equation.

Let us now write

$$-\varphi_c = \lambda = \text{unknown constant} \tag{2.102}$$

and introduce Ψ to be

$$\Psi = F(x, y, y') + \lambda G(x, y, y'). \tag{2.103}$$

Then we may state that along an optimal arc $y = y(x)$, $x_0 \leq x \leq x_1$, the Euler–Lagrange equation

$$\Psi_y - \frac{d}{dx}\Psi_{y'} = 0 \tag{2.104}$$

must hold. In addition we have the two boundary conditions

$$y(x_0) = y_0, \qquad y(x_1) = y_1, \tag{2.105}$$

and the auxiliary condition

$$\int_{x_0}^{x_1} G(x, y, y')\,dx = c. \tag{2.106}$$

Notice that we now have a two-point boundary-value problem to solve for the optimizing arc $y = y(x)$. The Euler–Lagrange differential equation now contains an unknown constant which must be adjusted so that the integral constraint is fulfilled. The unknown constant λ, the Lagrange multiplier, has a physical meaning which our derivation via dynamic programming makes clear. It is the negative of the limiting ratio of the change in minimum of the integral of F caused by a change in c, keeping the terminal point fixed. Later, numerical aspects will be discussed.

Let us show that the problem of maximizing the area

$$\int_{x_0}^{x_1} y\,dx \tag{2.107}$$

under the curve

$$y = y(x), \qquad y(x_0) = 0, \qquad y(x_1) = 0, \tag{2.108}$$

when the length is prescribed,

$$\int_{x_0}^{x_1} (1 + y'^2)^{1/2}\,dx = l, \tag{2.109}$$

leads to circular arcs. We form the function

$$\Psi = y + \lambda(1 + y'^2)^{1/2}, \tag{2.110}$$

where λ is a constant. The Euler–Lagrange equation is

$$1 - \frac{d}{dx}\left[\frac{\lambda y'}{(1 + y'^2)^{1/2}}\right] = 0. \tag{2.111}$$

Differentiation then shows that

$$\frac{y''}{(1 + y'^2)^{3/2}} = \lambda^{-1}, \tag{2.112}$$

so that the arcs have constant curvature, as asserted.

2.4. Differential Equation Auxiliary Conditions

Consider determining functions $x = x(t)$ and $y = y(t)$ which minimize the integral

$$I = \int_{t_0}^{t_1} F(t, x, y, \dot{x}, \dot{y})\,dt, \tag{2.113}$$

where now the differential equation constraint

$$G(t, x, y, \dot{x}, \dot{y}) = 0, \qquad t_0 \leq t \leq t_1, \tag{2.114}$$

is imposed, as well as the boundary conditions

$$x(t_0) = x_0, \qquad y(t_0) = y_0, \tag{2.115}$$

$$x(t_1) = x_1, \qquad y(t_1) = y_1. \tag{2.116}$$

We shall first make use of Bellman's principle of optimality and then derive the differential equations for the optimal arc.

Suppose that the process starts in the assigned fashion and continues optimally to the general terminal time τ and terminal state $x(\tau) = \xi$ and $y(\tau) = \eta$. We introduce the minimal cost function $\varphi = \varphi(\xi, \eta, \tau)$ to be the value of the integral I over that interval, the differential equation constraint $G = 0$ being observed on the interval $t_0 \leq t \leq \tau$. The function φ satisfies the relation

$$\varphi(\xi, \eta, \tau) = \min_{w,z} [F(\tau, \xi, \eta, w, z)h + \varphi(\xi - wh, \eta - zh, \tau - h) + o(h)], \tag{2.117}$$

where the minimization is over the terminal velocities w and z which must satisfy the constraint

$$G(\tau, \xi, \eta, w, z) = 0. \tag{2.118}$$

Passing to the limit with $h \to 0$, we find that the cost function φ satisfies the partial differential equation

$$\varphi_\tau = \min_{w,z} [F(\tau, \xi, \eta, w, z) - w\varphi_\xi - z\varphi_\eta], \tag{2.119}$$

where w and z are subject to the constraint given above. Using the standard Lagrange multiplier theory for a function of two variables subject to a constraining equation, we find that the above partial differential equation yields

$$F_w - \varphi_\xi + \lambda G_w = 0, \tag{2.120}$$

$$F_z - \varphi_\eta + \lambda G_z = 0, \tag{2.121}$$

and

$$\varphi_\tau = F - w\varphi_\xi - z\varphi_\eta, \tag{2.122}$$

together with the condition

$$G = 0. \tag{2.123}$$

Generally λ will vary from point to point.

First we see that along optimal curves

$$\frac{d}{d\tau}(F_w + \lambda G_w) = \frac{d}{d\tau}\varphi_\xi = \varphi_{\xi\xi}\dot{\xi} + \varphi_{\xi\eta}\dot{\eta} + \varphi_{\xi\tau}. \tag{2.124}$$

Then partial differentiation of the cost equation (2.122) with respect to ξ

shows that

$$\varphi_{\tau\xi} = F_\xi + F_w w_\xi + F_z z_\xi - w_\xi \varphi_\xi - w\varphi_{\xi\xi} - z_\xi \varphi_\eta - z\varphi_{\eta\xi}. \quad (2.125)$$

Differentiation of the constraint equation

$$G(\tau, \xi, \eta, w, z) = 0 \quad (2.126)$$

with respect to ξ shows that

$$G_\xi + G_w w_\xi + G_z z_\xi = 0, \quad (2.127)$$

or

$$\lambda(G_\xi + G_w w_\xi + G_z z_\xi) = 0. \quad (2.128)$$

Similarly,

$$\lambda(G_\eta + G_w w_\eta + G_z z_\eta) = 0. \quad (2.129)$$

It follows that adding equation (2.128) to (2.125)

$$\phi_{\tau\xi} = F_\xi - w\phi_{\xi\xi} - z\phi_{\eta\xi} + (F_w - \phi_\xi + \lambda G_w)w_\xi + (F_z - \phi_\eta + \lambda G_z)z_\xi + \lambda G_\xi \quad (2.130)$$

The terms in the parentheses are zero so that

$$\varphi_{\tau\xi} = F_\xi - w\varphi_{\xi\xi} - z\varphi_{\eta\xi} + \lambda G_\xi \quad (2.131)$$

or

$$\varphi_{\xi\xi}\dot{\xi} + \varphi_{\eta\xi}\dot{\eta} + \varphi_{\tau\xi} = F_\xi + \lambda G_\xi \quad (2.132)$$

since $w = \dot{\xi}$ and $z = \dot{\eta}$. The left side of equation (2.132) is equal to $(d/d\tau)\phi_\xi$ so that substituting equation (2.124) into (2.132)

$$\frac{d}{d\tau}(F_w + \lambda G_w) = F_\xi + \lambda G_\xi. \quad (2.133)$$

Similarly

$$\frac{d}{d\tau}(F_z + \lambda G_z) = F_\eta + \lambda G_\eta. \quad (2.134)$$

Lastly we must obtain a differential equation which shows how λ itself varies along an optimal trajectory. The equation $G = 0$ provides this relation. Now we summarize the result for our original problem.

Introduce the auxiliary function Ψ,

$$\Psi = F(t, x, y, \dot{x}, \dot{y}) + \lambda(t)G(t, x, y, \dot{x}, \dot{y}). \quad (2.135)$$

Then along optimal arcs the functions x, y, and λ must satisfy the three differential equations

$$\Psi_x - \frac{d}{dt}\Psi_{\dot{x}} = 0, \tag{2.136}$$

$$\Psi_y - \frac{d}{dt}\Psi_{\dot{y}} = 0, \tag{2.137}$$

$$G(t, x, y, \dot{x}, \dot{y}) = 0. \tag{2.138}$$

In addition, they must satisfy the given initial and terminal conditions.

If nothing is prescribed at $t = t_1$, we see that we must have

$$\varphi_{x_1}(t_1, x_1, y_1) = 0, \tag{2.139}$$

$$\varphi_{y_1}(t_1, x_1, y_1) = 0, \tag{2.140}$$

so that

$$(F_{\dot{x}} + \lambda G_{\dot{x}})|_{t=t_1} = 0, \tag{2.141}$$

$$(F_{\dot{y}} + \lambda G_{\dot{y}})|_{t=t_1} = 0. \tag{2.142}$$

In control theory we meet this problem in the form of determining the function $y = y(t)$, $t_0 \leq t \leq t_1$ so that we minimize the integral

$$I = \int_{t_0}^{t_1} F(x, y)\, dt \tag{2.143}$$

subject to the dynamical equation

$$\dot{x} = g(x, y), \tag{2.144}$$

and the initial condition

$$x(t_0) = c. \tag{2.145}$$

In this case

$$\Psi = F(x, y) + \lambda(t)[\dot{x} - g(x, y)]. \tag{2.146}$$

The differential equations and finite condition along the optimal curves become

$$F_x - \lambda g_x - \frac{d}{dt}\lambda = 0, \tag{2.147}$$

$$F_y - \lambda g_y = 0, \tag{2.148}$$

$$\dot{x} = g(x, y). \tag{2.149}$$

The initial condition on x at $t = t_0$ is

$$x(t_0) = c. \tag{2.150}$$

The terminal condition at $t = t_1$ is

$$\lambda(t_1) = 0. \tag{2.151}$$

Thus we again have a two-point boundary-value problem to be solved.

2.5. Pontryagin's Maximum Principle

Let $x(t)$, a scalar, denote the state of a system at time t, with $x(0) = c$, and suppose that its dynamical equation is

$$\dot{x} = f(x, y), \qquad 0 \leq t \leq T, \tag{2.152}$$

where $y = y(t)$ is a control function to be chosen so that we minimize the cost functional

$$C = \int_0^T g(x, y)\, dt \tag{2.153}$$

and

$$|y| \leq b, \qquad 0 \leq t \leq T, \tag{2.154}$$

where b is a constraint on the size of the control variable. We introduce the minimum cost function

$$\begin{aligned} \varphi(c, T) = {} & \text{the cost of a process starting in state } c, \\ & \text{of duration } T, \text{ and using an optimal control policy,} \\ & -\infty < c < +\infty,\ T \geq 0. \end{aligned} \tag{2.155}$$

Then the principle of optimality shows that the function φ satisfies the functional equation

$$\varphi(c, T + \Delta) = \min_{|w| \leq b} [g(c, w)\, \Delta + \varphi(c + f(c, T)\, \Delta, T) + o(\Delta)], \tag{2.156}$$

or in the limit

$$\varphi_T = \min_{|w| \leq b} [g(c, w) + \varphi_c f(c, w)], \qquad T \geq 0. \tag{2.157}$$

In addition we see that

$$\varphi(c, 0) = 0, \qquad -\infty < c < +\infty. \tag{2.158}$$

The initial-value problem in the two equations above can be integrated numerically, which yields both the minimum cost function $f(c, T)$ and the optimal control law $w = w(c, T)$. Here, though, we wish to obtain the differential system for a particular optimal process. When w is chosen optimally at each instant, we have

$$\varphi_T = g(c, w) + \varphi_c f, \tag{2.159}$$

so that

$$\varphi_{Tc} = g_c + g_w w_c + \varphi_{cc} f + \varphi_c[f_c + f_w w_c], \tag{2.160}$$

Rearranging the terms

$$\varphi_{Tc} = g_c + \varphi_{cc} f + \varphi_c f_c + w_c[g_w + \varphi_c f_w]. \tag{2.161}$$

On the other hand along an optimal trajectory we have

$$\begin{aligned}\frac{d}{dt}\varphi_c(c, T) &= \varphi_{cc}\frac{dc}{dt} + \varphi_{cT}\frac{dT}{dt}\\ &= \varphi_{cc}\frac{dc}{dt} - \varphi_{cT}\\ &= \varphi_{cc} f - \varphi_{cT}\end{aligned} \qquad (2.162$$

where $dT/dt = -1$.

Thus substituting equation (2.161) into (2.162)

$$\frac{d}{dt}\varphi_c(c, T) = -g_c - \varphi_c f_c - w_c[g_w + \varphi_c f_w]. \tag{2.163}$$

If the minimizing value of w occurs at a particular moment with $|w| < b$, we shall have

$$g_w + \varphi_c f_w = 0, \tag{2.164}$$

so that

$$\frac{d}{dt}\varphi_c(c, T) = -g_c - f_c\varphi_c. \tag{2.165}$$

The last relation is a differential equation for the function φ_c which must hold along an optimal trajectory at points for which $|w| < b$. We shall now show that it holds even at points for which $|w| = b$, points for which the control constraint is active.

Consider the expression μ

$$\mu = g(c, w) + \varphi_c(c, T)f(c, w) - \varphi_T(c, T). \tag{2.166}$$

For a fixed T^* and any c^*, the minimum of this expression is 0, which we now assume occurs at $w^* = b$. Now hold w^* fixed, as well as T^* and consider μ as a function of c. The minimum is 0, and it occurs at $c = c^*$. Therefore the derivative of μ with respect to c must be 0 at $c = c^*$. Thus

$$g_c + \varphi_{cc}f + \varphi_c f_c - \varphi_{Tc} = 0, \tag{2.167}$$

so that again we have

$$\frac{d}{dt}\varphi_c(c, T) = -g_c - \varphi_c f_c. \tag{2.168}$$

The Pontryagin equations can be expressed in the more familiar form by substituting $\lambda = \phi_c$, $F = g$, and $x = c$ into equation (2.165). Then, assuming there are no constraints on the control variable equation (2.165) becomes

$$\dot{\lambda} = \frac{\partial F}{\partial x} - \lambda\frac{\partial f}{\partial x} = -\frac{\partial H}{\partial x}, \tag{2.169}$$

where

$$H = F + \lambda f. \tag{2.170}$$

Furthermore, in view of equation (2.152), we have

$$\dot{x} = \frac{\partial H}{\partial \lambda}. \tag{2.171}$$

Substituting $\lambda = \phi_c$, $F = g$, and $y = w$ into equation (2.164) yields

$$\frac{\partial F}{\partial y} + \lambda\frac{\partial f}{\partial y} = 0 \tag{2.172}$$

or

$$\frac{\partial H}{\partial y} = 0. \tag{2.173}$$

The boundary condition is

$$\lambda(T) = 0. \tag{2.174}$$

2.6. Equilibrium of a Perfectly Flexible Inhomogeneous Suspended Cable

As an illustration of the theory for optimization with differential constraints, consider an inhomogeneous perfectly flexible cable suspended from two points. The weight per unit length of the cable at the distance (along the cable) s from the left end is $q(s)$, $0 \leq s \leq L$. In the equilibrium configuration the potential energy of the cable is a minimum. Let the x axis be horizontal and the y axis be directed vertically upward. Then the potential energy of the system is

$$V = \int_0^L q(s)y(s)\, ds. \tag{2.175}$$

The differential equation

$$\dot{x}^2 + \dot{y}^2 = 1 \tag{2.176}$$

expresses the fact that s is arc length along the cable.

Introducing the auxiliary function

$$\Psi = q(s)y(s) + \frac{\lambda(s)}{2} (\dot{x}^2 + \dot{y}^2 - 1), \tag{2.177}$$

we find the Euler–Lagrange equations to be

$$\frac{d}{ds} [\lambda(s)\dot{x}] = 0, \tag{2.178}$$

$$\frac{d}{ds} [\lambda(s)\dot{y}] = q(s), \qquad 0 \leq s \leq L. \tag{2.179}$$

In addition, of course, we have the differential equation

$$\dot{x}^2 + \dot{y}^2 = 1, \qquad 0 \leq s \leq L, \tag{2.180}$$

and the boundary conditions

$$x(0) = 0, \qquad y(0) = 0, \tag{2.181}$$

$$x(L) = x_1, \qquad y(L) = y_1, \tag{2.182}$$

where L is the length of the cable.

Now let us interpret these equations from the point of view of statics. Let $H = H(s)$ and $V = V(s)$ be the horizontal and vertical components of tension at s, $0 \leq s \leq L$. Equilibrium of forces of the section between s and $s + ds$ requires that

$$dH = 0, \tag{2.183}$$

$$dV = q\,ds \tag{2.184}$$

or

$$\dot{H} = 0, \tag{2.185}$$

$$\dot{V} = q. \tag{2.186}$$

Comparing these against the Euler–Lagrange equations given earlier we see that

$$H = \lambda\dot{x}, \tag{2.187}$$

$$V = \lambda\dot{y}, \tag{2.188}$$

so that

$$\lambda^2 = H^2 + V^2. \tag{2.189}$$

Thus the Lagrange multiplier function $\lambda = \lambda(s)$ is the tension at s, a concrete physical interpretation for the seemingly artificial Lagrange multiplier function.

For the case in which the cable has a uniform horizontal load, as in the case of suspension bridges, we can see that the cable hangs in the shape of a parabola. In this situation we have

$$q\,ds = \alpha\,dx \qquad (\alpha = \text{const}) \tag{2.190}$$

so that substituting $q(s)\,ds$ from equation (2.179) into (2.190) we obtain

$$d(\lambda\dot{y}) = \alpha\,dx. \tag{2.191}$$

Integrating equation (2.191) gives

$$\lambda\dot{y} = \alpha x + \beta. \tag{2.192}$$

Also from equation (2.178) we have

$$\lambda\dot{x} = \gamma, \tag{2.193}$$

which expresses the fact that the horizontal component of tension is constant. Thus dividing equation (2.192) by (2.193) we obtain

$$\frac{dy}{dx} = \frac{\alpha x + \beta}{\gamma}, \tag{2.194}$$

which establishes the assertion. We leave it to the reader to show that the shape is a catenary when $q(s) = q = \text{const}$.

2.7. New Approaches to Optimal Control and Filtering

By now it is well understood that many problems in the areas of optimal filtering and control can be reduced to problems in the calculus of variations. Suppose, for example, that a system is initially in state c and that the system undergoes a process described by the differential equation

$$\dot{x} = f(x, t), \qquad 0 \leq t \leq T. \tag{2.195}$$

Further suppose that noisy observations on the process have been made, so that

$$x(t) \cong y(t), \qquad 0 \leq t \leq T, \tag{2.196}$$

where $y(t)$ is the experimentally observed function. A basic task is to obtain a filtered estimate of the process which the system has actually undergone. This can be done by finding the function $x(t)$, $0 \leq t \leq T$, which minimizes the cost functional

$$I(x) = \int_0^T \{[\dot{x} - f(x, t)]^2 + \lambda(x - y)^2\}\, dt, \tag{2.197}$$

where

$$x(0) = c. \tag{2.198}$$

Observe also that frequently an estimate is desired of x at the fixed time t as T, the duration of the period of observation, increases. This focuses attention on the optimizer as a function of the interval.

As a typical control problem consider determining the control function $y(t)$, $0 \leq t \leq T$, which minimizes the cost functional

$$I(y) = \int_0^T [y^2 + x^2]\, dt, \tag{2.199}$$

where the system obeys the dynamical equation

$$\dot{x} = f(x, y), \qquad 0 \leq t \leq T, \tag{2.200}$$

and has the initial condition

$$x(0) = c. \tag{2.201}$$

By solving equation (2.200) for y in terms of x and $\dot{x}$ and substituting this into equation (2.199), it is seen that the task is to choose the function $x(t)$, $0 \leq t \leq T$, to minimize

$$I = \int_0^T F(t, x, \dot{x})\, dt, \tag{2.202}$$

where

$$x(0) = c, \tag{2.203}$$

and

$$x(T) = \text{free.} \tag{2.204}$$

Typically, this is viewed as a problem in the calculus of variations, which implies that a nonlinear two-point boundary-value problem is to be solved. Even with modern computers this is not a completely routine task. The Euler–Lagrange equation and boundary conditions are

$$F_x - (d/dt)F_{\dot{x}} = 0, \qquad 0 \leq t \leq T, \tag{2.205}$$

and

$$x(0) = c, \qquad F_{\dot{x}}|_{t=T} = 0. \tag{2.206}$$

Initial-value or Cauchy problems enjoy certain evident computational advantages over boundary-value problems. It is not surprising, therefore, that much attention has been given to the converting of such two-point boundary-value problems into initial-value problems. In particular, it is known how to convert the boundary-value problem described above into an initial-value problem (References 12, 13).

The purpose of this section is to present a new technique for transforming the minimization problem into an initial-value problem. Though the goal is to achieve a computational advance, in essence, a new conceptual approach to the calculus of variations is obtained. It does not, as will be seen, use Euler–Lagrange equations, Bellman's principle of optimality, or Pontryagin's maximum principle.

Primarily, the case in which F is quadratic is considered. This is a most important special case, and a particularly effective special technique is available for treating it, one that involves initial-value problems for ordinary differential equations only. The general case leads to partial differential equations, but is conceptually simpler (References 13–15).

Cauchy Problem for the Quadratic Case (Reference 16)

Consider minimizing I,

$$I(w) = \frac{1}{2}\int_a^T [\dot{w}^2 + g(t)w^2]\, dt, \tag{2.207}$$

where

$$w(a) = 1, \qquad w(T) = \text{free}. \tag{2.208}$$

Regard the lower limit, a, as a variable, and denote the optimizing choice of the function w on the interval $a \leq t \leq T$ (T fixed) by

$$w = x(t, a), \qquad a \leq t \leq T. \tag{2.209}$$

For $|a - T|$ sufficiently small, this problem will have a unique solution, the function $g(t)$ being sufficiently regular. Let ε be a small parameter and let $\varepsilon\eta = \varepsilon\eta(t, a)$, $a \leq t \leq T$, be an arbitrary variation about the minimizing function, x. The functions x and η must satisfy the relations

$$x(a, a) = 1$$

and

$$\eta(a, a) = 0 \tag{2.210}$$

at $t = a$.

In the usual manner it is observed that the first variation, F, must be zero:

$$F = \int_a^T [\dot{x}\dot{\eta} + gx\eta]\, dt = 0. \tag{2.211}$$

Integration by parts and use of the fundamental lemma would yield the Euler–Lagrange equation. We shall tread an entirely different path.

We observe that the first variation, F, is linear and homogeneous in the function x. It follows that if x is a solution of equation (2.211), Ax is also a solution, where A does not depend on t. Then using the uniqueness assumption concerning the optimizer it is readily seen that

$$x(t, a + \Delta) = \frac{x(t, a)}{x(a + \Delta, a)}, \qquad \Delta > 0, \qquad a + \Delta \leq t \leq T. \tag{2.212}$$

Introduce the function s by means of the definition

$$s(a) = \dot{x}(a, a), \qquad a \leq T. \tag{2.213}$$

In the limit as Δ tends to zero, equation (2.212) becomes the relation

$$x_a(t, a) = -s(a)x(t, a), \qquad a \leq t \leq T. \tag{2.214}$$

This is one of the basic ordinary differential equations. The dependent variable is x, the independent variable is a, and t is a fixed parameter. The initial condition at $a = t$ is

$$x(t, t) = 1. \tag{2.215}$$

Through differentiation of equation (2.212) with respect to t it is seen that

$$x(t, a + \Delta) = \frac{\dot{x}(t, a)}{x(a + \Delta, a)}, \qquad a + \Delta < T. \tag{2.216}$$

Expanding both sides of this equation in powers of Δ and equating the coefficients of the terms in the first power of Δ results in the ordinary differential equation in the independent variable a for the function $\dot{x}(t, a)$,

$$\dot{x}_a(t, a) = -s(a)\dot{x}(t, a), \qquad a \leq t. \tag{2.217}$$

The initial condition at $a = t$ is

$$\dot{x}(t, t) = s(t). \tag{2.218}$$

It remains to obtain an initial-value problem for the function s, which represents the missing initial slope of an optimal trajectory.

Return to equation (2.211) and put

$$\eta = t - a, \tag{2.219}$$

a special variation for which

$$\eta(a, a) = 0, \qquad \dot{\eta} = 1, \qquad \eta_a = -1. \tag{2.220}$$

Then differentiate equation (2.211) with respect to a to obtain the equation

$$-\dot{x}(a, a) + \int_a^T (\dot{x}_a + gx_a\eta - gx)\, dt = 0. \tag{2.221}$$

In view of equations (2.217) and (2.214) for $\dot{x}_a$ and x_a, and equation (2.211), we may write

$$\int_a^T (\dot{x}_a + gx_a\eta)\, dt = 0. \tag{2.222}$$

It follows that

$$s(a) = -\int_a^T g(t)x(t, a)\, dt, \qquad a \le T. \tag{2.223}$$

By differentiation this becomes

$$\begin{aligned} s_a(a) &= g(a)x(a, a) - \int_a^T g(t)x_a(t, a)\, dt, \\ &= g(a) + s(a)\int_a^T g(t)x(t, a)\, dt, \\ &= g(a) - s^2(a), \qquad a \le T, \end{aligned} \tag{2.224}$$

which is the desired ordinary differential equation for the function s. The initial condition at $a = T$ is, from equation (2.223),

$$s(T) = 0. \tag{2.225}$$

The requisite differential equations and initial conditions have now been obtained, and a summary is in order. In the derivation the key equations are equations (2.212) and (2.223), both of which are consequences of the disappearance of the first variation.

Statement of Cauchy Problem

Let $T > 0$ be fixed, and assume that the minimization problem of equations (2.207) and (2.208) possesses a unique solution for all values of the lower limit a for which $0 \le a \le T$. Furthermore, let t be a value such that $0 < t < T$. This is how the value of $u(t, 0)$ is to be obtained.

On the interval $t \le a \le T$ the Riccati differential equation

$$s_a(a) = g(a) - s^2(a) \tag{2.226}$$

and initial condition at $a = T$,

$$s(T) = 0, \tag{2.227}$$

determine the function s. At $a = t$ the additional ordinary differential equation

$$x_a(t, a) = -s(a)x(t, a) \tag{2.228}$$

is adjoined to the equation (2.226). The initial condition on $x(t, a)$ at $a = t$ is

$$x(t, t) = 1. \tag{2.229}$$

The differential equations for s and x are then integrated from $a = t$ to $a = 0$. In this manner the value of $u(t, 0)$ is obtained. If the value of $\dot{x}(t, 0)$ is desired, at $a = t$ the ordinary differential equation

$$\dot{x}_a(t, a) = -s(a)\dot{x}(t, a) \tag{2.230}$$

is also adjoined. Its initial condition at $a = t$ is

$$\dot{x}(t, t) = s(t). \tag{2.231}$$

The value of $s(t)$ is available from the integration of equations (2.226) and (2.227) from $a = T$ to $a = t$.

Observe that this technique is a one-sweep method. The method will be referred to in Chapter 3 as Kalaba's initial-value method.

Computational Aspects and Discussion

The procedure just described is easily instrumented for numerical solution. Some numerical results are given in Section 3.2.4 of Chapter 3 and Reference 17. Generally the value of x for several values of t, $t = t_1$, $t_2, \ldots, t_N$, would be desired. As the variable a passes each such value of t, another differential equation of the form of equation (2.214) is adjoined, along with an initial condition of the form (2.215). If values of $\dot{x}$ are also desired, appropriate equations (2.217) and (2.218) are added. In this way the values of $x(t_i, 0)$ and $\dot{x}(t_i, 0)$, $i = 1, 2, \ldots, N$, are obtained. Values of x and $\dot{x}$ for intermediate values of t might be obtained by using the Euler–Lagrange equation together with these complete sets of initial conditions.

Frequently, though, the Euler–Lagrange equation will prove to be unstable numerically. For example, if $g(t) = 1$, the Euler–Lagrange equation is

$$\ddot{x} = x. \tag{2.232}$$

On a long interval this is difficult to deal with numerically, since the solution is a linear combination of the functions $\exp(t)$ and $\exp(-t)$. Notice that the corresponding equations for the initial-value method are stable,

$$s_a = 1 - s^2, \qquad 0 \leq a \leq T, \tag{2.233}$$

$$s(T) = 0, \tag{2.234}$$

$$x_a = -s(a)x, \qquad 0 \leq a \leq t, \tag{2.235}$$

$$x(t, t) = 1. \tag{2.236}$$

Table 2.1. Summary of Equations Commonly Encountered in Optimal Control Problems

Performance index: $J = \min_{\mathbf{y}} \int_0^T F(t, \mathbf{x}, \mathbf{y})\, dt$

Auxiliary equation constraint: $\dot{\mathbf{x}} = \mathbf{f}(\mathbf{x}, \mathbf{y})$

Boundary conditions: $\mathbf{x}(0) = \mathbf{x}_0$; $\mathbf{x}(T) =$ unspecified

	Definitions	Equations	Boundary conditions
Euler–Lagrange equations	$\Phi = F + \boldsymbol{\lambda}^T(\mathbf{f} - \dot{\mathbf{x}})$	$\frac{\partial \Phi}{\partial \mathbf{x}} - \frac{d}{dt}\frac{\partial \Phi}{\partial \dot{\mathbf{x}}} = 0$ $\frac{\partial \Phi}{\partial \mathbf{y}} - \frac{d}{dt}\frac{\partial \Phi}{\partial \dot{\mathbf{y}}} = 0$	$\left.\frac{\partial \Phi}{\partial \dot{\mathbf{x}}}\right\vert_{t=T} = 0$ or $\boldsymbol{\lambda}(T) = 0$
Pontryagin's maximum principle	$H = F + \boldsymbol{\lambda}^T\mathbf{f}$	$\dot{\boldsymbol{\lambda}} = -\frac{\partial H}{\partial \mathbf{x}}$ $\dot{\mathbf{x}} = \frac{\partial H}{\partial \mathbf{x}}$ $\frac{\partial H}{\partial \mathbf{y}} = 0$	$\boldsymbol{\lambda}(T) = 0$
Dynamic programming	$f(\mathbf{c}, \tau) = \min J(y)$ [not to be confused with auxiliary constraint equation, $\mathbf{f}(\mathbf{x}, \mathbf{y})$] $\dot{\mathbf{c}} = d\mathbf{c}/d\tau$ $\dot{\mathbf{x}} = \mathbf{f}(\mathbf{x}, \mathbf{y})$ $\dot{\mathbf{c}} = \mathbf{f}(\mathbf{c}, \mathbf{z})$ or	$-\frac{\partial f}{\partial \tau} = \min_{\mathbf{z}}\left[F(\tau, \mathbf{c}, \mathbf{z}) + \left(\frac{\partial f}{\partial \mathbf{c}}\right)^T \dot{\mathbf{c}}\right]$	$f(\mathbf{c}, T) = 0$
	$f(\mathbf{c}, T) = \min J(\mathbf{y})$	$-\frac{\partial f}{\partial T} = \min_{\mathbf{z}}\left[F(T, \mathbf{c}, \mathbf{z}) + \left(\frac{\partial f}{\partial \mathbf{c}}\right)^T \dot{\mathbf{c}}\right]$	$f(\mathbf{c}, 0) = 0$

The optimal choice of the points $t_1, t_2, \ldots, t_N$ and the interpolation method are still open to discussion.

It might be thought that the derivations for the general case would be more difficult. Actually this is not so. The results are given in References 12, 13, and 15.

The method described here evolved from the theory of invariant imbedding (References 18). One of the primary goals of this theory is the transformation of two-point boundary-value problems, integral equations, and even multipoint boundary-value problems into initial-value problems. It has now been shown that general classes of variational problems can also be directly transformed into initial-value problems.

2.8. Summary of Commonly Used Equations

Table 2.1 gives a summary of equations commonly encountered in optimal control problems in the remainder of this text.

Exercises

1. Seemingly the problem of the cable with a uniform horizontal load would be formulated as

$$\int_{x_0}^{x_1} \alpha y \, dx = \min$$

subject to the condition

$$\int_{x_0}^{x_1} (1 + y'^2)^{1/2} \, dx = L.$$

But, as we know, the solution to this problem leads to circular and not parabolic area. Find the error.

2. The total downward force on the cable is $\int_0^L q(s)\, ds$. Show that this is automatically balanced by the upward forces at the ends by integrating equation (2.179). Also note that automatically

$$\int_0^L (\dot{x}^2 + \dot{y}^2)^{1/2} \, ds = \int_0^L ds = L.$$

3. Given the system

$$\dot{x} = a_1 x + y, \qquad x(0) = 1,$$

derive the two-point boundary-value equations for minimizing the cost functional

$$I(y) = \frac{1}{2}\int_0^T (y^2 + x^2)\,dt.$$

Use the Euler–Lagrange equation method.

4. Use Pontryagin's maximum principle for deriving the two-point boundary-value equations in Exercise 3.

5. Use dynamic programming to derive the initial-value equations for minimizing the cost functional in Exercise 3.

6. Use Kalaba's initial-value method, described in Section 2.7, to derive the initial-value equations for minimizing the cost functional in Exercise 3.

7. Given the system

$$\dot{x} = y, \qquad x(0) = 1.$$

Use Kalaba's initial-value method to derive the initial-value equations for minimizing the cost functional

$$I(y) = \frac{1}{2}\int_0^T (y^2 + x^2)\,dt.$$

8. Derive the Euler–Lagrange equation (2.232).

3

Numerical Solutions for Linear Two-Point Boundary-Value Problems

Linear systems with quadratic performance criteria typically lead to two-point boundary-value problems. Analytical solutions for these problems can only be obtained for simple systems. Thus in general numerical methods must be used. The numerical methods considered in this chapter are the matrix Riccati equation, the method of complementary functions, and invariant imbedding. While it would be desirable to be able to use only one of the numerical methods exclusively, this is not always possible since there are advantages and disadvantages to each method. The numerical solutions of two-point boundary-value problems are sometimes difficult because of round-off and truncation errors. In such cases the method of invariant imbedding should be used. Examples are given in this chapter and the results are compared on the basis of numerical accuracy.

3.1. Numerical Solution Methods

A typical optimal control problem consists of finding the control function $\mathbf{u}(t)$, $0 \leq t \leq T_f$, which minimizes the quadratic performance index

$$I = \frac{1}{2} \int_0^{T_f} (\mathbf{x}^T Q \mathbf{x} + \mathbf{u}^T R \mathbf{u})\, dt, \tag{3.1}$$

where the system is characterized by the linear dynamical equation

$$\dot{\mathbf{x}} = A\mathbf{x} + B\mathbf{u} \tag{3.2}$$

and the initial condition

$$x(0) = \mathbf{c}. \tag{3.3}$$

In general, $\mathbf{x}$, $\mathbf{u}$, A, B, Q, and R, are functions of time, where $\mathbf{x}(t)$ is an n-component state vector, $\mathbf{u}(t)$ is an r-component control vector, $A(t)$ is an $n \times n$ matrix, $B(t)$ is an $n \times r$ matrix, $Q(t)$ is an $n \times n$ positive semidefinite matrix, and $R(t)$ is an $r \times r$ positive definite matrix. The problem equations can be formulated via any of the several methods discussed in Chapter 2. Solution of the problem via Pontryagin's maximum principle leads to a two-point boundary-value problem.

The solution of the two-point boundary-value problem is considered using the following four methods:

1. matrix Riccati equation (References 1 and 2);
2. method of complementary functions (Reference 3);
3. invariant imbedding (References 4 and 5);
4. analytical solution.

The first method involves the numerical integration of the matrix Riccati equation from which the initial conditions for the integration of $\dot{\mathbf{x}}$ can be obtained. The second method entails the numerical integration of a set of homogeneous differential equations and the numerical solution of a system of linear algebraic equations. The third method entails the conversion of the two-point boundary-value equations into initial-value equations which are then integrated numerically. The last method entails assuming a certain form for the solution of the two-point boundary-value problem, analytically calculating the eigenvalues, and solving numerically for the constant coefficients of the solution of the homogeneous equations.

Pontryagin's maximum principle involves maximizing the Hamiltonian function

$$H = \tfrac{1}{2}\mathbf{x}^T Q \mathbf{x} + \tfrac{1}{2}\mathbf{u}^T R \mathbf{u} + \boldsymbol{\lambda}^T A \mathbf{x} + \boldsymbol{\lambda}^T B \mathbf{u}. \tag{3.4}$$

The costate vector $\boldsymbol{\lambda}(t)$ is the solution of the vector differential equation

$$\dot{\boldsymbol{\lambda}} = -\frac{\partial H}{\partial \mathbf{x}} = -Q\mathbf{x} - A^T\boldsymbol{\lambda}. \tag{3.5}$$

The vector $\mathbf{u}(t)$ that maximizes H is obtained as follows:

$$\frac{\partial H}{\partial \mathbf{u}} = \mathbf{0} = R\mathbf{u} + B^T\boldsymbol{\lambda}, \tag{3.6}$$

$$\mathbf{u} = -R^{-1}B^T\boldsymbol{\lambda}. \tag{3.7}$$

Substituting equation (3.7) into equation (3.2) gives

$$\dot{\mathbf{x}} = A\mathbf{x} - BR^{-1}B^T\boldsymbol{\lambda}. \tag{3.8}$$

Equations (3.5) and (3.8) yield the linear two-point boundary-value equation

$$\begin{bmatrix} \dot{\mathbf{x}} \\ \dot{\boldsymbol{\lambda}} \end{bmatrix} = \begin{bmatrix} A & -BR^{-1}B^T \\ -Q & -A^T \end{bmatrix} \begin{bmatrix} \mathbf{x} \\ \boldsymbol{\lambda} \end{bmatrix} \tag{3.9}$$

with boundary conditions

$$\mathbf{x}(0) = \mathbf{c}, \qquad \boldsymbol{\lambda}(T_f) = \mathbf{0}. \tag{3.10}$$

The initial condition, $\mathbf{x}(0)$ is given, while the boundary condition $\boldsymbol{\lambda}(T_f)$ is derived from the transversality conditions for a fixed-time, free-endpoint problem.

3.1.1. Matrix Riccati Equation

Solution by the matrix Riccati equation assumes that

$$\boldsymbol{\lambda}(t) = P(t)\mathbf{x}(t), \tag{3.11}$$

where $P(t)$ is an $n \times n$ matrix. Substituting equation (3.11) into equation (3.8) we obtain

$$\dot{\mathbf{x}} = A\mathbf{x} - BR^{-1}B^TP\mathbf{x}. \tag{3.12}$$

Also, from equations (3.11) and (3.5) we have

$$\dot{\boldsymbol{\lambda}} = \dot{P}\mathbf{x} + P\dot{\mathbf{x}} = -Q\mathbf{x} - A^TP\mathbf{x}. \tag{3.13}$$

Substituting equation (3.12) into equation (3.13) gives

$$(\dot{P} + PA + A^TP - PBR^{-1}B^TP + Q)\mathbf{x} = 0. \tag{3.14}$$

This equation must hold for all nonzero $\mathbf{x}(t)$, thus the term multiplying $\mathbf{x}(t)$ must be zero. Therefore we have

$$\dot{P} = -PA - A^TP + PBR^{-1}B^TP - Q. \tag{3.15}$$

Equation (3.15) is the matrix Riccati equation. The terminal condition is obtained from equations (3.10) and (3.11):

$$P(T_f) = 0. \tag{3.16}$$

Substituting equation (3.11) into equation (3.8) gives

$$\dot{\mathbf{x}} = (A - BR^{-1}B^TP)\mathbf{x}. \tag{3.17}$$

To obtain the optimal trajectory equation (3.15) is integrated backward from time T_f to 0. The initial condition, $P(0) = K$, is thus obtained. Equation (3.17) is then integrated forward with the initial conditions $\mathbf{x}(0) = \mathbf{c}$ and $P(0) = K$.

3.1.2. Method of Complementary Functions

The solution of equation (3.9) can be expressed in terms of the transition matrix by the equation

$$\begin{bmatrix} \mathbf{x}(t) \\ \boldsymbol{\lambda}(t) \end{bmatrix} = \begin{bmatrix} H_{11}(t) & H_{12}(t) \\ H_{21}(t) & H_{22}(t) \end{bmatrix} \begin{bmatrix} \mathbf{x}(0) \\ \boldsymbol{\lambda}(0) \end{bmatrix}, \tag{3.18}$$

where

$$H(t) = \begin{bmatrix} H_{11}(t) & H_{12}(t) \\ H_{21}(t) & H_{22}(t) \end{bmatrix} \tag{3.19}$$

is the $2n \times 2n$ partitioned transition matrix. Then

$$\mathbf{x}_B(t) = H(t)\mathbf{x}_B(0), \tag{3.20}$$

where

$$\mathbf{x}_B(t) = \begin{bmatrix} \mathbf{x}(t) \\ \boldsymbol{\lambda}(t) \end{bmatrix}. \tag{3.21}$$

Now equation (3.9) can be expressed in the form

$$\dot{\mathbf{x}}_B(t) = F\mathbf{x}_B(t), \tag{3.22}$$

where

$$F = \begin{bmatrix} A & -BR^{-1}B^T \\ -Q & -A^T \end{bmatrix}. \tag{3.23}$$

Differentiating equation (3.20) and substituting into equation (3.22) gives

$$\dot{H}\mathbf{x}_B(0) = FH\mathbf{x}_B(0);$$

and simplifying yields

$$\dot{H} = FH \tag{3.24}$$

with initial condition

$$H(0) = I. \tag{3.25}$$

The initial condition is easily derived from the identity equation obtained when t is set equal to zero in equation (3.20).

The initial condition vector, $\mathbf{x}_B(0)$, is determined from the equation

$$\begin{bmatrix} \mathbf{x}(0) \\ \boldsymbol{\lambda}(T_f) \end{bmatrix} = \begin{bmatrix} H_{11}(0) & H_{12}(0) \\ H_{21}(T_f) & H_{22}(T_f) \end{bmatrix} \begin{bmatrix} \mathbf{x}(0) \\ \boldsymbol{\lambda}(0) \end{bmatrix}, \tag{3.26}$$

where the boundary conditions are given by equation (3.10). Inverting equation (3.26) gives

$$\begin{bmatrix} \mathbf{x}(0) \\ \boldsymbol{\lambda}(0) \end{bmatrix} = \begin{bmatrix} H_{11}(0) & H_{12}(0) \\ H_{21}(T_f) & H_{22}(T_f) \end{bmatrix}^{-1} \begin{bmatrix} \mathbf{x}(0) \\ \boldsymbol{\lambda}(T_f) \end{bmatrix}. \tag{3.27}$$

The optimal solution is obtained by integrating equation (3.24) from time $t = 0$ to time $t = T_f$ and storing the values of $H(t)$ at the boundaries. The vector $\mathbf{x}_B(0)$ is then evaluated from equation (3.27) in order to obtain the full set of initial conditions. Equation (3.22) is integrated from time $t = 0$ to $t = T_f$ and the optimal control function is obtained from equation (3.7). Alternately the optimal trajectory can be obtained by storing the values of $H(t)$ at the instants where a solution is desired and substituting into equation (3.20).

For a two-component state vector, the transition matrix, H, is a 4×4 matrix. Then we have

$$\dot{H} = FH = \begin{bmatrix} a_1 & b_1 & c_1 & d_1 \\ a_2 & b_2 & c_2 & d_2 \\ a_3 & b_3 & c_3 & d_3 \\ a_4 & b_4 & c_4 & d_4 \end{bmatrix} \begin{bmatrix} h_{11} & h_{21} & h_{31} & h_{41} \\ h_{12} & h_{22} & h_{32} & h_{42} \\ h_{13} & h_{23} & h_{33} & h_{43} \\ h_{14} & h_{24} & h_{34} & h_{44} \end{bmatrix} \tag{3.28}$$

with the initial condition

$$H(0) = I. \tag{3.29}$$

The coefficients of the F matrix in equation (3.24) are obtained by comparing the coefficients with the coefficients in equation (3.23).

In expanded form, the homogeneous differential equations are

$$
\begin{aligned}
\dot{h}_{11} &= a_1 h_{11} + b_1 h_{12} + c_1 h_{13} + d_1 h_{14}, & h_{11}(0) &= 1,\\
\dot{h}_{12} &= a_2 h_{11} + b_2 h_{12} + c_2 h_{13} + d_2 h_{14}, & h_{12}(0) &= 0,\\
\dot{h}_{13} &= a_3 h_{11} + b_3 h_{12} + c_3 h_{13} + d_3 h_{14}, & h_{13}(0) &= 0,\\
\dot{h}_{14} &= a_4 h_{11} + b_4 h_{12} + c_4 h_{13} + d_4 h_{14}, & h_{14}(0) &= 0,\\[6pt]
\dot{h}_{21} &= a_1 h_{21} + b_1 h_{22} + c_1 h_{23} + d_1 h_{24}, & h_{21}(0) &= 0,\\
\dot{h}_{22} &= a_2 h_{21} + b_2 h_{22} + c_2 h_{23} + d_2 h_{24}, & h_{22}(0) &= 1,\\
\dot{h}_{23} &= a_3 h_{21} + b_3 h_{22} + c_3 h_{23} + d_3 h_{24}, & h_{23}(0) &= 0,\\
\dot{h}_{24} &= a_4 h_{21} + b_4 h_{22} + c_4 h_{23} + d_4 h_{24}, & h_{24}(0) &= 0,\\[6pt]
\dot{h}_{31} &= a_1 h_{31} + b_1 h_{32} + c_1 h_{33} + d_1 h_{34}, & h_{31}(0) &= 0,\\
\dot{h}_{32} &= a_2 h_{31} + b_2 h_{32} + c_2 h_{33} + d_2 h_{34}, & h_{32}(0) &= 0,\\
\dot{h}_{33} &= a_3 h_{31} + b_3 h_{32} + c_3 h_{33} + d_3 h_{34}, & h_{33}(0) &= 1,\\
\dot{h}_{34} &= a_4 h_{31} + b_4 h_{32} + c_4 h_{33} + d_4 h_{34}, & h_{34}(0) &= 0,\\[6pt]
\dot{h}_{41} &= a_1 h_{41} + b_1 h_{42} + c_1 h_{43} + d_1 h_{44}, & h_{41}(0) &= 0,\\
\dot{h}_{42} &= a_2 h_{41} + b_2 h_{42} + c_2 h_{43} + d_2 h_{44}, & h_{42}(0) &= 0,\\
\dot{h}_{43} &= a_3 h_{41} + b_3 h_{42} + c_3 h_{43} + d_3 h_{44}, & h_{43}(0) &= 0,\\
\dot{h}_{44} &= a_4 h_{41} + b_4 h_{42} + c_4 h_{43} + d_4 h_{44}, & h_{44}(1) &= 1,
\end{aligned}
\tag{3.30}
$$

where the h's are all functions of time.

The solution is given by the linear combination

$$
\begin{bmatrix} x_1 \\ x_2 \\ \lambda_1 \\ \lambda_2 \end{bmatrix} = A_1 \begin{bmatrix} h_{11} \\ h_{12} \\ h_{13} \\ h_{14} \end{bmatrix} + A_2 \begin{bmatrix} h_{21} \\ h_{22} \\ h_{23} \\ h_{24} \end{bmatrix} + A_3 \begin{bmatrix} h_{31} \\ h_{32} \\ h_{33} \\ h_{34} \end{bmatrix} + A_4 \begin{bmatrix} h_{41} \\ h_{42} \\ h_{43} \\ h_{44} \end{bmatrix}, \tag{3.31}
$$

where A_1, A_2, A_3, and A_4 are constants to be determined by the boundary conditions and are equal to the components of the initial condition vector $\mathbf{x}_B(0)$. Utilizing the boundary conditions at $t = 0$ and $t = T_f$, we readily obtain four linear algebraic equations in terms of A_1, A_2, A_3, and A_4:

$$
\begin{aligned}
x_1(0) &= A_1 h_{11}(0) + A_2 h_{21}(0) + A_3 h_{31}(0) + A_4 h_{41}(0),\\
x_2(0) &= A_1 h_{12}(0) + A_2 h_{22}(0) + A_3 h_{32}(0) + A_4 h_{42}(0),\\
\lambda_1(T_f) &= A_1 h_{13}(T_f) + A_2 h_{23}(T_f) + A_3 h_{33}(T_f) + A_4 h_{43}(T_f),\\
\lambda_2(T_f) &= A_1 h_{14}(T_f) + A_2 h_{24}(T_f) + A_3 h_{34}(T_f) + A_4 h_{44}(T_f),
\end{aligned}
\tag{3.32}
$$

where

$$x_1(0) = c_1, \qquad x_2(0) = c_2, \qquad \lambda_1(T_f) = \lambda_2(T_f) = 0. \tag{3.33}$$

The optimal trajectory is obtained by integrating equation (3.28) with initial condition, equation (3.29), from $t = 0$ to $t = T_f$, storing the values of $H(t)$ at the end points and at the instants where a solution is desired. The linear algebraic equations (3.32) are then solved for A_1, A_2, A_3, and A_4 and the solution is obtained from equations (3.31).

3.1.3. Invariant Imbedding

The invariant imbedding method converts the two-point boundary-value problem into an initial-value problem. The two-point boundary-value equation (3.9) and the boundary conditions, equation (3.10), can be expressed in the form

$$\dot{\mathbf{x}} = A\mathbf{x} + B\boldsymbol{\lambda}, \qquad \mathbf{x}(a) = \mathbf{c}, \tag{3.34}$$

$$\dot{\boldsymbol{\lambda}} = D\mathbf{x} + E\boldsymbol{\lambda}, \qquad \boldsymbol{\lambda}(T_f) = \mathbf{0}, \tag{3.35}$$

where $\mathbf{x}$ and $\boldsymbol{\lambda}$ are n-dimensional column vectors; A, B, D, and E are $n \times n$ matrices; and a is the initial time and is regarded as a variable. It should be noted that the matrix B is different here from the definition given in equation (3.2). To derive the initial-value equations, let W and Z be $n \times n$ matrices:

$$\dot{W} = AW + BZ, \qquad W(a) = I, \tag{3.36}$$

$$\dot{Z} = DW + EZ, \qquad Z(T_f) = 0, \tag{3.37}$$

where

$$\mathbf{x} = W\mathbf{c}, \tag{3.38}$$

$$\boldsymbol{\lambda} = Z\mathbf{c}, \tag{3.39}$$

then

$$\begin{aligned} \dot{\mathbf{x}} &= \dot{W}\mathbf{c} = (AW + BZ)\mathbf{c} \\ &= A(W\mathbf{c}) + B(Z\mathbf{c}) \\ &= A\mathbf{x} + B\boldsymbol{\lambda}. \end{aligned} \tag{3.40}$$

Similarly, we have

$$\begin{aligned} \dot{\boldsymbol{\lambda}} &= \dot{Z}\mathbf{c} = (DW + EZ)\mathbf{c} \\ &= D\mathbf{x} + E\boldsymbol{\lambda}. \end{aligned} \tag{3.41}$$

To indicate the dependence of the matrices W and Z upon t and a, write

$$W = W(t, a), \tag{3.42}$$

$$Z = Z(t, a). \tag{3.43}$$

Differentiating equations (3.36) and (3.37) with respect to a gives

$$\dot{W}_a = AW_a + BZ_a, \qquad \dot{W}(a, a) + W_2(a, a) = 0, \tag{3.44}$$

$$\dot{Z}_a = DW_a + EZ_a, \qquad Z_a(T_f) = 0, \tag{3.45}$$

where the total derivative of $W(a, a)$ is given by the derivative of W with respect to its first argument plus the derivative of W with respect to its second argument. Now, since the functional forms of equations (3.36) and (3.37) are the same as (3.44) and (3.45), and

$$W(a, a) = W(a) = I, \tag{3.46}$$

$$W_2(a, a) = W_2(a) = -\dot{W}(a, a), \tag{3.47}$$

it follows that[†]

$$W_a(t, a) = W(t, a)[-\dot{W}(a, a)], \tag{3.48}$$

$$Z_a(t, a) = Z(t, a)[-\dot{W}(a, a)]. \tag{3.49}$$

But

$$\begin{aligned} \dot{W}(a, a) &= AW(a, a) + BZ(a, a) \\ &= A + BR(a), \end{aligned} \tag{3.50}$$

where

$$R(a) = Z(a, a). \tag{3.51}$$

[†] This is due to the system linearity and is easily illustrated by an example. Let

$$x(t) = Hx(0), \qquad x(0) = c,$$

$$y(t) = Hy(0), \qquad y(0) = \alpha c,$$

where α is a constant multiplier. Then we have

$$\begin{aligned} y(t) &= H(\alpha c) = \alpha Hc \\ &= \alpha H[H^{-1}x(t)] \\ &= \alpha x(t). \end{aligned}$$

Thus we have

$$W_a(t, a) = -W(t, a)[A + BR(a)], \qquad W(t, t) = I, \tag{3.52}$$

and

$$Z_a(t, a) = -Z(t, a)[A + BR(a)], \qquad Z(t, t) = R(t). \tag{3.53}$$

Furthermore, differentiating $R(a)$ in equation (3.51) with respect to a gives

$$\begin{aligned} R_a(a) &= \dot{Z}(a, a) + Z_2(a, a) \\ &= D + ER(a) - R(a)[A + BR(a)]. \end{aligned} \tag{3.54}$$

Thus we obtain

$$R_a = D + ER - RA - RBR, \qquad R(T_f) = 0. \tag{3.55}$$

Equations (3.52), (3.53), and (3.55) form the initial-value equations.

To obtain the optimal trajectory, matrix equation (3.55) is integrated from $a = T_f$ to $a = t$. At $a = t$ matrix equation (3.52) is adjoined and both differential equations are integrated from $a = t$ to $a = 0$. Since the optimal trajectory is generally desired at several values of t, (i.e., $t = t_1, t_2, \ldots, t_N$) as the variable a passes each such value of t, another matrix differential equation of the form of equation (3.52) is adjoined. Note that by integrating the initial-value problem for R, the missing initial condition of Z for the original two-point boundary-value problem is obtained. This converts the original boundary-value problem into an initial-value problem with data on W and Z given at $t = a$.

3.1.4. Analytical Solution

An analytical solution to the linear two-point boundary-value problem can be obtained if the order of the system is small and if the equations are relatively simple. The analytical solution is often useful for checking numerical results.

Consider the system of linear differential equations with constant coefficients

$$\begin{aligned} \dot{x}_1 &= a_{11}x_1 + a_{12}x_2 + \cdots + a_{1n}x_n, \\ \dot{x}_2 &= a_{21}x_1 + a_{22}x_2 + \cdots + a_{2n}x_n, \\ &\vdots \\ \dot{x}_n &= a_{n1}x_1 + a_{n2}x_2 + \cdots + a_{nn}x_n, \end{aligned} \tag{3.56}$$

and boundary conditions

$$x_i(0) = b_i, \qquad i = 1 \text{ to } r, \tag{3.57}$$

$$x_i(T_f) = b_i, \qquad i = r + 1 \text{ to } n. \tag{3.58}$$

The equations can be expressed in the matrix form

$$\dot{\mathbf{x}} = F\mathbf{x}. \tag{3.59}$$

The characteristic equation of the matrix F is obtained by setting the determinant of $mI - F$ equal to zero:

$$| mI - F | = 0, \tag{3.60}$$

where I is the identity matrix and m is a scalar. The roots of the characteristic equation are the eigenvalues of the matrix F.

Assume that the eigenvalues are $m_1, m_2, \ldots, m_n$. The general form of the solution for $x_1(t)$, assuming the eigenvalues are distinct, is

$$x_1(t) = c_1 e^{m_1 t} + c_2 e^{m_2 t} + \cdots + c_n e^{m_n t}. \tag{3.61}$$

If the matrix F is sparse, then it is relatively simple to obtain expressions for $x_2(t), x_3(t), \ldots, x_n(t)$ in terms of $c_1, c_2, \ldots, c_n$ and $m_1, m_2, \ldots, m_n$. Then utilizing the boundary conditions at $t = 0$ and $t = T_f$, n linear algebraic equations in terms of the c_i's are obtained. Solving for the c_i's, the optimal trajectory is then given by equation (3.61).

If any of the eigenvalues is complex, i.e.,

$$m_1 = a + jb, \qquad m_2 = a - jb, \tag{3.62}$$

then

$$c_1 e^{m_1 t} + c_2 e^{m_2 t} = e^{at}(K_1 \cos bt + K_2 \sin bt). \tag{3.63}$$

For two equal eigenvalues, the solution is of the form

$$c_1 e^{m_1 t} + c_2 t e^{m_2 t}. \tag{3.64}$$

3.2. An Optimal Control Problem for a First-Order System

Consider the optimal control problem (Reference 6) of finding the control function $y(t)$, $0 \leq t \leq T_f$, which minimizes the cost functional

$$I(y) = \frac{1}{2} \int_0^{T_f} (y^2 + x^2)\, dt, \tag{3.65}$$

where the system is characterized by the first-order dynamical equation

$$\dot{x} = a_1 x + y, \qquad 0 \leq t \leq T_f \tag{3.66}$$

and the initial condition

$$x(0) = 1. \tag{3.67}$$

This is the state-regulator problem (References 1, 2) in which the cost functional is such that the state variable $x(t)$ is to be kept near zero without excessive expenditure of control energy.

The methods for the formulation of the problem equations from Chapter 2 are briefly reviewed and the equations are derived for the given example via the following four methods:

1. Euler–Lagrange equations (Reference 7);
2. Pontryagin's maximum principle (References 1, 2);
3. dynamic programming (Reference 8);
4. Kalaba's initial-value method (References 9–11).

The first two methods entail solution of two-point boundary-value problems, and the numerical solutions are obtained using the method of complementary functions and the analytical solution. The third method entails solution of an initial-value problem involving the Riccati equation. The fourth method entails solution of an initial-value problem but has computational advantages over the solution of the corresponding problem by dynamic programming. The analytical solution for this example is easily derived, but this becomes increasingly difficult as the dimensions of the problem increase.

Substituting from equation (3.66) into equation (3.65) yields the cost functional

$$I(x) = \frac{1}{2} \int_0^{T_f} [\dot{x}^2 - 2a_1 x\dot{x} + (1 + a_1^2)x^2]\, dt. \tag{3.68}$$

3.2.1. The Euler–Lagrange Equations

Equation (3.68) has the general form

$$I(x) = \int_0^{T_f} F(x, \dot{x}, t)\, dt, \tag{3.69}$$

where

$$F(x, \dot{x}, t) = \tfrac{1}{2}[\dot{x}^2 - 2a_1 x\dot{x} + (1 + a_1^2)x^2]. \tag{3.70}$$

The general form of the Euler–Lagrange equation is

$$\frac{\partial F}{\partial x} - \frac{d}{dt}\left(\frac{\partial F}{\partial \dot{x}}\right) = 0, \qquad 0 \leq t \leq T \tag{3.71}$$

with associated boundary conditions

$$x(0) = c, \qquad \left.\frac{\partial F}{\partial \dot{x}}\right|_{t=T_f} = 0. \tag{3.72}$$

Differentiating equation (3.70) with respect to x and $\dot{x}$ yields

$$\frac{\partial F}{\partial x} = -a_1\dot{x} + (1 + a_1^2)x, \tag{3.73}$$

$$\begin{aligned}\frac{d}{dt}\left(\frac{\partial F}{\partial \dot{x}}\right) &= \frac{d}{dt}(\dot{x} - a_1 x) \\ &= \ddot{x} - a_1\dot{x}.\end{aligned} \tag{3.74}$$

Then equation (3.71) becomes

$$\begin{gathered}-a_1\dot{x} + (1 + a_1^2)x - (\ddot{x} - a_1 x) = 0, \\ \ddot{x} = (1 + a_1^2)x.\end{gathered} \tag{3.75}$$

The boundary conditions are

$$x(0) = 1, \tag{3.76}$$

$$\left.\frac{\partial F}{\partial \dot{x}}\right|_{t=T_f} = \dot{x}(T_f) - a_1 x(T_f) = 0. \tag{3.77}$$

Equation (3.75) is a second-order linear differential equation with two-point boundary-value conditions. The solution can be obtained numerically using the method of complementary functions. First express equation (3.75) in the form of a system of first-order linear differential equations. Let $x = x_1$; then

$$\dot{x}_1 = x_2, \qquad x_1(0) = 1, \tag{3.78}$$

$$\dot{x}_2 = (1 + a_1^2)x_1, \qquad x_2(T_f) - a_1 x_1(T_f) = 0. \tag{3.79}$$

Equations (3.78) and (3.79) can be expressed in the form

$$\dot{x}_1 = ax_1 + bx_2, \tag{3.80}$$

$$\dot{x}_2 = dx_1 + ex_2, \tag{3.81}$$

where

$$\begin{aligned} a &= e = 0, \\ b &= 1, \\ d &= (1 + a_1^2). \end{aligned} \tag{3.82}$$

The associated homogeneous differential equations are

$$\begin{aligned} \dot{h}_{11} &= ah_{11} + bh_{12}, \qquad & h_{11}(0) &= 1, \\ \dot{h}_{12} &= dh_{11} + eh_{12}, & h_{12}(0) &= 0, \\ \dot{h}_{21} &= ah_{21} + bh_{22}, & h_{21}(0) &= 0, \\ \dot{h}_{22} &= dh_{21} + eh_{22}, & h_{22}(0) &= 1. \end{aligned} \tag{3.83}$$

Comparing the coefficients in equations (3.78) and (3.79) with (3.83), or using the definitions given in equations (3.82), it is obvious that

$$\begin{aligned} \dot{h}_{11}(t) &= h_{12}(t), & h_{11}(0) &= 1, \\ \dot{h}_{12}(t) &= (1 + a_1^2)h_{11}(t), \qquad & h_{12}(0) &= 0, \\ \dot{h}_{21}(t) &= h_{22}(t), & h_{21}(0) &= 0, \\ \dot{h}_{22}(t) &= (1 + a_1^2)h_{21}(t), & h_{22}(0) &= 1. \end{aligned} \tag{3.84}$$

The solution can be represented by the linear combinations

$$\begin{aligned} x_1(t) &= Ah_{11}(t) + Bh_{21}(t), \\ x_2(t) &= Ah_{12}(t) + Bh_{22}(t), \end{aligned} \tag{3.85}$$

where A and B are constants to be determined by use of the boundary conditions. Utilizing the boundary conditions at $t = 0$ and $t = T_f$, two linear algebraic equations in terms of A and B are readily obtained:

$$\begin{aligned} x_1(0) = 1 &= Ah_{11}(0) + Bh_{21}(0), \\ x_2(T_f) - a_1x_1(T_f) = 0 &= A[h_{12}(T_f) - a_1h_{11}(T_f)] \\ &\quad + B[h_{22}(T_f) - a_1h_{21}(T_f)]. \end{aligned} \tag{3.86}$$

Homogeneous equations (3.84) are integrated from $t = 0$ to $t = T_f$ and the values of $h_{11}(t)$, $h_{12}(t)$, $h_{21}(t)$, and $h_{22}(t)$ are stored as the integration process moves forward. The linear algebraic equations are then solved for A and B and the solution is obtained from equation (3.85). It is not necessary to store all the values of the solutions of the homogeneous equations; it suffices to store values at the end points and at the instants where a solution printout is desired.

3.2.2. Pontryagin's Maximum Principle

Pontryagin's maximum principle involves maximizing the Hamiltonian function. The Hamiltonian function[†] is formed by the equation

$$H = \sum_{i=1}^{n} \lambda_i f_i - F(t, \mathbf{x}, \mathbf{y}), \tag{3.87}$$

where

$$\dot{x}_i = f_i(\mathbf{x}, \mathbf{y}), \qquad i = 1, 2, \ldots, n, \tag{3.88}$$

$$\mathbf{x} = (x_1, x_2, \ldots, x_n)^T \quad \text{is a state vector,} \tag{3.89}$$

$$\mathbf{y} = (y_1, y_2, \ldots, y_n)^T \quad \text{is a control vector.} \tag{3.90}$$

The functions $\lambda_i(t)$ satisfy the system of differential equations

$$\dot{\lambda}_i = -\frac{\partial H}{\partial x_i}, \qquad i = 1, 2, \ldots, n, \tag{3.91}$$

and the original system of differential equations can be expressed as

$$\dot{x}_i = \frac{\partial H}{\partial \lambda_i}, \qquad i = 1, 2, \ldots, n \tag{3.92}$$

with boundary conditions

$$x_i(0) = c_i, \qquad \lambda_i(T_f) = 0, \qquad i = 1, 2, \ldots, n. \tag{3.93}$$

The initial conditions, $x_i(0)$, are given, while boundary conditions, $\lambda_i(T_f)$, are derived from the transversality conditions for a fixed time, free-end-point problem. The optimal control is the vector $\mathbf{y}(t)$ that maximizes the Hamiltonian function.

For the given example, the Hamiltonian function is

$$H = \lambda(a_1 x + y) - \tfrac{1}{2}(x^2 + y^2) \tag{3.94}$$

and the system of differential equations (3.91) and (3.92) is

$$\dot{x} = a_1 x + y, \tag{3.95}$$

$$\dot{\lambda} = x - a_1 \lambda. \tag{3.96}$$

[†] In this example an alternative definition of the Hamiltonian function is given in which F is subtracted instead of added as in Equation (3.4). The Hamiltonian is then maximized instead of minimized for the minimization problem.

The function $y(t)$ that maximizes H is obtained from

$$\frac{\partial H}{\partial y} = \lambda(t) - y = 0, \tag{3.97}$$

$$y(t) = \lambda(t). \tag{3.98}$$

Then substituting equation (3.98) into equation (3.95) the differential equations are

$$\dot{x} = a_1 x + \lambda, \tag{3.99}$$

$$\dot{\lambda} = x - a_1 \lambda, \tag{3.100}$$

with boundary conditions

$$x(0) = 1, \qquad \lambda(T_f) = 0. \tag{3.101}$$

Solution of the differential equations (3.99) and (3.100) together with satisfaction of boundary conditions (3.101) is a two-point boundary-value problem which can be solved numerically by the method of complementary functions. The corresponding homogeneous equations are

$$\begin{aligned} \dot{h}_{11}(t) &= a_1 h_{11}(t) + h_{12}(t), & h_{11}(0) &= 1, \\ \dot{h}_{12}(t) &= h_{11}(t) - a_1 h_{12}(t), & h_{12}(0) &= 0, \\ \dot{h}_{21}(t) &= a_1 h_{21}(t) + h_{22}(t), & h_{21}(0) &= 0, \\ \dot{h}_{22}(t) &= h_{21}(t) - a_1 h_{22}(t), & h_{22}(0) &= 1, \end{aligned} \tag{3.102}$$

and the associated linear algebraic equations are

$$\begin{aligned} x(0) &= 1 = A h_{11}(0) + B h_{21}(0), \\ \lambda(T_f) &= 0 = A h_{12}(T_f) + B h_{22}(T_f). \end{aligned} \tag{3.103}$$

The method of solution follows the same procedure as discussed for the Euler–Lagrange equation case.

3.2.3. Dynamic Programming

Solution of the two-point boundary-value problem can be avoided by the use of dynamic programming. Thus, in general the function $f(c, T_f)$ is defined as

$$f(c, T_f) = \text{minimum cost of the process of duration } T_f \text{ and initial state } x(0) = c. \tag{3.104}$$

This function is *not* to be confused with $f_i(x, y)$ as defined earlier. Utilizing the principle of optimality, it is readily shown that for the given example

$$\frac{\partial f}{\partial T_f} = \min_z \left[F(T_f, c, z) + \dot{c}\frac{\partial f}{\partial c} \right] \tag{3.105}$$

$$= \min_z \left[\frac{1}{2}(c^2 + z^2) + (a_1 c + z)\frac{\partial f}{\partial c} \right], \tag{3.106}$$

where

$$z_{\min} = z_{\min}(c, T_f) \tag{3.107}$$

is the optimal control law. Equation (3.106) is a nonlinear partial differential equation and the associated initial condition is $f(c, 0) = 0$. A candidate for the minimum z is obtained by differentiating the right-hand side of the equation (3.106) with respect to z and equating the results to zero. Thus we have

$$z_{\min} = -\frac{\partial f}{\partial c}. \tag{3.108}$$

Substituting from equation (3.108) into equation (3.106) gives

$$\frac{\partial f}{\partial T_f} = \frac{1}{2}c^2 + a_1 c\frac{\partial f}{\partial c} - \frac{1}{2}\left(\frac{\partial f}{\partial c}\right)^2. \tag{3.109}$$

Introducing the technique of solving the above partial differential equation by the method of separation of variables leads to solution of an ordinary differential equation. Thus, let us assume

$$f(c, T_f) = \psi(T_f)c^2. \tag{3.110}$$

Substituting equation (3.110) into equation (3.109) gives

$$c^2\frac{d\psi(T_f)}{dT_f} = \frac{1}{2}c^2 + 2a_1 c^2\psi(T_f) - 2\psi^2(T_f)c^2. \tag{3.111}$$

Dividing both sides of equation (3.111) by c^2 yields

$$\frac{d\psi}{dT_f} = \frac{1}{2} + 2a_1\psi(T_f) - 2\psi^2(T_f). \tag{3.112}$$

The initial condition is obtained by setting $T_f = 0$ in equation (3.110). Then we have

$$\psi(0) = 0. \tag{3.113}$$

Differentiating equation (3.110) and substituting into equation (3.108) gives

$$y_{\text{opt}} = -2\psi(T_f)c. \tag{3.114}$$

Since $x(t)$ is the current state of the system with time remaining $T_f - t$ the optimal input can be expressed in the form

$$y_{\text{opt}} = -2\psi(T_f - t)x(t). \tag{3.115}$$

Substituting equation (3.115) into equation (3.66) yields

$$\dot{x} = [a_1 - 2\psi(T_f - t)]x(t) \tag{3.116}$$

with initial condition

$$x(0) = c = 1. \tag{3.117}$$

To evaluate $x(t)$ numerically by solving (3.116) subject to satisfaction of (3.117), equation (3.112) is first integrated from 0 to the terminal time T_f and the values of $\psi(T_f)$ are stored. Equation (3.116) is then integrated from time $t = 0$ to time $t = T_f$ utilizing the stored values $\psi(T_f - t)$. An alternative procedure is to integrate equation (3.112) from 0 to T_f, storing only the final value of $\psi(T_f)$, i.e., $\psi(T_f) = k$. Then integrate equation (3.116) from $t = 0$ to $t = T_f$, using the initial conditions

$$\begin{aligned} x(0) &= c = 1, \\ \psi(T_f - t)|_{t=0} &= k. \end{aligned} \tag{3.118}$$

This is the procedure which is utilized here to obtain the numerical results. The accuracy of these results depends somewhat upon the degree to which the numerical values for $\psi(T_f)$ obtained by integrating from T_f to 0 are the same as the numerical values obtained by integrating from 0 to T_f.

Equation (3.112) is a Riccati equation. It can be shown that, in general, the matrix Riccati method and dynamic programming are the same for linear quadratic problems (Reference 12).

3.2.4. Kalaba's Initial-Value Method

Utilizing the new approach to optimal control described in Section 2.7 of Chapter 2 the problem can be converted into an initial-value problem and can be evaluated by a one-sweep computational procedure. As in

Section 2.1 of Chapter 2, replace $x(t)$ by

$$x(t) + \varepsilon\eta(t), \tag{3.119}$$

where $x(t)$ is the function that minimizes $I(x)$, $\eta(t)$ is a variation in $x(t)$, and ε is a small number. Then equation (3.69) satisfies the inequality

$$\int_0^{T_f} F(x, \dot{x}, t)\, dt \leq \int_0^{T_f} F(x + \varepsilon\eta, \dot{x} + \varepsilon\dot{\eta}, t)\, dt. \tag{3.120}$$

Expanding the right-hand side in a Taylor series and letting ε be sufficiently small, the equation for the first variation is obtained:

$$0 = \int_0^{T_f} [\eta(t)F_x(x, \dot{x}, t) + \dot{\eta}(t)F_{\dot{x}}(x, \dot{x}, t)]\, dt. \tag{3.121}$$

The first variation of the given example satisfies the equation

$$0 = \int_a^{T_f} [(\dot{x} - a_1 x)\dot{\eta} + (-a_1\dot{x} + x + a_1^2 x)\eta]\, d\tau, \tag{3.122}$$

where

$$x(a) = 1, \tag{3.123}$$

$$x(T_f) = \text{free}. \tag{3.124}$$

The lower limit, a, is regarded as a variable. To obtain a Fredholm integral equation for x, the arbitrary variation η is chosen to be

$$\eta = k(t, \tau, a), \tag{3.125}$$

where

$$k = \begin{cases} t - a, & t \leq \tau \leq T_f \\ \tau - a, & a \leq \tau \leq t \end{cases} \tag{3.126}$$

or

$$k = \min(t - a, \tau - a), \qquad a \leq t, \qquad \tau \leq T_f. \tag{3.127}$$

Then taking the derivative of k with respect to τ we have

$$\dot{\eta}(\tau) = \begin{cases} 1, & a \leq \tau < t \qquad (3.128) \\ 0, & t < \tau \leq T_f. \qquad (3.129) \end{cases}$$

To simplify the notation express equation (3.122) in the form

$$0 = \int_a^{T_f} [(\dot{x} - a_1 x)\dot{\eta} + g(x, \dot{x})\eta]\, d\tau, \tag{3.130}$$

where

$$g(x, \dot{x}) = -a_1\dot{x} + x + a_1^2 x^2. \tag{3.131}$$

Then making use of equations (3.128) and (3.129) we obtain

$$\begin{aligned}\int_a^{T_f} (\dot{x} - a_1 x)\dot{\eta}\, d\tau &= \int_a^t (\dot{x} - a_1 x)\dot{\eta}\, d\tau + \int_t^{T_f} (\dot{x} - a_1 x)\dot{\eta}\, d\tau \\ &= \int_a^t (\dot{x} - a_1 x)\, d\tau.\end{aligned} \tag{3.132}$$

Integrating the first term on the right-hand side of equation (3.132) gives

$$\begin{aligned}\int_a^t (\dot{x} - a_1 x)\, d\tau &= x(t, a) - x(a, a) - \int_a^t a_1 x\, d\tau \\ &= x(t, a) - 1 - \int_a^t a_1 x\, d\tau.\end{aligned} \tag{3.133}$$

Substituting equation (3.133) into equation (3.130), the Fredholm integral equation is obtained:

$$x(t, a) = 1 - \int_a^{T_f} g(x, \dot{x})k(t, \tau, a)\, d\tau + \int_a^t a_1 x\, d\tau. \tag{3.134}$$

Differentiating equation (3.134) with respect to a and noting from equation (3.126) that

$$\eta_a = -1, \tag{3.135}$$

$$\eta(a) = 0 \tag{3.136}$$

yields

$$\begin{aligned}x_a(t, a) &= g(x, \dot{x})k(t, \tau, a)\Big|_{\tau=a} - \int_a^{T_f} g(x, \dot{x})k_a(t, \tau, a)\, d\tau \\ &\quad - \int_a^{T_f} g(x_a, \dot{x}_a)k(t, \tau, a)\, d\tau - a_1 x(\tau, a)\Big|_{\tau=a} + \int_a^t a_1 x_a\, d\tau \\ &= -a_1 + \int_a^{T_f} g(x, \dot{x})\, d\tau - \int_a^{T_f} g(x_a, \dot{x}_a)k(t, \tau, a)\, d\tau + \int_a^t a_1 x_a\, d\tau.\end{aligned} \tag{3.137}$$

Comparing the "forcing" terms of the integral equations (3.134) and (3.137), it is seen that[†]

$$x_a(t, a) = -S(a)x(t, a), \qquad 0 \le a \le t, \tag{3.138}$$

where

$$S(a) = a_1 - \int_a^{T_f} g(x, \dot{x})\, d\tau \tag{3.139}$$

$$a_1 - \int_a^{T_f} (-a_1\dot{x} + x + a_1^2 x)\, d\tau. \tag{3.140}$$

Differentiating equation (3.140) with respect to a yields the Riccati differential equation

$$\begin{aligned} S_a(a) &= g(x, \dot{x})\Big|_{\tau=a} - \int_a^{T_f} g(x_a, \dot{x}_a)\, d\tau \\ &= -a_1\dot{x}(a, a) + x(a, a) + a_1^2 x(a, a) \\ &\quad - \int_a^{T_f} (-a_1\dot{x}_a + x_a + a_1^2 x_a)\, d\tau \\ &= -a_1 S(a) + 1 + a_1^2 + S(a)\int_0^{T_f} (-a_1\dot{x} + x + a_1^2 x)\, d\tau \\ &= -a_1 S(a) + 1 + a_1^2 + S(a)[a_1 - S(a)] \\ &= 1 + a_1^2 - S^2(a), \end{aligned} \tag{3.141}$$

where use has been made of equations (3.123), (3.138), (3.140), and

$$\dot{x}(a, a) = S(a). \tag{3.142}$$

The initial condition for the Riccati equation is obtained by setting $a = T_f$ in equation (3.140):

$$S(T_f) = a_1. \tag{3.143}$$

Equations (3.141) and (3.143) and the equation

$$x_a(t, a) = -S(a)x(t, a), \qquad 0 \le a \le t \tag{3.144}$$

[†] This is due to the system linearity. Consider the two linear functional equations

$$Ax = b \quad \text{and} \quad Ay = 2b$$

where b is the forcing function. Then

$$y = A^{-1}(2b) = 2A^{-1}b = 2x.$$

with initial condition

$$x(t, t) = 1 \tag{3.145}$$

form a set of initial-value ordinary differential equations. Equation (3.141) is integrated from $a = T_f$ to $a = t$. Equation (3.144) is then adjoined at $a = t$ and both differential equations are integrated from $a = t$ to $a = 0$. Since the optimal trajectory is generally desired at several values of t, (i.e., $t = t_1, t_2, \ldots, t_N$), as the variable a passes each such value of t, another equation of the form (3.144) is adjoined.

3.2.5. Analytical Solution

The analytical solution is easily derived from the Euler–Lagrange equations and boundary conditions. Expressing equations (3.78) and (3.79) in matrix form yields

$$\begin{bmatrix} \dot{x}_1 \\ \dot{x}_2 \end{bmatrix} = \begin{bmatrix} 0 & 1 \\ 1 + a_1^2 & 0 \end{bmatrix} \begin{bmatrix} x_1 \\ x_2 \end{bmatrix}, \qquad \begin{aligned} x_1(0) &= 1, & (3.146) \\ x_2(T_f) - a x_1(T_f) &= 0, & (3.147) \end{aligned}$$

$$\dot{\mathbf{x}} = F\mathbf{x}. \tag{3.148}$$

The characteristic equation of the F matrix is

$$| mI - F | = \begin{vmatrix} m & -1 \\ -(1 + a_1^2) & m \end{vmatrix} = 0, \tag{3.149}$$

$$m^2 - (1 + a_1^2) = 0. \tag{3.150}$$

The eigenvalues are roots of the characteristic equation

$$m = \pm(1 + a_1^2)^{1/2}. \tag{3.151}$$

Assume that the solution of Equation (3.75) is given by

$$x(t) = k_1 e^{-\alpha t} + k_2 e^{\alpha t}, \tag{3.152}$$

where

$$\alpha = (1 + a_1^2)^{1/2}. \tag{3.153}$$

In order to satisfy the boundary conditions, we set

$$x(0) = 1 = k_1 + k_2, \tag{3.154}$$

$$\dot{x}(T_f) - a_1 x(T_f) = -\alpha k_1 e^{-\alpha T_f} + \alpha k_2 e^{\alpha T_f} - a_1(k_1 e^{-\alpha T_f} + k_2 e^{\alpha T_f}). \tag{3.155}$$

Table 3.1. $x(t)$ as a Function of Time[a]

T_f	Time T	x: Euler–Lagrange equations	x: Pontryagin's maximum principle	x: Dynamic programming	x: Invariant imbedding	x: Exact solution
1	1	0.28197	0.28197	0.28197	0.28197	0.28197
	0.9	0.313085	0.313085	0.313085	0.313085	0.313085
	0.8	0.350473	0.350473	0.350473	0.350473	0.350473
	0.7	0.394881	0.394881	0.394881	0.394881	0.394881
	0.6	0.447201	0.447201	0.447201	0.447201	0.447201
	0.5	0.508479	0.508479	0.508479	0.508479	0.508479
	0.4	0.579944	0.579944	0.579944	0.579944	0.579944
	0.3	0.663027	0.663027	0.663027	0.663027	0.663027
	0.2	0.759393	0.759393	0.759393	0.759393	0.759393
	0.1	0.870972	0.870972	0.870972	0.870972	0.870972
	0	1	1	1	1	1
5	5	9.96590 E−4	9.94921 E−4	1.00227 E−3	9.95046 E−4	9.95047 E−4
	4.5	1.79434 E−3	1.79398 E−3	1.79794 E−3	1.79438 E−3	1.79438 E−3
	4	3.52907 E−3	3.52907 E−3	3.53067 E−3	3.52891 E−3	3.52892 E−3

	3.5	7.10249 E−3	7.10267 E−3	7.10353 E−3	7.10266 E−3	7.10267 E−3
	3	1.43783 E−2	1.43782 E−2	1.43786 E−2	1.43782 E−2	1.43782 E−2
	2.5	2.91474 E−2	2.91475 E−2	2.91476 E−2	2.91474 E−2	2.91474 E−2
	2	5.91078 E−2	5.91078 E−2	5.91079 E−2	5.91078 E−2	5.91078 E−2
	1.5	0.119874	0.119874	0.119874	0.119874	0.119874
	1	0.243117	0.243117	0.243117	0.243117	0.243117
	0.5	0.493069	0.493069	0.493069	0.493069	0.493069
	0	1	1	1	1	1
10	10	−2.92969 E−3	−1.22070 E−3	0.012099	8.45118 E−7	8.45119 E−7
	9	−2.44141 E−4	0	2.94427 E−3	2.99720 E−6	2.99720 E−6
	8	6.10352 E−5	−4.57764 E−5	7.27284 E−4	1.22118 E−5	1.22118 E−5
	7	3.05176 E−5	4.95911 E−5	2.24048 E−4	5.02018 E−5	5.02018 E−5
	6	1.86920 E−4	2.13623 E−4	2.48751 E−4	2.06486 E−4	2.06486 E−4
	5	8.49724 E−4	8.51393 E−4	8.59601 E−4	8.49325 E−4	8.49326 E−4
	4	3.49307 E−3	3.49414 E−3	3.49599 E−3	3.49349 E−3	3.49349 E−3
	3	1.43695 E−2	1.43698 E−2	1.43702 E−2	1.43696 E−2	1.43696 E−2
	2	5.91057 E−2	5.91058 E−2	5.91059 E−2	5.91057 E−2	5.91057 E−2
	1	0.243117	0.243117	0.243117	0.243117	0.243117
	0	1	1	1	1	1

[a] Numbers following E denote powers of ten. Underlined digits are those not agreeing with the exact solution.

Solving equations (3.154) and (3.155) simultaneously for k_1 and k_2 and substituting the resulting expressions into equation (3.152), we obtain the analytical solution:

$$x(t) = \frac{1}{e^{-\alpha T_f} + \left(\dfrac{\alpha - a_1}{\alpha + a_1}\right) e^{\alpha T_f}} \left[\left(\frac{\alpha - a_1}{\alpha + a_1}\right) e^{-\alpha(t - T_f)} + e^{\alpha(t - T_f)}\right]. \tag{3.156}$$

3.2.6. Numerical Results

The numerical results are shown tabulated for $a_1 = -1$ and for three different values of the terminal time T_f (Table 3.1). A fourth-order Runge–Kutta integration method with grid intervals of 1/100 sec was utilized for both boundary-value and initial-value solutions. The exact solution was obtained from the explicit analytic solution.

The results show that for $T_f = 1$ sec, all four methods yield six-digit accuracy or better. For $T_f = 5$ sec, five-digit accuracy or better is obtained by utilizing the invariant imbedding approach; however, as t approaches T_f, only two-digit accuracy is obtained by utilizing the Euler–Lagrange equation approach or the Pontryagin maximum principle approach. Utilizing the dynamic programming approach, none of the digits agree with the exact solution at time $t = T_f$. The digits which do not agree with the exact solution are shown underlined in the table. For $T_f = 10$ sec, five-digit accuracy or better is still obtained at all values of time by utilizing the invariant imbedding approach, whereas the other three methods give poor results at values of time near T_f.

The reason for the poor results utilizing the Euler–Lagrange equation approach or the Pontryagin maximum principle approach is that, as is frequently the case, the boundary-value equations obtained by these methods are numerically unstable. The dynamic programming results can be improved if enough values of the solution of the Riccati equation are stored. The initial-value equations obtained by the invariant imbedding approach are stable and do not require storage.

3.3. An Optimal Control Problem for a Second-Order System

Consider the optimal control problem (Reference 13) of minimizing the integral

$$I = \frac{1}{2} \int_0^{T_f} (x_1^2 + u^2)\, dt, \tag{3.157}$$

where

$$\dot{x}_1 = x_2, \qquad x_1(0) = 1, \tag{3.158}$$

$$\dot{x}_2 = u, \qquad x_2(0) = 0. \tag{3.159}$$

Comparing these equations with equations (3.1) and (3.2) it is obvious that

$$Q = \begin{bmatrix} 1 & 0 \\ 0 & 0 \end{bmatrix}, \qquad R = 1, \tag{3.160}$$

$$A = \begin{bmatrix} 0 & 1 \\ 0 & 0 \end{bmatrix}, \qquad B = \begin{bmatrix} 0 \\ 1 \end{bmatrix}. \tag{3.161}$$

The two-point boundary-value equation (3.9) then becomes

$$\begin{bmatrix} \dot{x}_1 \\ \dot{x}_2 \\ \hline \dot{\lambda}_1 \\ \dot{\lambda}_2 \end{bmatrix} = \left[\begin{array}{cc|cc} 0 & 1 & 0 & 0 \\ 0 & 0 & 0 & 1 \\ \hline 1 & 0 & 0 & 0 \\ 0 & 0 & -1 & 0 \end{array}\right] \begin{bmatrix} x_1 \\ x_2 \\ \hline \lambda_1 \\ \lambda_2 \end{bmatrix} \tag{3.162}$$

with the boundary conditions

$$\mathbf{x}(0) = \begin{bmatrix} 1 \\ 0 \end{bmatrix}, \qquad \boldsymbol{\lambda}(T_f) = \begin{bmatrix} 0 \\ 0 \end{bmatrix}. \tag{3.163}$$

3.3.1. Numerical Methods

The two-point boundary-value problem for the second-order system is evaluated numerically using the three methods; the matrix Riccati equation, the method of complementary functions, and invariant imbedding. Once the matrices A, B, Q, and R are defined, the matrix Riccati equation (3.15) can be integrated backward from time T_f to 0 and the optimal trajectory obtained by integrating equation (3.17) forward.

The two-component state vector for the second-order system yields a 4×4 transition matrix H in the method of complementary functions. The equations for a 4×4 transition matrix were given in expanded form in Section 3.1.2. Comparing the coefficients in equation (3.28) with the coefficients in equation (3.162), it is obvious that

$$\begin{aligned} &b_1 = d_2 = a_3 = 1, \\ &c_4 = -1, \\ &\text{all the rest of the coefficients} = 0. \end{aligned} \tag{3.164}$$

The solution is then obtained as described in Section 3.1.2.

Using the invariant imbedding method, equations (3.34) and (3.35) can be expressed in terms of the parameters given in equation (3.162). Then we have

$$\begin{bmatrix} \dot{x}_1 \\ \dot{x}_2 \end{bmatrix} = \begin{bmatrix} 0 & 1 \\ 0 & 0 \end{bmatrix} \begin{bmatrix} x_1 \\ x_2 \end{bmatrix} + \begin{bmatrix} 0 & 0 \\ 0 & 1 \end{bmatrix} \begin{bmatrix} \lambda_1 \\ \lambda_2 \end{bmatrix}$$

$$= A\mathbf{x} + B\boldsymbol{\lambda} \tag{3.165}$$

and

$$\begin{bmatrix} \dot{\lambda}_1 \\ \dot{\lambda}_2 \end{bmatrix} = \begin{bmatrix} 1 & 0 \\ 0 & 0 \end{bmatrix} \begin{bmatrix} x_1 \\ x_2 \end{bmatrix} + \begin{bmatrix} 0 & 0 \\ -1 & 0 \end{bmatrix} \begin{bmatrix} \lambda_1 \\ \lambda_2 \end{bmatrix}$$

$$= D\mathbf{x} + E\boldsymbol{\lambda}, \tag{3.166}$$

where $\mathbf{x}$ and $\boldsymbol{\lambda}$ are two-dimensional column vectors and A, B, D, and E are 2×2 matrices. The solution is then obtained as described in Section 3.1.3.

3.3.2. Analytical Solution

The two-point boundary-value problem given by equations (3.162) and (3.163) can be evaluated analytically as follows. Expressing equation (3.162) in the form

$$\dot{\mathbf{x}} = F\mathbf{x}, \tag{3.167}$$

where

$$\mathbf{x} = \begin{bmatrix} x_1 \\ x_2 \\ \lambda_1 \\ \lambda_2 \end{bmatrix}, \tag{3.168}$$

the eigenvalues can be calculated from the determinant

$$| mI - F | = \begin{vmatrix} m & -1 & 0 & 0 \\ 0 & m & 0 & -1 \\ -1 & 0 & m & 0 \\ 0 & 0 & 1 & m \end{vmatrix}. \tag{3.169}$$

The eigenvalues are

$$m_1 = a_1 + jb_1 = \frac{1+j}{(2)^{1/2}}, \qquad m_3 = a_2 + jb_2 = \frac{-1+j}{(2)^{1/2}}, \tag{3.170}$$

$$m_2 = a_1 - jb_1 = \frac{1-j}{(2)^{1/2}}, \qquad m_4 = a_2 - jb_2 = \frac{-1-j}{(2)^{1/2}}. \tag{3.171}$$

The solution is then of the form

$$
\begin{aligned}
x_1(t) &= C_1e^{a_1t}\cos b_1t + C_2e^{a_1t}\sin b_1t + C_3e^{a_2t}\cos b_2t \\
&\quad + C_4e^{a_2t}\sin b_2t, \\
x_2(t) &= \dot{x}_1 = C_1e^{a_1t}(a_1\cos b_1t - b_1\sin b_1t) \\
&\quad + C_2e^{a_1t}(a_1\sin b_1t + b_1\cos b_1t) \\
&\quad + C_3e^{a_2t}(a_2\cos b_2t - b_2\sin b_2t) \\
&\quad + C_4e^{a_2t}(a_2\sin b_2t + b_2\cos b_2t), \\
\lambda_1(t) &= -\dot{\lambda}_2 = 2a_1b_1C_1e^{a_1t}(a_1\sin b_1t + b_1\cos b_1t) \\
&\quad -2a_1b_1C_2e^{a_1t}(a_1\cos b_1t - b_1\sin b_1t) \\
&\quad +2a_2b_2C_3e^{a_2t}(a_2\sin b_2t + b_2\cos b_2t) \\
&\quad -2a_2b_2C_4e^{a_2t}(a_2\cos b_2t - b_2\sin b_2t), \\
\lambda_2(t) &= \dot{x}_2 = -2a_1b_1C_1e^{a_1t}\sin b_1t + 2a_1b_1C_2e^{a_1t}\cos b_1t \\
&\quad -2a_2b_2C_3e^{a_2t}\sin b_2t + 2a_2b_2C_4e^{a_2t}\cos b_2t.
\end{aligned}
\tag{3.172}
$$

Utilizing the boundary conditions at $t = 0$ and $t = T_f$, four linear algebraic equations in terms of C_1, C_2, C_3, and C_4 are obtained:

$$
\begin{aligned}
x_1(0) &= C_1 + (0)C_2 + C_3 + (0)C_4, \\
x_2(0) &= a_1C_1 + b_1C_2 + a_2C_3 + b_2C_4, \\
\lambda_1(T_f) &= 2a_1b_1C_1e^{a_1T_f}(a_1\sin b_1T_f + b_1\cos b_1T_f) \\
&\quad -2a_1b_1C_2e^{a_1T_f}(a_1\cos b_1T_f - b_1\sin b_1T_f) \\
&\quad +2a_2b_2C_3e^{a_2T_f}(a_2\sin b_2T_f + b_2\cos b_2T_f) \\
&\quad -2a_2b_2C_4e^{a_2T_f}(a_2\cos b_2T_f - b_2\sin b_2T_f), \\
\lambda_2(T_f) &= -2a_1b_1C_1e^{a_1T_f}\sin b_1T_f + 2a_1b_1C_2e^{a_1T_f}\cos b_1T_f \\
&\quad -2a_2b_2C_3e^{a_2T_f}\sin b_2T_f + 2a_2b_2C_4e^{a_2T_f}\cos b_2T_f,
\end{aligned}
\tag{3.173}
$$

where

$$
\begin{aligned}
x_1(0) &= 1, \\
x_2(0) &= \lambda_1(T_f) = \lambda_2(T_f) = 0.
\end{aligned}
\tag{3.174}
$$

The optimal trajectory is obtained by solving linear algebraic equations (3.173) for C_1, C_2, C_3, and C_4. The solution is then given by the equation

$$
\begin{aligned}
x_1(t) &= C_1e^{a_1t}\cos b_1t + C_2e^{a_1t}\sin b_1t + C_3e^{a_2t}\cos b_2t \\
&\quad + C_4e^{a_2t}\sin b_2t.
\end{aligned}
\tag{3.175}
$$

3.3.3. Numerical Results and Discussion

The numerical results are shown tabulated in Table 3.2 for two different values of the terminal time T_f. A fourth-order Runge–Kutta method with grid intervals of 1/100 sec was utilized for numerical integration.

The results show that for $T_f = 1$ sec, all four methods yield six-digit accuracy or better. For $T_f = 5$ sec, some degradation in the numerical accuracy of the matrix Riccati equation and complementary function methods is observed, whereas the invariant imbedding and the analytical solution methods still yield six-digit accuracy or better. The digits which do not agree with the exact solution are shown underlined in the table. The exact solution is assumed to be given by the invariant imbedding and the analytical solution methods, since the two widely different approaches yield the same results, digit by digit. Furthermore, it is well known that the numerical solution of two-point boundary-value problems yields grossly inaccurate results for long terminal times for all except the invariant imbedding approach.

The table also shows that the analytical solution takes the least computer computation time followed in order of increasing computer time by the complementary function, the matrix Riccati equation, and the invariant imbedding methods.

The analytical solution was obtained on the digital computer by solving the linear algebraic equations (3.173) for the constant coefficients of the solution of the homogeneous equations. Since the eigenvalues were known, the optimal trajectory, $x_1(t)$, was easily obtained for any value of t, $0 \leq t \leq T_f$, from equation (3.175). This method required the least computer time since numerical integration was not required. In general, however, the determination of eigenvalues for large systems of linear differential equations with constant coefficients is not trivial. Even when the eigenvalues can be accurately determined, the solving of linear algebraic equations for the constant multipliers of the solution of the homogeneous equations may be difficult because of ill conditioning (Reference 4). Ill conditioning occurs in the given example when some of the coefficients of C_1, C_2, C_3, and C_4 in equations (3.173) become excessively large in magnitude as the terminal time, T_f, increases.

The matrix Riccati equation method required only the integration of a 2×2 matrix equation, $\dot{P}$ [equation (3.15)], from $t = T_f$ to $t = 0$ and the integration of $\dot{P}$ and the two-dimensional vector, $\dot{\mathbf{x}}$ [equation (3.17)], from $t = 0$ to $t = T_f$. However, the method yields inaccurate results for long terminal times, since the integration of $\dot{P}$ from 0 to T_f is not completely

Table 3.2. Optimal Trajectories $x = x_1(t)^a$

	Time T	Matrix Riccati equation	Complementary functions	Invariant imbedding	Analytical solution
		x	x	x	x
$T_f = 1$ sec	1	0.884476	0.884476	0.884476	0.884476
	0.8	0.915144	0.915144	0.915144	0.915144
	0.6	0.944974	0.944974	0.944974	0.944974
	0.4	0.971782	0.971782	0.971782	0.971782
	0.2	0.991873	0.991873	0.991873	0.991873
	0	1	1	1	1
Computer computation time (sec)		29	21	37	2
		x	x	x	x
$T_f = 5$ sec	5	−0.107244	−0.107245	−0.107242	−0.107242
	4	−5.89635 E−2	−5.89633 E−2	−5.89624 E−2	−5.89624 E−2
	3	4.02167 E−2	4.02169 E−2	4.02171 E−2	4.02171 E−2
	2	0.282831	0.282831	0.282831	0.282831
	1	0.697249	0.697249	0.697249	0.697249
	0	1	1	1	1
Computer computation time (sec)		136	88	184	2

[a] Numbers following E denote powers of ten. Underlined digits are those not agreeing with the exact solution.

identical to the integration of $\dot{P}$ from T_f to 0. Accuracy can be improved by storing the values of P as $\dot{P}$ is integrated from T_f to 0 and then utilizing these stored values as $\dot{\mathbf{x}}$ is integrated from 0 to T_f. This method is undesirable, however, when the storage capacity of the computer is limited.

The method of complementary functions requires the integration of matrix equation (3.28) from $t = 0$ to $t = T_f$ and the numerical solution of the system of linear algebraic equations (3.32). This method also yields inaccurate results for long terminal times because of ill conditioning.

The invariant imbedding method requires the integration of matrix equation (3.55) from $a = T_f$ to $a = 0$ with the adjoining of equations of the form of equation (3.52) as the variable a passes each value of t where a solution printout is desired. This method does not yield inaccurate results for long terminal times.

Exercises

1. Given the linear time-varying system,

$$\dot{\mathbf{x}} = A(t)x, \qquad \mathbf{x}(0) = \mathbf{x}_0,$$

where $\mathbf{x}$ is an n-component state vector. Let

$$\mathbf{x}(t) = H(t)\mathbf{z}(t).$$

Show that the $n \times n$ transition matrix $H(t)$ satisfies the equation

$$\mathbf{x}(t) = H(t)\mathbf{x}(0),$$

where

$$\dot{H}(t) = A(t)H(t), \qquad H(0) = I.$$

2. Derive the dynamic programming equations in Section 3.2.3 using the initial value a as an imbedding parameter instead of T_f.

3. Show that equation (3.142),

$$\dot{x}(a, a) = S(a),$$

is correct.

4. Derive the equations obtained for the first-order system example using Kalaba's initial-value method in Section 3.2.4 by the method of invariant imbedding.

5. Derive the two-point boundary-value equations for the example given in Section 3.3 without the use of matrix notation.

6. Given the system

$$\ddot{x} + b\dot{x} + ax = y(t),$$

where

$$a = 4, \qquad b = 2,$$

$$x(0) = 1, \qquad x(0) = 0,$$

find the control, $y(t)$, such that

$$I = \int_0^T (x^2 + y^2)\, dt$$

is minimized. Derive the two-point boundary-value equations using (a) Euler–Lagrange equations, (b) Pontryagin's maximum principle.

4

Numerical Solutions for Nonlinear Two-Point Boundary-Value Problems

Optimal control problems for linear systems with quadratic performance criteria involve the solution of linear two-point boundary-value problems. The numerical solutions of these problems were discussed in the previous chapter. For nonlinear optimal control problems, iterative methods must be used to obtain the numerical solutions of the nonlinear two-point boundary-value problems. Two iterative methods for the numerical solutions are discussed in this chapter, (i) the method of quasilinearization and (ii) the Newton–Raphson method. These methods can also be used for linear or nonlinear optimal control problems subject to integral constraints.

Nonlinear two-point boundary-value problems also occur in problems involving the determination of the buckling load of columns and post buckling beam configurations. Numerical solutions for these problems are discussed using integral equation and imbedding methods.

4.1. Numerical Solution Methods

Consider the problem of extremizing the cost functional

$$J = \int_0^T F_1(\mathbf{x}, \mathbf{y}, t)\, dt \tag{4.1}$$

subject to the nonlinear differential constraint and initial condition

$$\dot{\mathbf{x}} = \mathbf{f}(\mathbf{x}, \mathbf{y}, t), \qquad \mathbf{x}(0) = \mathbf{b}, \tag{4.2}$$

where $\mathbf{x}$ is the n-dimensional state vector, $\mathbf{y}$ is an r-dimensional control vector, and $\mathbf{f}$ is an n-dimensional constraint vector. The methods of Chapter 2, such as the Euler–Lagrange equations or Pontryagin's maximum principle, yield a nonlinear two-point boundary-value problem. For example the Euler–Lagrange equations are

$$\frac{\partial F}{\partial \mathbf{x}} - \frac{d}{dt}\frac{\partial F}{\partial \dot{\mathbf{x}}} = 0, \tag{4.3}$$

$$\frac{\partial F}{\partial \mathbf{y}} - \frac{d}{dt}\frac{\partial F}{\partial \dot{\mathbf{y}}} = 0 \tag{4.4}$$

with boundary conditions

$$\left.\frac{\partial F}{\partial \dot{\mathbf{x}}}\right|_{t=T} = 0, \qquad \left.\frac{\partial F}{\partial \dot{\mathbf{y}}}\right|_{t=T} = 0, \tag{4.5}$$

where

$$F = F_1(\mathbf{x}, \mathbf{y}, t) + \boldsymbol{\lambda}^T(t)[\mathbf{f}(\mathbf{x}, \mathbf{y}, t) - \dot{\mathbf{x}}]. \tag{4.6}$$

Equations (4.2)–(4.5) yield the nonlinear two-point boundary-value problem and associated boundary conditions:

$$\begin{bmatrix} \dot{\mathbf{x}} \\ \dot{\boldsymbol{\lambda}} \end{bmatrix} = \begin{bmatrix} \mathbf{f}(\mathbf{x}, \boldsymbol{\lambda}, t) \\ \mathbf{g}(\mathbf{x}, \boldsymbol{\lambda}, t) \end{bmatrix}, \qquad \begin{bmatrix} \mathbf{x}(0) \\ \boldsymbol{\lambda}(T) \end{bmatrix} = \begin{bmatrix} \mathbf{b} \\ \mathbf{0} \end{bmatrix} \tag{4.7}$$

Equation (4.7) can be expressed in the form

$$\dot{\mathbf{x}} = \mathbf{f}(\mathbf{x}, t), \tag{4.8}$$

$$x_i(0) = b_i, \qquad i = 1, 2, \ldots, n, \tag{4.9}$$

$$x_j(T) = \lambda_i(T) = 0, \qquad j = n + 1, \ldots, 2n, \tag{4.10}$$

where $\mathbf{x}$ and $\mathbf{f}$ have been redefined (with increased dimensions) to include the Lagrange multipliers, i.e.,

$$\mathbf{x} = (x_1, x_2, \ldots, x_n, \lambda_1, \lambda_2, \ldots, \lambda_n)^T, \tag{4.11}$$

$$\mathbf{f} = (f_1, f_2, \ldots, f_{2n})^T. \tag{4.12}$$

The solution of these nonlinear equations is the subject of this chapter. The solution is obtained using the method of quasilinearization (Reference 1) or the Newton–Raphson method (Reference 2).

4.1.1. Quasilinearization

Using the method of quasilinearization, differential equation (4.8) is first linearized about the kth approximation by expanding the function $\mathbf{f}$ in a Taylor series and retaining only the linear terms. The $(k+1)$ approximation is

$$\dot{\mathbf{x}}_{k+1} = \mathbf{f}(\mathbf{x}_k, t) + J_k(\mathbf{x}_{k+1} - \mathbf{x}_k), \tag{4.13}$$

where the Jacobian matrix, J_k, is defined by

$$J_k = J_k[\mathbf{f}(\mathbf{x}_k, t)] \tag{4.14}$$

with the ij element

$$J_k|_{ij} = \left.\frac{\partial f_i(x)}{\partial x_j}\right|_{\mathbf{x}=\mathbf{x}_k(t)}. \tag{4.15}$$

Equation (4.13) can be expressed in the form

$$\dot{\mathbf{x}}_{k+1} = A_k(t)\mathbf{x}_{k+1} + \mathbf{u}_k(t), \tag{4.16}$$

where

$$A_k(t) = J_k, \tag{4.17}$$

$$\mathbf{u}_k(t) = \mathbf{f}(\mathbf{x}_k, t) - J_k\mathbf{x}_k. \tag{4.18}$$

The numerical solution to linear equation (4.16) is obtained using the method of complementary functions. The solution is given by the linear combination

$$\mathbf{x}_{k+1}(t) = \mathbf{p}_{k+1}(t) + H_{k+1}(t)\mathbf{c}_A, \tag{4.19}$$

where the $2n \times 1$ vector, $\mathbf{p}_{k+1}$, and the $2n \times 2n$ matrix, H_{k+1}, are the solutions of the initial-value equations

$$\dot{\mathbf{p}}_{k+1}(t) = A_k(t)\mathbf{p}_{k+1}(t) + \mathbf{u}_k(t), \qquad \mathbf{p}_{k+1}(0) = 0, \tag{4.20}$$

$$\dot{H}_{k+1}(t) = A_k(t)H_{k+1}(t), \qquad H_{k+1}(0) = I. \tag{4.21}$$

The vector of constants

$$\mathbf{c}_A = (\gamma_1, \gamma_2, \ldots, \gamma_{2n})^T \tag{4.22}$$

is obtained by substituting the boundary conditions into equation (4.19) and solving for $\mathbf{c}_A$.

To obtain a numerical solution, an initial guess is made for the vector $\mathbf{x}(t)$, i.e., assume $\mathbf{x}(t) = \mathbf{x}_0(t)$. Equations (4.20) and (4.21) are integrated from time $t = 0$ to $t = T_f$. The values of $p(t)$ and $H(t)$ at the boundaries are utilized to evaluate the vector $\mathbf{c}_A$ from equation (4.19). The full set of initial conditions of the first approximation, $\mathbf{x}_1(t)$, is equal to $\mathbf{c}_A$. Equation (4.16) is then integrated from time $t = 0$ to $t = T_f$ to obtain $\mathbf{x}_1(t)$. The values of $\mathbf{x}_1(t)$ are stored and the above sequence is repeated to obtain the second approximation, etc. This method requires the integration of $4n(n + 1)$ differential equations for each approximation.

4.1.2. Newton–Raphson Method

Using the Newton–Raphson method, the unknown initial conditions of equations (4.7) are assumed to be given by the n-dimensional vector

$$\boldsymbol{\lambda}(0) = \mathbf{c}. \tag{4.23}$$

Expanding the boundary conditions

$$\boldsymbol{\lambda}(T, \mathbf{c}) = \mathbf{0} \tag{4.24}$$

at terminal time T in a Taylor series around the kth approximation and retaining only the linear terms, we obtain

$$\boldsymbol{\lambda}(T, \mathbf{c}_k) + \Lambda_c(\mathbf{c}_{k+1} - \mathbf{c}_k) = 0, \tag{4.25}$$

where Λ_c is an $n \times n$ matrix whose i, j element is

$$\Lambda_c|_{ij} = \frac{\partial \lambda_i(T, \mathbf{c}_k)}{\partial c_j}. \tag{4.26}$$

Solving for the $k + 1$ approximation gives

$$\mathbf{c}_{k+1} = \mathbf{c}_k - \Lambda_c^{-1}\boldsymbol{\lambda}(T, \mathbf{c}_k). \tag{4.27}$$

The equations for the components of matrix Λ_c are obtained by differentiating equation (4.7) with respect to the components of $\mathbf{c}$:

$$\begin{aligned} \dot{\mathbf{x}}_{c_1} &= \mathbf{f}_{c_1}(\mathbf{x}, \boldsymbol{\lambda}, t), & \dot{\boldsymbol{\lambda}}_{c_1} &= \mathbf{g}_{c_1}(\mathbf{x}, \boldsymbol{\lambda}, t), \\ \dot{\mathbf{x}}_{c_2} &= \mathbf{f}_{c_2}(\mathbf{x}, \boldsymbol{\lambda}, t), & \dot{\boldsymbol{\lambda}}_{c_2} &= \mathbf{g}_{c_2}(\mathbf{x}, \boldsymbol{\lambda}, t), \\ &\vdots \\ \dot{\mathbf{x}}_{c_n} &= \mathbf{f}_{c_n}(\mathbf{x}, \boldsymbol{\lambda}, t), & \dot{\boldsymbol{\lambda}}_{c_n} &= \mathbf{g}_{c_n}(\mathbf{x}, \boldsymbol{\lambda}, t). \end{aligned} \tag{4.28}$$

The initial conditions are

$$\begin{gathered} x_{ic_j}(0) = 0, \qquad \underset{i \neq j}{\lambda_{ic_j}(0)} = 0, \qquad \lambda_{ic_i}(0) = 1, \\ i = 1, 2, \ldots, n, \qquad j = 1, 2, \ldots, n. \end{gathered} \tag{4.29}$$

Equations (4.29) are obtained by differentiating equation (4.23) with respect to **c**.

To obtain a numerical solution, an initial guess is made for initial conditions (4.23). Initial-value equations (4.7) and (4.28) are then integrated from time $t = 0$ to $t = T$. A new value for **c** is calculated from equation (4.27) and the above sequence is repeated to obtain the second approximation, etc. This method requires the integration of $2n(n + 1)$ differential equations for each approximation.

4.2. Examples of Problems Yielding Nonlinear Two-Point Boundary-Value Problems

Nonlinear two-point boundary-value problems obviously occur in optimization problems where the system equations are nonlinear. They may also occur, however, for optimization problems with integral constraints. In this section examples are given for the optimization of a first-order nonlinear optimal control problem and for the optimization of functionals subject to integral or energy constraints. The numerical solutions are obtained via the method of quasilinearization or the Newton–Raphson method.

4.2.1. A First-Order Nonlinear Optimal Control Problem

Consider the simple nonlinear optimal control problem with cost functional

$$J = \frac{1}{2} \int_0^T (x^2 + y^2)\, dt \tag{4.30}$$

subject to the differential equation constraint and initial condition

$$\dot{x} = -x^2 + y, \qquad x(0) = 1. \tag{4.31}$$

Utilizing the Lagrange multipliers, the cost functional can be expressed in the form

$$J' = \int_0^T F\,dt, \tag{4.32}$$

where

$$F = \tfrac{1}{2}(x^2 + y^2) + \lambda(-x^2 + y - \dot{x}). \tag{4.33}$$

The Euler–Lagrange equations and associated transversality conditions are

$$\frac{\partial F}{\partial x} - \frac{d}{dt}\frac{\partial F}{\partial \dot{x}} = 0, \qquad \left.\frac{\partial F}{\partial \dot{x}}\right|_{t=T} = 0, \tag{4.34}$$

$$x - 2\lambda x + \dot{\lambda} = 0, \qquad \lambda(T) = 0, \tag{4.35}$$

and

$$\frac{\partial F}{\partial y} - \frac{d}{dt}\frac{\partial F}{\partial \dot{y}} = 0, \qquad \left.\frac{\partial F}{\partial \dot{y}}\right|_{t=T} = 0, \tag{4.36}$$

$$y + \lambda = 0, \qquad 0 = 0. \tag{4.37}$$

These equations augmented by equation (4.31) yield the nonlinear two-point boundary-value equations and associated boundary conditions

$$\dot{x} = -x^2 - \lambda, \qquad x(0) = 1, \tag{4.38}$$

$$\dot{\lambda} = -x + 2\lambda x, \qquad \lambda(T) = 0, \tag{4.39}$$

where

$$y = -\lambda. \tag{4.40}$$

The two-point boundary-value equations (4.38) and (4.39) are to be solved via quasilinearization and the Newton–Raphson method.

Quasilinearization

The Jacobian matrix for the given example is defined by the equation (4.14)

$$J_k = \begin{bmatrix} \dfrac{\partial f_1}{\partial x} & \dfrac{\partial f_1}{\partial \lambda} \\ \dfrac{\partial f_2}{\partial x} & \dfrac{\partial f_2}{\partial \lambda} \end{bmatrix}, \tag{4.41}$$

where f_1 and f_2 are defined at the kth approximation by the equations

(4.38) and (4.39):

$$f_1 = f_1(x_k, \lambda_k, t) = \dot{x}_k = -x_k^2 - \lambda_k, \tag{4.42}$$

$$f_2 = f_2(x_k, \lambda_k, t) = \dot{\lambda}_k = -x_k - 2\lambda_k x_k. \tag{4.43}$$

The quasilinear equations are then defined by equation (4.16). Alternatively, equations (4.42) and (4.43) can be expanded directly into a Taylor series:

$$\dot{x}_{k+1} = f_1 + \frac{\partial f_1}{\partial x}(x_{k+1} - x_k) + \frac{\partial f_1}{\partial \lambda}(\lambda_{k+1} - \lambda_k), \tag{4.44}$$

$$\dot{\lambda}_{k+1} = f_2 + \frac{\partial f_2}{\partial x}(x_{k+1} - x_k) + \frac{\partial f_2}{\partial \lambda}(\lambda_{k+1} - \lambda_k). \tag{4.45}$$

Performing the above operations we obtain

$$\begin{aligned} \dot{x}_{k+1} &= -x_k^2 - \lambda_k + (-2x_k)(x_{k+1} - x_k) + (-1)(\lambda_{k+1} - \lambda_k) \\ &= (-2x_k)x_{k+1} - \lambda_{k+1} + x_k^2 \end{aligned} \tag{4.46}$$

$$\begin{aligned} \dot{\lambda}_{k+1} &= -x_k + 2\lambda_k x_k + (-1 + 2\lambda_k)(x_{k+1} - x_k) + (2x_k)(\lambda_{k+1} - \lambda_k) \\ &= (-1 + 2\lambda_k)x_{k+1} + (2x_k)\lambda_{k+1} - 2\lambda_k x_k. \end{aligned} \tag{4.47}$$

The quasilinearized equations for boundary-value equations (4.38) and (4.39) are then

$$\begin{bmatrix} \dot{x}_{k+1} \\ \dot{\lambda}_{k+1} \end{bmatrix} = \begin{bmatrix} -2x_k & -1 \\ -1 + \lambda_k & 2x_k \end{bmatrix} \begin{bmatrix} x_{k+1} \\ \lambda_{k+1} \end{bmatrix} + \begin{bmatrix} x_k^2 \\ -2x_k\lambda_k \end{bmatrix} \tag{4.48}$$

with boundary conditions

$$\begin{bmatrix} x_{k+1}(0) \\ \lambda_{k+1}(T) \end{bmatrix} = \begin{bmatrix} 1 \\ 0 \end{bmatrix}. \tag{4.49}$$

Equation (4.48) can be expressed in the form

$$\dot{\mathbf{x}}_{k+1} = A_k(t)\mathbf{x}_{k+1} + \mathbf{u}_k(t). \tag{4.50}$$

The numerical solution is obtained utilizing the method of complementary functions,

$$\dot{\mathbf{p}}_{k+1} = A_k(t)\mathbf{p}_{k+1} + \mathbf{u}_k(t), \qquad \mathbf{p}_{k+1}(0) = \mathbf{0}, \tag{4.51}$$

$$\dot{H}_{k+1} = A_k(t)H_{k+1}, \qquad H_{k+1}(0) = I, \tag{4.52}$$

where

$$\mathbf{p}_{k+1} = \begin{bmatrix} p_1^{k+1} \\ p_2^{k+1} \end{bmatrix}, \tag{4.53}$$

$$H_{k+1} = \begin{bmatrix} h_{11}^{k+1} & h_{21}^{k+1} \\ h_{12}^{k+1} & h_{22}^{k+1} \end{bmatrix}. \tag{4.54}$$

The solution for each approximation is

$$\begin{bmatrix} x(t) \\ \lambda(t) \end{bmatrix} = \begin{bmatrix} p_1(t) \\ p_2(t) \end{bmatrix} + \gamma_1 \begin{bmatrix} h_{11} \\ h_{12} \end{bmatrix} + \gamma_2 \begin{bmatrix} h_{21} \\ h_{22} \end{bmatrix}, \tag{4.55}$$

where the superscripts, $k + 1$, have been dropped and γ_1 and γ_2 are obtained by the simultaneous solution of the equations

$$1 = p_1(0) + \gamma_1 h_{11}(0) + \gamma_2 h_{21}(0), \tag{4.56}$$

$$0 = p_2(T) + \gamma_1 h_{12}(T) + \gamma_2 h_{22}(T). \tag{4.57}$$

Newton–Raphson Method

The Newton–Raphson method is initiated by guessing the unknown boundary conditions at time $t = 0$. The boundary-value equations (4.38) and (4.39) then become

$$\dot{x} = -x^2 - \lambda, \qquad x(0) = 1, \tag{4.58}$$

$$\dot{\lambda} = -x + 2\lambda x, \qquad \lambda(0) = c, \tag{4.59}$$

where c is the initial guess. The Newton–Raphson equations are obtained by expanding the boundary condition of the scalar

$$\lambda(T, c) = 0 \tag{4.60}$$

in a Taylor series and truncating after the first-order terms

$$\lambda(T, c_k) + (c_{k+1} - c_k)\lambda_c(T, c_k) = 0, \tag{4.61}$$

where c_k is the current approximation of c and

$$\lambda_c = \frac{\partial \lambda}{\partial c}. \tag{4.62}$$

Solving for c_{k+1} gives

$$c_{k+1} = c_k - \frac{\lambda(T, c_k)}{\lambda_c(T, c_k)}. \tag{4.63}$$

Differentiating equations (4.58) and (4.59) with respect to c, we have

$$\dot{x}_c = -2xx_c - \lambda_c, \qquad x_c(0) = 0, \tag{4.64}$$

$$\dot{\lambda}_c = -x_c + 2\lambda x_c + 2\lambda_c x, \qquad \lambda_c(0) = 1. \tag{4.65}$$

The four initial-value equations (4.58), (4.59), (4.64), and (4.65) are integrated from time $t = 0$ to $t = T$ with the initial approximation $\lambda(0) = c$. The new value of c is the calculated from equation (4.63) and the initial-value equations (4.58), (4.59), (4.64), and (4.65) are integrated again, etc.

Numerical Results

Numerical results are shown in Table 4.1 for the quasilinearization method. A fourth-order Runge–Kutta method with grid intervals of 1/100

Table 4.1. Quasilinearization[a]

Approximation number	Time T	x	y
1	0	1	−8.62883 E−2
	0.02	0.978867	−6.96205 E−2
	0.04	0.958892	−0.052692
	0.06	0.940035	−3.54688 E−2
	0.08	0.922258	−1.79165 E−2
	0.1	0.905526	2.32831 E−10
2	0	1	−0.086557
	0.02	0.97886	−6.98666 E−2
	0.04	0.958862	−5.28703 E−2
	0.06	0.939953	−0.035564
	0.08	0.922084	−1.79425 E−2
	0.1	0.905212	2.32831 E−10
3	0	1	−8.65571 E−2
	0.02	0.97886	−6.98666 E−2
	0.04	0.958862	−5.28703 E−2
	0.06	0.939953	−0.035564
	0.08	0.922084	−1.79425 E−2
	0.1	0.905212	1.16415 E−10
4	0	1	−8.65571 E−2
	0.02	0.97886	−6.98666 E−2
	0.04	0.958862	−5.28703 E−2
	0.06	0.939953	−0.035564
	0.08	0.922084	−1.79425 E−2
	0.1	0.905212	2.32831 E−10

[a] Numbers following E denote powers of ten.

sec was utilized with terminal time $T = 0.1$ sec. The initial approximation was obtained by assuming that $x(t)$ and $\lambda(t)$ equal their known boundary conditions throughout the time interval 0 to T, i.e.,

$$x(t) = 1, \qquad \lambda(t) = 0, \qquad 0 \leq t \leq T. \tag{4.66}$$

Table 4.1 shows that convergence has been obtained within two approximations except for the control function, $y(t)$, at time $t = 0$, and there the change is only in the fifth and sixth digits. The control function appears to be changing at time $t = T$; however, $y(t)$ is approximately zero at that time, so that the change is insignificant. Similar results are obtained using the Newton–Raphson method.

4.2.2. Optimization of Functionals Subject to Integral Constraints

The classical method for optimizing a functional subject to an integral constraint is to introduce the Lagrange multiplier and apply the Euler–Lagrange equations to the augmented integrand. The Lagrange multiplier is a constant whose value is selected such that the integral constraint is satisfied. This value is frequently an eigenvalue of the boundary-value problem and is determined by a trial-and-error procedure. An improved method for solving this isoperimetric problem is presented in this section (Reference 3). The Lagrange multiplier is introduced as a state variable and evaluated simultaneously with the optimum solution. A numerical example is given and is shown to have a large region of convergence.

Introduction

Consider the optimization problem of extremizing a functional J subject to an integral constraint. This is called the isoperimetric problem. The classical method for solving this problem is to introduce a constant Lagrange multiplier q, and extremize the sum of J and q times the integral constraint (Reference 4). The magnitude of the Lagrange multiplier q must be selected such that the integral constraint is satisfied. Unfortunately, this is a trial-and-error procedure. The optimization problem is solved for several values of q and the value of q for which the integral constraint is satisfied is selected.

In this section an improved method for solving the isoperimetric problem is formulated. The method utilizes the constant Lagrange multiplier as a state variable with differential constraint $dq/dt = 0$. The Euler–Lagrange equations and associated boundary conditions form a nonlinear two-point boundary-value problem, the solution of which simultaneously yields the state variables which extremize J and the value of q for which the integral constraint is satisfied.

The contents of this section are as follows. First the classical method for extremizing a functional subject to an integral constraint is briefly reviewed. The improved method is discussed, followed by a simple example. The method of solution via the Newton–Raphson method is given, followed by an analytical solution and the numerical results and discussion.

The example shows that convergence is obtained for a wide range of initial estimates of the Lagrange multiplier, q. Thus the improved method should have broad application to a number of optimization problems, both linear and nonlinear.

Classical Method

Consider the problem of finding the function $x(t)$ that extremizes the integral

$$J = \int_0^T F(t, x, \dot{x})\, dt \tag{4.67}$$

subject to the integral constraint

$$E = \int_0^T G(t, x, \dot{x})\, dt \tag{4.68}$$

and with initial condition

$$x(0) = c. \tag{4.69}$$

For simplicity we shall consider only the case where the state variable, $x(t)$, is a scalar. The derivations which follow can easily be extended to the n-dimensional case, however. The optimization problem can be solved by the extremization of the integral

$$J_1 = \int_0^T \left[F + q\left(G - \frac{E}{T}\right)\right] dt \tag{4.70}$$

where q is the Lagrange multiplier and is equal to a constant.

The Euler–Lagrange equation is

$$\frac{\partial}{\partial x}(F + qG) - \frac{d}{dt}\frac{\partial}{\partial \dot{x}}(F + qG) = 0 \tag{4.71}$$

subject to boundary conditions

$$x(0) = c, \tag{4.72}$$

$$\frac{\partial}{\partial \dot{x}} (F + qG)\Big|_{t=T} = 0. \tag{4.73}$$

The magnitude of the constant Lagrange multiplier q must be selected such that the integral constraint is satisfied. This is a trial-and-error procedure.

An Improved Method

The solution of equation (4.70) is unchanged if expressed in the form

$$J_1 = \int_0^T (F + qG)\, dt - q(T)E. \tag{4.74}$$

Now the value of the Lagrange multiplier q for which integral constraint (4.68) is satisfied can be obtained along with the optimum $x(t)$ by adjoining the differential constraint

$$\dot{q}(t) = 0 \tag{4.75}$$

with unknown initial condition and introducing the Lagrange multiplier, $\lambda(t)$. The parameter q is treated as a state variable. Thus we have

$$J_2 = \int_0^T (F + qG + \lambda\dot{q})\, dt - q(T)E. \tag{4.76}$$

The Euler–Lagrange equations are

$$\frac{\partial}{\partial x} (F + qG) - \frac{d}{dt} \frac{\partial}{\partial \dot{x}} (F + qG) = 0, \tag{4.77}$$

$$\frac{\partial}{\partial q} (qG) - \frac{d}{dt} \frac{\partial}{\partial \dot{q}} (\lambda\dot{q}) = G - \dot{\lambda} = 0 \tag{4.78}$$

subject to the boundary conditions

$$x(0) = c, \tag{4.79}$$

$$\frac{\partial}{\partial \dot{x}} (F + qG)\Big|_{t=T} = 0, \tag{4.80}$$

$$\frac{\partial}{\partial \dot{q}} (\lambda\dot{q})\Big|_{t=0} = \lambda(0) = 0, \tag{4.81}$$

$$\frac{\partial}{\partial \dot{q}} (\lambda\dot{q})\Big|_{t=T} + \frac{d}{dt} (-qE)\Big|_{t=T} = \lambda(T) - E = 0. \tag{4.82}$$

The above approach has effectively adjoined the following equation to the classical Euler–Lagrange equations:

$$\dot{\lambda}(t) = G \tag{4.83}$$

with boundary conditions

$$\lambda(0) = 0, \qquad \lambda(T) = E. \tag{4.84}$$

From equations (4.83) and (4.84) we have

$$\lambda(T) = \int_0^T G\, dt = E. \tag{4.85}$$

Example. A simple example is utilized to illustrate the procedure for this approach. Assume it is desired to minimize the functional

$$J = \int_0^T \dot{x}^2\, dt \tag{4.86}$$

subject to the integral constraint

$$E = \int_0^T x^2\, dt, \tag{4.87}$$

where

$$x(0) = c. \tag{4.88}$$

Equation (4.76) becomes

$$J_2 = \int_0^T (\dot{x}^2 + qx^2 + \lambda\dot{q})\, dt - q(T)E. \tag{4.89}$$

The Euler–Lagrange equations are

$$\ddot{x} - qx = 0, \tag{4.90}$$

$$x^2 - \dot{\lambda} = 0, \tag{4.91}$$

$$\dot{q} = 0, \tag{4.92}$$

with boundary conditions

$$x(0) = c, \qquad \dot{x}(T) = 0, \tag{4.93}$$

$$\lambda(0) = 0, \qquad \lambda(T) = E. \tag{4.94}$$

These equations form a nonlinear two-point boundary-value problem whereas the classical method for this example would yield a linear two-point boundary-value problem for a given value of q.

Introduce the notation

$$p_1(t) = \dot{x}(t), \qquad p_2(t) = \lambda(t). \tag{4.95}$$

Then the Euler–Lagrange equations and boundary conditions become

$$\dot{x} = p_1, \qquad x(0) = c, \tag{4.96}$$

$$\dot{q} = 0, \qquad p_2(0) = 0, \tag{4.97}$$

$$\dot{p}_1 = qx, \qquad p_1(T) = 0, \tag{4.98}$$

$$\dot{p}_2 = x^2, \qquad p_2(T) = E. \tag{4.99}$$

The unknown initial conditions are $p_1(0)$ and $q(0)$.

Solution via Newton–Raphson Method

Using the Newton–Raphson method, the unknown initial conditions are assumed to be given by

$$p_1(0) = c_1, \tag{4.100}$$

$$q(0) = c_2. \tag{4.101}$$

Expanding the boundary conditions

$$p_1(T, c_1, c_2) = 0, \tag{4.102}$$

$$p_2(T, c_1, c_2) = E \tag{4.103}$$

in a Taylor series around the kth approximation and retaining only the linear terms, we obtain

$$p_1(T, c_1^k, c_2^k) + (c_1^{k+1} - c_1^k)p_{1c_1} + (c_2^{k+1} - c_2^k)p_{1c_2} = 0, \tag{4.104}$$

$$p_2(T, c_1^k, c_2^k) + (c_1^{k+1} - c_1^k)p_{2c_1} + (c_2^{k+1} - c_2^k)p_{2c_2} = E, \tag{4.105}$$

where

$$p_{ic_j} = \frac{\partial p_i}{\partial c_j}. \tag{4.106}$$

Then

$$\begin{bmatrix} c_1^{k+1} \\ c_2^{k+1} \end{bmatrix} = \begin{bmatrix} c_1^k \\ c_2^k \end{bmatrix} - \begin{bmatrix} p_{1c_1} & p_{1c_2} \\ p_{2c_1} & p_{2c_2} \end{bmatrix}^{-1} \begin{bmatrix} p_1(T, c_1^k, c_2^k) \\ p_2(T, c_1^k, c_2^k) - E \end{bmatrix} \tag{4.107}$$

or

$$\mathbf{c}^{k+1} = \mathbf{c}^k - P_c^{-1}\mathbf{p}(T, \mathbf{c}^k). \tag{4.108}$$

The equations for the components of matrix P_c are obtained by differentiating equations (4.96)–(4.99) with respect to c_1 and c_2:

$$\dot{x}_{c_1} = p_{1c_1}, \qquad x_{c_1}(0) = 0, \tag{4.109}$$

$$\dot{q}_{c_1} = 0, \qquad q_{c_1}(0) = 0, \tag{4.110}$$

$$\dot{p}_{1c_1} = q_{c_1}x + qx_{c_1}, \qquad p_{1c_1}(0) = 1, \tag{4.111}$$

$$\dot{p}_{2c_1} = 2xx_{c_1}, \qquad p_{2c_2}(0) = 0, \tag{4.112}$$

and

$$\dot{x}_{c_2} = p_{1c_2}, \qquad x_{c_2}(0) = 0, \tag{4.113}$$

$$\dot{q}_{c_2} = 0, \qquad q_{c_2}(0) = 1, \tag{4.114}$$

$$\dot{p}_{1c_2} = q_{c_2}x + qx_{c_2}, \qquad p_{1c_2}(0) = 0, \tag{4.115}$$

$$\dot{p}_{2c_2} = 2xx_{c_2}, \qquad p_{2c_2}(0) = 0. \tag{4.116}$$

To obtain a numerical solution, an initial guess is made for initial conditions (4.100) and (4.101). Initial-value equations (4.96)–(4.99) and (4.109)–(4.116) are then integrated from time $t = 0$ to $t = T$. A new value for $\mathbf{c}$ is calculated from equation (4.108) and the above sequence is repeated to obtain the second approximation, etc. Twelve differential equations are integrated for each approximation.

Analytical Solution (Reference 5)

The analytical solution for the given example is easily derived in this case using the classical approach. The function $x(t)$ that minimizes the integral J is

$$x(t) = c\,\frac{\cosh q^{1/2}(t - T)}{\cosh q^{1/2}T}. \tag{4.117}$$

The derivative, $\dot{x}(t)$, is

$$\dot{x}(t) = q^{1/2}c\,\frac{\sinh q^{1/2}(t - T)}{\cosh q^{1/2}T}. \tag{4.118}$$

The integral constraint

$$E = \int_0^T x^2\,dt \tag{4.119}$$

as a function of Lagrange multiplier q is given by

$$E'(q) = \frac{c^2}{\cosh^2 q^{1/2}T}\left(\frac{1}{4q^{1/2}}\sinh 2q^{1/2}T + \frac{T}{2}\right) \tag{4.120}$$

and the cost functional

$$J = \int_0^T \dot{x}^2 \, dt \tag{4.121}$$

as a function of q is given by

$$J'(q) = \frac{qc^2}{\cosh^2 q^{1/2}T} \left(\frac{1}{4q^{1/2}} \sinh 2q^{1/2}T - \frac{T}{2} \right). \tag{4.122}$$

The integral constraint, $E'(q)$, and the cost, $J'(q)$, are shown in Figures 4.1 and 4.2 as a function of q for initial condition

$$x(0) = c = 1 \tag{4.123}$$

and terminal time

$$T = 1. \tag{4.124}$$

The maximum value of $E'(q)$ is obtained when the Lagrange multiplier, q, is equal to zero:

$$E'(q)_{\max} = c^2T = 1, \qquad q = 0. \tag{4.125}$$

As q increases, less control is exerted, and $E'(q)$ decreases. The corresponding cost, $J'(q)$, increases.

For a given value of E less than $E'(q)_{\max}$, for example

$$E = 0.276332, \tag{4.126}$$

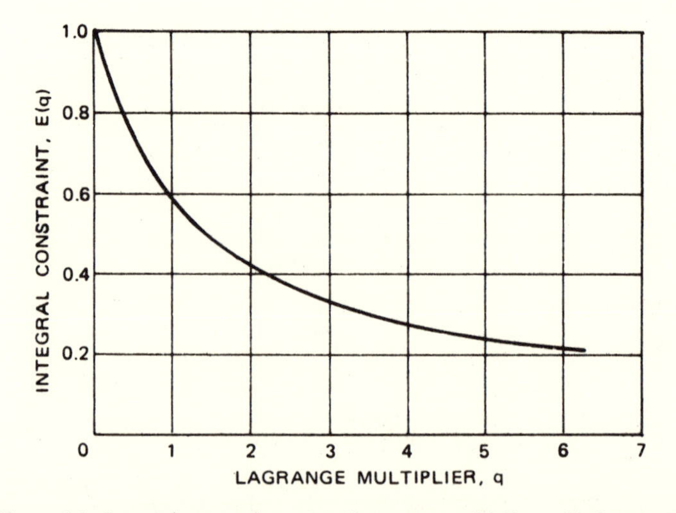

Figure 4.1. Integral constraint versus Lagrange multiplier q (Reference 3).

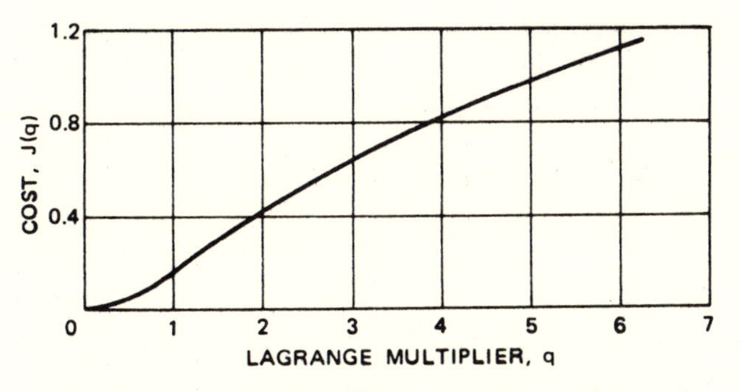

Figure 4.2. Cost versus Lagrange multiplier q (Reference 3).

there is a value of q for which equation (4.119) is satisfied. Figure 4.1 shows that the integral constraint is satisfied for $q = 4$. It is the purpose of the new approach to determine simultaneously the function $x(t)$ that minimizes cost equation (4.121) and the value of q that satisfies equation (4.119) for a given value of E.

Numerical Results and Discussion

Numerical results were obtained for the new approach using the Newton–Raphson method. A fourth-order Runge–Kutta method was utilized for integration with grid intervals of 1/100 sec. It was assumed that

$$T = 1 \quad \text{and} \quad x(0) = 1$$

with integral constraint

$$E = 0.276332.$$

To initialize the algorithm, it was assumed that

$$p_1(0) = c_1 = 0,$$
$$q(0) = c_2 = 5.$$

From the analytical solution, equations (4.118) and (4.120), it can be shown that

$$\left.\begin{aligned} E &= 0.276332 \\ \dot{x}(0) = p_1(0) &= -1.92806 \end{aligned}\right\} \quad \text{for } q(t) = \text{const} = 4$$

It is desired to determine the rate of convergence of the algorithm for this example and the accuracy of the converged solution.

Table 4.2. Convergence of q and $p_1(0)$ Using the New Approach. Initial $q(0) = 5$, $p_1(0) = 0$

Iteration number	q	$p_1(0)$
0	5	0
1	1.92019	0.0274530
2	1.2031	−0.629421
3	1.94918	−1.35058
4	3.54181	−1.85434
5	4.00459	−1.92776
6	3.99998	−1.92805
7	4	−1.92806

Table 4.2 shows the convergence of q and $p_1(0)$. After seven iterations the solution is exact to within six digits. The rate of convergence can be increased by using the classical Euler–Lagrange equations first to obtain $x(t)$ and $p_1(t)$ for a given value of q. The new approach is then used with the derived initial condition, $p_1(0)$, as shown in Table 4.3.

The region of convergence of the algorithm is also of interest. Often the selection of $\dot{x}(0)$ can be made on the basis of physical considerations, but no guideline is available for the selection of the Lagrange multiplier q. Table 4.4 shows the convergence of q and $p_1(0)$ with the algorithm initialized by

$$p_1(0) = -2, \qquad q(0) = 20.$$

Convergence is obtained in nine iterations. Table 4.5 shows the convergence

Table 4.3. Convergence of q and $p_1(0)$ Using the Classical Method Followed by the New Approach. Initial $q(0) = 5$, $p_1(0) = 0$

Iteration number	q	$p_1(0)$	Comments
0	5	0	Classical Euler–
1	5	−2.18556	Lagrange equations
0	5	−2.18556	New approach
1	3.79379	−1.895	
2	4.0009	−1.928	
3	4	−1.92805	

Table 4.4. Convergence of q and $p_1(0)$ Using the New Approach. Initial $q(0) = 20$, $p_1(0) = -2$

Iteration number	q	$p_1(0)$
0	20	−2
1	6.75286	0.67191
2	2.29085	1.08057
3	0.802321	0.299127
4	0.825438	−0.663266
5	2.10961	−1.46807
6	3.81964	−1.90974
7	3.99853	−1.92695
8	3.99999	−1.92805
9	4	−1.92806

of q and $p_1(0)$ with $p_1(0)$ reset to -2 at the end of the first two iterations. Convergence is obtained in five iterations in this case.

Table 4.6 shows the convergence of q and $p_1(0)$ with the algorithm initialized by

$$p_1(0) = -2, \qquad q(0) = 100$$

and $p_1(0)$ reset to -2 at the end of the first two iterations. Convergence is obtained in eight iterations. These results show that the initial choice of $q(0)$ is not critical.

The large region of convergence for the Lagrange multiplier q is consistent with the results in the literature. In Reference 6 rapid convergence was obtained in spite of the poor initial choice of Lagrange multiplier λ.

Table 4.5. Convergence of q and $p_1(0)$ Using the New Approach with $p_1(0)$ Kept Constant for Two Iterations. Initial $q(0) = 20$, $p_1(0) = -2$

Iteration number	q	$p_1(0)$
0	20	−2
1	6.75286	−2
2	3.67059	−2
3	3.52566	−1.79835
4	3.96374	−1.922
5	4.00003	−1.92806

Table 4.6. Convergence of q and $p_1(0)$ Using the New Approach with $p_1(0)$ Kept Constant for Two Iterations. Initial $q(0) = 100$, $p_1(0) = -2$

Iteration number	q	$p_1(0)$
0	100	-2
1	33.3332	-2 (20.0)[a]
2	11.1225	-2 (3.40849)
3	4.18492	-0.895274
4	2.26352	-0.970177
5	2.40412	-1.43458
6	3.54516	-1.84137
7	4.00176	-1.92855
8	4	-1.92806

[a] Parentheses indicate values $p_1(0)$ would have assumed had they been allowed to change.

In example 1 of References 7 and 8 convergence was obtained even though the initial choices of the multipliers were orders of magnitude greater than the converged values.

4.2.3. Design of Linear Regulators with Energy Constraints

In the usual design of linear quadratic optimal control systems, the regulator performance is obtained for several different values of the constant Lagrange multiplier q. The Lagrange multiplier determines the amount of control energy expended. If the energy is to be constrained, then the value of q must be found such that the energy constraint is satisfied. In this section a method is described for determining simultaneously the optimal trajectory and the value of q which satisfies the energy constraint (Reference 9). The method makes use of the improved method for the optimization of functionals subject to integral constraints described in the previous section.

Introduction

The linear regulator problem is discussed in several papers and texts on optimal control theory (References 10–13). The regulator problem is, given the linear time-varying system

$$\dot{x} = Ax + Bu, \qquad x(0) = c, \tag{4.127}$$

find the control which minimizes the cost functional J_1,

$$J_1 = \frac{1}{2} \int_0^T (x^T Q x + u^T R u)\, dt, \tag{4.128}$$

where $x(t)$ is the n-dimensional state vector, $u(t)$ is an r-dimensional control vector, $A(t)$ is an $n \times n$ matrix, $B(t)$ is an $n \times r$ matrix, $Q(t)$ is an $n \times n$ positive semidefinite matrix, and $R(t)$ is an $r \times r$ positive definite matrix. The state of the system is to be kept near zero without the expenditure of excessive control energy. The method for selecting the weighting matrices, Q and R, is not always clear, however. In the examples given in Sections 3.2 and 3.3, the components of matrices Q and R are assumed to be equal to either zero or one.

Several curves are given in Reference 11 which show the effect of a constant scalar R on the response of a first-order system optimal regulator. Presumably one could select the value of R to obtain a desired response by solving the problem for several different values of R. A more direct approach, however, is possible.

The direct approach can be used where it is desired to limit the input energy. The linear regulator problem can then be rephrased as the following isoperimetric problem. Find the control which minimizes the cost functional J,

$$J = \frac{1}{2} \int_0^T x^T Q x\, dt, \tag{4.129}$$

subject to the energy constraint

$$E = \int_0^T u^T u\, dt. \tag{4.130}$$

The problem is solved by minimizing the integral

$$J_1 = \frac{1}{2} \int_0^T [x^T Q x + q(u^T u - E/T)]\, dt, \tag{4.131}$$

where q is the Lagrange multiplier and is equal to a constant. The value of q must be selected to satisfy equation (4.130). If the problem is solved for several different values of q, then the method of solution is essentially the same as for equation (4.128). If, however, the Lagrange multiplier is considered as a state variable with differential constraint $dq/dt = 0$, then the energy constraint E can be specified such that the solution simultaneously yields the state variables which minimize J_1 and the value of q which satisfies the energy constraint (4.130). This method requires the solution of a nonlinear two-point boundary-value problem.

In a related work, a method for selecting the weighting matrices in linear-quadratic optimal control designs is discussed in References 14 and 15. The asymptotic behavior of the eigenvalues and eigenvectors is used to provide a unique specification of the weighting matrices. The cost functional is

$$J_1 = \int_0^T (x^T Q x + \varrho u^T R u)\, dt, \tag{4.132}$$

where the modal properties of the regulator are a function of the weights Q and R. The parameter ϱ is a scalar which controls the tradeoff between quadratic regulator performance and control effort expended. Again however, the solution must be obtained for several values of ϱ in order to determine the actual control energy. Thus the direct approach described in this section can be used in problems of the form of equation (4.132) when the energy constraint is specified.

The derivation of the equations for obtaining the optimal trajectory simultaneously with the value of q which satisfies the energy constraint is given in the next section. This is followed by an example and a discussion of the method of solution via the Newton–Raphson method. An analytical solution for the example is given, followed by a discussion of the numerical results.

Derivation of Equations

Cost functional (4.131) is unchanged if expressed in the form

$$J_1 = \frac{1}{2}\int_0^T (x^T Q x + q u^T u)\, dt - q(T)E/2. \tag{4.133}$$

The Lagrange multiplier q, for which energy constraint (4.130) is satisfied, is obtained along with the optimal control by adjoining the differential constraint

$$\dot{q}(t) = 0 \tag{4.134}$$

with unknown initial condition, $q(0)$.

Utilizing Pontryagin's maximum principle, the Hamiltonian function for cost functional (4.133) subject to differential constraints (4.127) and (4.134) is

$$H = \tfrac{1}{2}(x^T Q x + q u^T u) + p^T(Ax + Bu) + p_{n+1} \times 0, \tag{4.135}$$

where $p_{n+1}(t)$ is the augmented scalar costate variable due to differential constraint (4.134). The costate vector $p(t)$ and the scalar $p_{n+1}(t)$ are the solutions of the differential equations

$$\dot{p} = -\frac{\partial H}{\partial x} = -Qx - A^T p, \tag{4.136}$$

$$\dot{p}_{n+1} = -\frac{\partial H}{\partial q} = -\frac{u^T u}{2}. \tag{4.137}$$

The vector $u(t)$ that minimizes H and the boundary conditions are

$$\frac{\partial H}{\partial u} = qu + B^T p, \qquad u = -\frac{1}{q} B^T p, \tag{4.138}$$

$$p(T) = 0, \qquad p_{n+1}(0) = 0, \tag{4.139}$$

$$p_{n+1}(T) = \frac{\partial}{\partial q}\left[-\frac{qE}{2}\right]_{t=T} = -\frac{E}{2}. \tag{4.140}$$

The nonlinear two-point boundary-value equations and associated boundary conditions are then given by

$$\dot{x} = Ax - \frac{1}{q} BB^T p, \qquad x(0) = c, \tag{4.141}$$

$$\dot{q} = 0, \qquad p_{n+1}(0) = 0, \tag{4.142}$$

$$\dot{p} = -Qx - A^T p, \qquad p(T) = 0, \tag{4.143}$$

$$\dot{p}_{n+1} = -\frac{1}{2q^2} p^T BB^T p, \qquad p_{n+1}(T) = -\frac{E}{2}. \tag{4.144}$$

Example. Consider the scalar linear dynamic system given by

$$\dot{x} = -ax + u, \qquad x(0) = c \tag{4.145}$$

with cost functional

$$J = \frac{1}{2}\int_0^T x^2\, dt \tag{4.146}$$

subject to the energy constraint

$$E = \int_0^T u^2\, dt. \tag{4.147}$$

The problem is solved by minimizing the integral

$$J_1 = \frac{1}{2}\int_0^T (x^2 + qu^2)\,dt - q(T)E/2. \qquad (4.148)$$

The Hamiltonian function is

$$H = \tfrac{1}{2}(x^2 + qu^2) + p_1(-ax + u) + p_2 \times 0. \qquad (4.149)$$

The two-point boundary-value equations and associated boundary conditions are

$$\dot{x} = -ax - \frac{1}{q}p_1, \qquad x(0) = c, \qquad (4.150)$$

$$\dot{q} = 0, \qquad p_2(0) = 0, \qquad (4.151)$$

$$\dot{p}_1 = -x + ap_1, \qquad p_1(T) = 0, \qquad (4.152)$$

$$\dot{p}_2 = -\frac{1}{2q^2}p_1^2, \qquad p_2(T) = -\frac{E}{2}. \qquad (4.153)$$

Since the equations are nonlinear, the solution, in general, must be obtained by an iterative procedure such as the Newton–Raphson method. The unknown initial conditions are $p_1(0)$ and $q(0)$. The optimal control is

$$u = -\frac{1}{q}p_1. \qquad (4.154)$$

Solution via Newton–Raphson Method

Using the Newton–Raphson method, the unknown initial conditions are assumed to be given by

$$p_1(0) = c_1, \qquad (4.155)$$

$$q(0) = c_2. \qquad (4.156)$$

If we expand the boundary conditions

$$p_1(T, c_1, c_2) = 0, \qquad (4.157)$$

$$p_2(T, c_1, c_2) = -E/2 \qquad (4.158)$$

in a Taylor series around the kth approximation and retain only the linear terms

$$p_1(T, c_1^k, c_2^k) + (c_1^{k+1} - c_1^k)p_{1c_1} + (c_2^{k+1} - c_2^k)p_{1c_2} = 0, \qquad (4.159)$$

$$p_2(T, c_1^k, c_2^k) + (c_1^{k+1} - c_1^k)p_{2c_1} + (c_2^{k+1} - c_2^k)p_{2c_2} = -E/2, \qquad (4.160)$$

where

$$p_{ic_j} = \frac{\partial p_i}{\partial c_j}, \tag{4.161}$$

then we have

$$\begin{bmatrix} c_1^{k+1} \\ c_2^{k+1} \end{bmatrix} = \begin{bmatrix} c_1^k \\ c_2^k \end{bmatrix} - \begin{bmatrix} p_{1c_1} & p_{1c_2} \\ p_{2c_1} & p_{2c_2} \end{bmatrix}^{-1} \begin{bmatrix} p_1(T, c_1^k, c_2^k) \\ p_2(T, c_1^k, c_2^k) + E/2 \end{bmatrix} \tag{4.162}$$

or

$$\mathbf{c}^{k+1} = \mathbf{c}^k - P_c^{-1}\mathbf{p}(T, \mathbf{c}^k). \tag{4.163}$$

The equations for the components of matrix P_c are obtained by differentiating equations (4.150)–(4.153) with respect to c_1 and c_2:

$$\dot{x}_{c_1} = -ax_{c_1} + q_{c_1}q^{-2}p_1 - q^{-1}p_{1c_1}, \qquad x_{c_1}(0) = 0, \tag{4.164}$$

$$\dot{q}_{c_1} = 0, \qquad q_{c_1}(0) = 0, \tag{4.165}$$

$$\dot{p}_{1c_1} = -x_{c_1} + ap_{1c_1}, \qquad p_{1c_1}(0) = 1, \tag{4.166}$$

$$\dot{p}_{2c_1} = q_{c_1}q^{-3}p_1^2 - q^{-2}p_{1c_1}p_1, \qquad p_{2c_1}(0) = 0, \tag{4.167}$$

and

$$\dot{x}_{c_2} = -ax_{c_2} + q_{c_2}q^{-2}p_1 - q^{-1}p_{1c_2}, \qquad x_{c_2}(0) = 0, \tag{4.168}$$

$$\dot{q}_{c_2} = 0, \qquad q_{c_2}(0) = 1, \tag{4.169}$$

$$\dot{p}_{1c_2} = -x_{c_2} + ap_{1c_2}, \qquad p_{1c_2}(0) = 0, \tag{4.170}$$

$$\dot{p}_{2c_2} = q_{c_2}q^{-3}p_1^2 - q^{-2}p_{1c_2}p_1, \qquad p_{2c_2}(0) = 0. \tag{4.171}$$

To obtain a numerical solution, an initial guess is made for initial conditions (4.155) and (4.156). Initial-value equations (4.150)–(4.153) and (4.164)–(4.171) are then integrated from time $t = 0$ to $t = T$. A new value for $\mathbf{c}$ is calculated from equation (4.163) and the above sequence is repeated to obtain the second approximation, etc. Twelve differential equations are integrated for each approximation.

Analytical Solution

An analytical solution for the given example can be derived in this case using the classical approach (Reference 11). For $a = 1$, dynamic system (4.145) becomes

$$\dot{x} = -x + u, \qquad x(0) = c. \tag{4.172}$$

For cost functional

$$J_1 = \frac{1}{2}\int_0^T (x^2 + qu^2)\,dt \tag{4.173}$$

the optimal control is

$$u(t) = -(1/q)k(t)x(t). \tag{4.174}$$

The scalar $k(t)$ is the solution of the Riccati equation

$$\dot{k} = 2k + (1/q)k^2 - 1, \qquad k(T) = 0. \tag{4.175}$$

The solution of (4.175) is

$$k(t) = q[-1 + w\tanh(-wt + \xi)], \tag{4.176}$$

where

$$w = (1/q + 1)^{1/2}, \tag{4.177}$$

$$\xi = \tanh^{-1}(1/w) + wT. \tag{4.178}$$

The optimal trajectory is the solution of the differential equation

$$\dot{x}(t) = -[1 + (1/q)k(t)]x(t). \tag{4.179}$$

Then we have

$$x(t) = x(0)e^{\alpha}, \tag{4.180}$$

where

$$\alpha = \int_0^t -\left[1 + \frac{1}{q}k(\tau)\right] d\tau. \tag{4.181}$$

Upon substituting (4.176) into (4.181) and integrating, (4.180) becomes

$$x(t) = \frac{x(0)\cosh(-wt + \xi)}{\cosh \xi}. \tag{4.182}$$

When (4.176) and (4.182) are substituted into (4.174), $u(t)$ becomes

$$u(t) = \frac{x(0)}{\cosh \xi}[1 - w\tanh(-wt + \xi)]\cosh(-wt + \xi). \tag{4.183}$$

Table 4.7. Energy Constraint $E'(q)$ versus q. $a = 1$; $x(0) = 1$; $T = 1$

q	$E'(q)$
0.2	0.395044
0.4	0.164223
0.6	0.0902612
0.8	0.0570906
1	0.0393639

The energy constraint as a function of q is then

$$\begin{aligned} E'(q) &= \int_0^T u^2(t)\,dt \\ &= -\frac{x^2(0)}{4w\cosh^2\xi}\,[(w^2+1)\sinh 2(-wT+\xi) \\ &\quad - 2w\cosh 2(-wT+\xi) - 2wT(1-w^2) \\ &\quad - (w^2+1)\sinh 2\xi + 2w\cosh 2\xi] \end{aligned} \tag{4.184}$$

where w is given by equation (4.177).

The energy constraint, $E'(q)$, is given in Table 4.7 as a function of q for initial condition and terminal time

$$x(0) = c = 1 \quad \text{and} \quad T = 1. \tag{4.185}$$

For a given value of $E'(q)$, for example,

$$E = 0.0393639, \tag{4.186}$$

there is a value of q for which equation (4.147) is satisfied. From Table 4.7 the value of q for which the energy constraint is satisfied is $q = 1$. The purpose of the approach described in this section is to determine simultaneously the optimal trajectory and the value of q which satisfies the energy constraint for a given value of E.

Numerical Results

Numerical results were obtained using the Newton–Raphson method and a fourth-order Runge–Kutta integration method with grid intervals of 1/100 sec. It was assumed that

$$T = 1, \qquad x(0) = 1, \qquad \text{and } a = 1 \tag{4.187}$$

with energy constraint

$$E = 0.0393639. \tag{4.188}$$

To initialize the algorithm, it was assumed that

$$p_1(0) = c_1 = 0, \tag{4.189}$$

$$q(0) = c_2 = 2. \tag{4.190}$$

From the analytical solution of the previous section it is known that the converged value of q for the given energy constraint is

$$q(t) = \text{const} = 1. \tag{4.191}$$

Table 4.8 shows the convergence of q and $p_1(0)$. The value of q is within 97% of its final value in eight iterations.

Table 4.8. Convergence of q and $p_1(0)$

Iteration number	$q(0) = 2;\ p_1(0) = 0$		$q(0) = 2;\ p_1(0) = 0.7$	
	q	$p_1(0)$	q	$p_1(0)$
0	2	0	2	0.7
1	0.306856	0.434253	1.15683	0.39776
2	0.233475	0.330893	0.995926	0.389147
3	0.244818	0.29811	0.998699	0.385701
4	0.356458	0.300179	0.999856	0.385785
5	0.481532	0.331212	0.999985	0.385815
6	0.676733	0.353426	0.999999	0.385818
7	0.865759	0.372026	1	0.385819
8	0.969872	0.381616		
9	0.996015	0.385005		
10	0.999547	0.385715		
11	0.999951	0.385807		
12	0.999995	0.385817		
13	1	0.385818		

The initial selection of $p_1(0)$, for $q(0) = 2$, can be improved by the following considerations. Assume that the absolute value of $u(t)$ decreases linearly from $u_{\max}$ at $t = 0$ to zero at $t = 1$. Then in order to satisfy the energy constraint, $E \doteq 0.04$, the value of $u_{\max}$ is approximately 0.35. Then from equation (4.154), the value of $p_1(0) \doteq 0.7$ when $q(0) = 2$.

Table 4.8 shows the convergence of q and $p_1(0)$ for initial conditions

$$p_1(0) = c_1 = 0.7, \tag{4.192}$$

$$q(0) = c_2 = 2. \tag{4.193}$$

The table shows that the value of q is within 98% of its final value in three iterations.

4.3. Examples Using Integral Equation and Imbedding Methods

Nonlinear two-point boundary-value problems occur not only in optimization problems, but in the determination of the buckling load of columns, magnetohydrodynamics, post buckling beam configurations, nonlinear filtering, etc. In this section, examples of these nonlinear two-point boundary-value problems are given with numerical solutions obtained via integral equation and imbedding methods.

4.3.1. Integral Equation Method for Buckling Loads

An initial-value method for the integral equation of the column is presented for determining the buckling load of columns (Reference 16). The differential equation of the column is reduced to a Fredholm integral equation. An initial-value problem is derived for this integral equation, which is reduced to a set of ordinary differential equations with prescribed initial conditions in order to find the Fredholm resolvent. The singularities of the resolvent occur at the eigenvalues. Integration of the equations proceeds until the integrals become excessively large, indicating that a critical load has been reached. Numerical results are given for two examples for which the critical load is well known. One is the Euler load of a simply supported beam and the other case is the buckling load of a cantilever beam under its own weight. The advantage of this initial-value method is that it can be applied easily to solve other nonlinear problems for which the critical loads are unknown.

Introduction

An important problem in the theory of stability is the determination of the critical load of columns. At the critical load the column will buckle sidewise. The critical load may be found by determining the smallest eigenvalue of the differential equation of the elastic curve. Many methods exist that permit determination of the buckling load of columns by solving a boundary-value problem as described in References 17 and 18. The disadvantage of all these methods is that they cannot be used for nonlinear buckling problems. A method is described in this section that permits the determination of the buckling loads of linear as well as nonlinear problems.

The method consists of reducing the differential equation of the column to a Fredholm integral equation. This integral equation is reduced to an initial-value problem as described in Reference 19. The Fredholm resolvent is obtained by integrating a set of first-order differential equations with given initial conditions. The singularities of the resolvent occur at the eigenvalues.

Thus the smallest eigenvalue or the critical load may be found by integrating the equations for the resolvent from zero up to the point where integrals become excessively large or overflow. The point at which overflow occurs is equal to the smallest eigenvalue. The method is applied to determine the buckling load for two well-known cases as described below. The same method can be used to solve nonlinear problems.

Example 1. Consider the case of a simply supported beam, as shown in Figure 4.3a. The differential equation of the deflected beam (derived initially by the great mathematician, Leonhart Euler in 1757), see Reference 17, is

$$y''(s) + \lambda y(s) = 0, \qquad 0 \leq s \leq 1, \tag{4.194}$$

$$\lambda = Pl^2/EI, \qquad s = x/l, \tag{4.195}$$

where l is the span of the column, E is Young's modulus, I is the moment of inertia of the column, x is the abscissa, and s is the nondimensional abscissa. The boundary conditions for the differential equation (4.194) are

$$s = 0, \qquad y(0) = 0, \tag{4.196a}$$

$$s = 1, \qquad y(1) = 0. \tag{4.196b}$$

In order to reduce differential equation (4.194) to an integral equation,

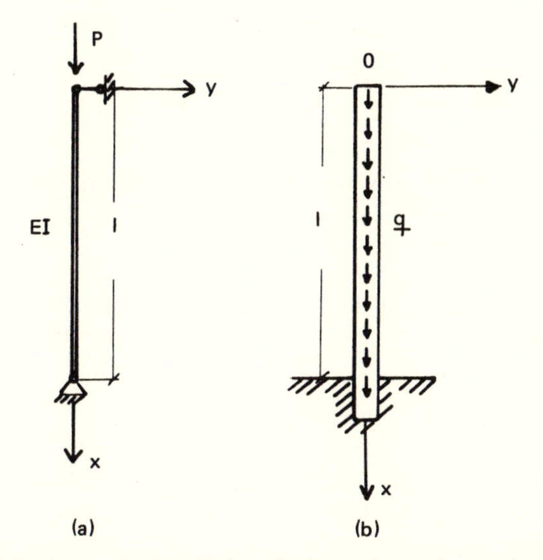

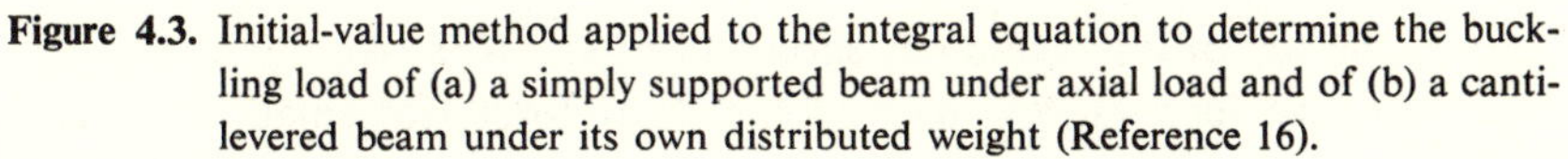

Figure 4.3. Initial-value method applied to the integral equation to determine the buckling load of (a) a simply supported beam under axial load and of (b) a cantilevered beam under its own distributed weight (Reference 16).

let us integrate equation (4.194) twice with respect to s; integrating by parts yields

$$y(s) + A + Bs + \lambda \int_0^s (s - t)y(t)\, dt = 0, \tag{4.197}$$

where A and B are constants to be determined from the boundary conditions (4.196). Making use of (4.196a) and (4.196b) in equation (4.197), we obtain

$$A = 0, \tag{4.198a}$$

$$B = -\lambda \int_0^1 (1 - t)y(t)\, dt. \tag{4.198b}$$

Substituting (4.198a) and (4.198b) into equation (4.197) gives

$$y(s) = \lambda s \int_0^1 (1 - t)y(t)\, dt - \lambda \int_0^s (s - t)y(t)\, dt. \tag{4.199}$$

Equation (4.199) can be written as

$$y(s) = \lambda \int_0^1 k(s, t)y(t)\, dt, \tag{4.200}$$

where the kernel $k(s, t)$ is given by

$$k(s,t) = \begin{cases} s(1-t) & \text{for } s \leq t \leq 1 \quad (4.201a) \\ t(1-s) & \text{for } 0 \leq t \leq s. \quad (4.201b) \end{cases}$$

The critical load for this case was derived for the first time by Euler and is

$$\lambda = \pi^2.$$

It will be shown that the imbedding method gives the same value of λ but has the advantage that it can be extended to determine the eigenvalues of nonlinear problems.

Example 2. Consider the case of buckling of a cantilevered column under its own weight as shown in Figure 4.3b. Such a case arises in the design of columns, high-rise constructions, and chimneys. To analyze the stability of the column under its own weight, let us choose the coordinate system at the top of the column as shown in Figure 4.3b. At any section x, the compressive force is $N = qx$ and the shear force (for small deformations) is

$$Q = N\frac{dy}{dx}. \tag{4.202a}$$

Recalling the differential relationship for the columns

$$EI\frac{d^3y}{dx^3} = -Q, \tag{4.202b}$$

substitution of equation (4.202a) into (4.202b) gives

$$EI\frac{d^3y}{dx^3} + qx\frac{dy}{dx} = 0. \tag{4.203}$$

By introducing the nondimensional coordinates

$$s = \frac{x}{l}, \qquad w = \frac{dy}{ds} \tag{4.204}$$

and calling

$$a^2 = ql^3/EI, \tag{4.205}$$

equation (4.203) reduces to

$$\frac{d^2w}{ds^2} + a^2sw = 0, \qquad 0 \leq s \leq 1. \tag{4.206}$$

The boundary conditions for equation (4.206) are

$$x = 0, \qquad M = 0, \text{ or } EI\frac{d^2y}{dx^2} = 0 \tag{4.207a}$$

$$x = l, \qquad \frac{dy}{dx} = 0. \tag{4.207b}$$

In nondimensional form, the boundary conditions (4.207a) and (4.207b) are expressed as

$$s = 0, \qquad \frac{dw}{ds} = 0, \tag{4.208a}$$

$$s = 1, \qquad w = 0. \tag{4.208b}$$

Integrating the differential equation (4.206) twice with respect to s, integrating by parts, and making use of the boundary conditions (4.208a) and (4.208b), we obtain

$$w(s) = a^2 \int_0^1 (1 - t)tw(t)\,dt - a^2 \int_0^s (s - t)tw(t)\,dt. \tag{4.209}$$

Equation (4.209) is an integral equation, which can be rewritten as

$$w(s) = \lambda \int_0^1 k(s, t)w(t)\,dt, \tag{4.210}$$

where the kernel $k(s, t)$ is given by

$$k(s, t) = \begin{cases} (1 - t)t, & s \le t \le 1 \\ (1 - s)t, & 0 \le t \le s. \end{cases} \tag{4.211}$$

The solution to equations (4.206) and (4.208) is given in terms of the Bessel functions.

Initial-Value Method

Integral equations (4.200) and (4.210) have the form of the inhomogeneous Fredholm integral equation

$$y(x) = g(x) + \lambda \int_0^L k(x, t')y(t')\,dt', \qquad 0 \le x \le L. \tag{4.212}$$

The solution of (4.212) is given in terms of the resolvent, K, by the equation

$$y(x) = g(x) + \lambda \int_0^L K(x, t', \lambda)g(t')\,dt', \tag{4.213}$$

where the resolvent kernel itself satisfies the Fredholm integral equation

$$K(x, t, \lambda) = k(x, t) + \lambda \int_0^L k(x, t')K(t', t, \lambda)\, dt', \qquad 0 \leq x, \qquad t \leq L. \tag{4.214}$$

The initial-value method for the resolvent is discussed in References 19 and 20 for the case where L is a variable. Differentiating equation (4.214) with respect to λ, it can be shown that the initial-value equation for the case where λ is a variable is given by

$$K_\lambda(x, t, \lambda) = \int_0^L K(x, t', \lambda)K(t', t, \lambda)\, dt', \qquad 0 \leq x, \qquad t \leq L, \tag{4.215}$$

where

$$K_\lambda(x, t, \lambda) = \frac{\partial K}{\partial \lambda}. \tag{4.216}$$

The initial condition on the resolvent at $\lambda = 0$ from equation (4.214) is

$$K(x, t, 0) = k(x, t). \tag{4.217}$$

The method of computation requires the use of normalized variables in place of x and t. However, since $L = 1$ in the examples, the variables are already normalized.

Method of Computation. To evaluate the resolvent, equation (4.215) is approximated in the interval (0, 1) by a quadrature formula of the form

$$\int_0^1 f(t')\, dt' \doteq \sum_{j=1}^N f(r_j)w_j, \tag{4.218}$$

where w_j are weights (for example from Simpson's rule). Introducing the nomenclature

$$K_{ij}(\lambda) = K(r_i, r_j, \lambda), \qquad i, j = 1, 2, \ldots, N, \tag{4.219}$$

equation (4.215) is replaced by the approximation

$$\frac{dK_{ij}(\lambda)}{d\lambda} = \sum_{m=1}^N K_{im}(\lambda)K_{mj}(\lambda)w_m, \qquad i, j = 1, 2, \ldots, N \tag{4.220}$$

with initial conditions on K_{ij} at $\lambda = 0$ given by

$$K_{ij}(0) = k(r_i, r_j). \tag{2.221}$$

Numerical Results and Discussion

Numerical results were obtained for Examples 1 and 2. The quadrature formula utilized for approximating the integral in equation (4.215) was Simpson's rule with $N = 7$, i.e., six equal intervals. The differential equations were integrated using a fourth-order Runge–Kutta method with grid intervals, $\Delta\lambda = 1/20$.

Results obtained for Example 1 are shown in Tables 4.9 and 4.10. Table 4.9 shows the resolvent obtained by the initial-value method for the case of $\lambda = 2$.

The resolvent kernel satisfies the linear integral equation

$$K(x, t, \lambda) = k(x, t) + \lambda \int_0^1 k(x, t')K(t', t, \lambda)\, dt',$$

which can be approximated by

$$K_{ij}(\lambda) = k(r_i, r_j) + \lambda \sum_{m=1}^{N} k(r_i, r_m)K_{mj}(\lambda)w_m.$$

This equation for the resolvent can be solved by the standard method of linear algebraic equations for each value of λ. A solution was obtained at $\lambda = 2$. The initial-value method and the solution of the linear algebraic equations agree up to six digits, for $\lambda = 2$. The exact analytical expression

Table 4.9. Initial-Value Solution for $\lambda = 2$: Resolvent for Kernel

$$k(x, t) = \begin{cases} x(1-t), & x \le t \le 1 \\ t(1-x), & 0 \le t \le x \end{cases}$$

$$N = 7, \quad \Delta\lambda = 1/20$$

x \ t	0	1/6	1/3	1/2	2/3	5/6	1
0	0	0	0	0	0	0	0
1/6	0	0.157536	0.136737	0.110873	0.076796	0.039875	0
1/3	0	0.136737	0.263345	0.213533	0.147903	0.076796	0
1/2	0	0.110873	0.213533	0.308284	0.213533	0.110873	0
2/3	0	0.076796	0.147903	0.213533	0.263345	0.136737	0
5/6	0	0.039875	0.076796	0.110873	0.136737	0.157536	0
1	0	0	0	0	0	0	0

Table 4.10. Resolvent Element K_{ij} for $i = j = 4$ and for Same Kernel, N, and $\Delta\lambda$ as for Table 4.9.

λ	Linear algebraic equation solution	Initial-value solution	Analytical solution
1	0.275796	0.275796	0.273151
2	0.308284	0.308284	0.302115
5	0.48975	0.48975	0.459655
8	1.36004	1.36004	1.11972
9	3.7000	3.69999	2.35024
9.45	18.0584	18.0508	4.8169
9.5	31.9797	31.8803	5.46171
9.55	140.509	121.66	6.30828
9.6	−58.5377	5.17169×10^{6}	7.46884
9.65	−24.1953	1.29203×10^{40}	9.15786

for the resolvent is given by

$$K(x, t, \lambda) = \begin{cases} \dfrac{1}{\sqrt{\lambda}}\left[\cos\sqrt{\lambda}t - \dfrac{\sin\sqrt{\lambda}t}{\tan\sqrt{\lambda}}\right]\sin\sqrt{\lambda}x, & 0 \le x \le t \\[2ex] \dfrac{1}{\sqrt{\lambda}}\left[\cos\sqrt{\lambda}x - \dfrac{\sin\sqrt{\lambda}x}{\tan\sqrt{\lambda}}\right]\sin\sqrt{\lambda}t, & 0 \le t \le x. \end{cases}$$

The maximum error for the resolvent, obtained via the initial-value method or by linear algebraic equations, is less than 2.1% (from the analytical solution) for the case of $\lambda = 2$. Table 4.10 shows the resolvent element, K_{ij}, for $i = j = 4$ as a function of λ. The linear algebraic equation solution, the initial-value solution, and the analytical solution are shown for comparison in the same table. The analytical solution approaches infinity at $\lambda = 9.87$ ($\doteq \pi^2$). The linear algebraic equation solution becomes negative at $\lambda = 9.6$, but gives no other indication that the solution is invalid. The initial-value solution becomes excessively large at $\lambda = 9.6$ and thus gives a critical value which is in error by less than 2.8%. This is so in spite of the fact that the resolvent obtained by the initial-value method differs from the true value by a considerable amount for λ greater than 9.

Similar results are obtained for Example 2. The analytical solution for the eigenvalue is given in terms of Bessel functions (see Reference 17) and has a critical value $\lambda = 7.84$. The linear algebraic equation solution becomes negative at $\lambda = 7.75$, but again gives no other indication that the solution is invalid. The initial-value solution becomes excessively large at $\lambda = 7.75$.

The initial-value solution thus gives a critical value of about 7.75, which is in error by less than 1.2% (compared to the exact solution).

Additional results were obtained for Example 1 utilizing Simpson's rule with $N = 5$ instead of 7. The maximum error of any of the matrix elements K_{ij} at $\lambda = 2$ is found to be approximately twice as large for $N = 5$ as it is for $N = 7$. The initial-value solution becomes excessively large at $\lambda = 9.2$, which differs by only 6.8% from the exact solution.

4.3.2. An Imbedding Method for Buckling Loads

An imbedding method for determining the critical length of buckling columns is presented (Reference 21). The method can be applied to a large class of problems and requires the integration of only one initial-value equation. Integration proceeds until the integral becomes excessively large, indicating that the critical length has been reached. Alternately the reciprocal of the equation can be integrated, and the critical length determined by the zero crossing. Utilizing this method the critical length for an example is shown to be obtained to an accuracy of greater than 1/10,000 of one percent. Computation time is less than approximately 3 sec.

Introduction

The critical load of columns may be determined by solving two-point boundary-value problems. As described in references 17 and 18, however, the solution of these boundary-value problems is far from trivial. In the previous section it was shown that these problems can be solved by an initial-value approach to the integral equation method. Utilizing this method, the differential equation of a column is expressed in the form of a Fredholm integral equation. Initial-value equations are then derived which require the integration of a set of ordinary differential equations to find the Fredholm resolvent. Since the singularities of the resolvent occur at the eigenvalues, the equations are integrated until the integrals become excessively large, indicating that a critical load has been reached. While this approach is valid for both linear and nonlinear problems, it has the disadvantage that the accuracy of the solution is dependent upon the number of differential equations which are integrated. Another method for solving for the critical load of columns is discussed in this section. This method is an imbedding method (References 22 and 23) and requires the integration of only one initial-value equation. Integration proceeds until the integral becomes excessively large, indicating that the critical length has been reached. The critical load is then easily calculated from the critical length. Only linear problems are considered; however, the method can easily be extended to nonlinear problems.

The Imbedding Method

Consider the two-point boundary-value equations

$$\begin{aligned} \dot{u}(t) &= au(t) + bv(t), \\ \dot{v}(t) &= cu(t) + dv(t), \qquad 0 \leq t \leq x \end{aligned} \tag{4.222}$$

with boundary conditions

$$u(0) = 0, \qquad u(x) = x. \tag{4.223}$$

In general a, b, c, and d may be functions of t, and u and v may be considered to be vectors. For simplicity, however, the following derivations are given for the scalar equations. The reason for choosing the terminal boundary condition, $u(x) = x$, will become clear when the examples are discussed. It is desired to convert the boundary-value problem into an initial-value problem. The imbedding parameter utilized to derive the initial-value equations is the interval length, x. Rewriting equations (4.222) and (4.223) to explicitly indicate the dependence of the solution upon x, we obtain

$$\begin{aligned} \dot{u}(t, x) &= au(t, x) + bv(t, x) \\ \dot{v}(t, x) &= cu(t, x) + dv(t, x) \\ u(0, x) &= 0, \qquad u(x, x) = x. \end{aligned} \tag{4.224}$$

Differentiating equations (4.224) with respect to x gives

$$\begin{aligned} \dot{u}_x(t, x) &= au_x(t, x) + bv_x(t, x), \\ \dot{v}_x(t, x) &= cu_x(t, x) + dv_x(t, x), \\ u_x(0, x) &= 0, \qquad \dot{u}(x, x) + u_x(x, x) = 1. \end{aligned} \tag{4.225}$$

Here a dot represents differentiation with respect to t, while $(\ )_x$ represents differentiation with respect to x. Comparing equations (4.224) and (4.225), it is seen that

$$
\begin{aligned}
u_x(t, x) &= \left[\frac{1 - \dot{u}(x, x)}{x}\right]u(t, x), \\
v_x(t, x) &= \left[\frac{1 - \dot{u}(x, x)}{x}\right]v(t, x).
\end{aligned}
\tag{4.226}
$$

Define

$$
r(x) = v(x, x). \tag{4.227}
$$

Then we have

$$
\begin{aligned}
1 - \dot{u}(x, x) &= 1 - [au(x, x) + bv(x, x)] \\
&= 1 - ax - br(x).
\end{aligned}
\tag{4.228}
$$

Substituting equation (4.228) into equations (4.226) gives

$$
\begin{aligned}
u_x(t, x) &= \left[\frac{1 - ax - br(x)}{x}\right]u(t, x), \\
v_x(t, x) &= \left[\frac{1 - ax - br(x)}{x}\right]v(t, x).
\end{aligned}
\tag{4.229}
$$

Differentiating equation (4.227) with respect to x

$$
\begin{aligned}
r'(x) &= \dot{v}(x, x) + v_x(x, x) \\
&= cx + dr(x) + \left[\frac{1 - ax - br(x)}{x}\right]r(x).
\end{aligned}
\tag{4.230}
$$

The initial condition for equation (4.230) is obtained by assuming $r'(x)$ remains finite as x approaches zero. Thus as x approaches zero the numerator of the bracketed term in equation (4.230) approaches zero, requiring that

$$
r(0) = 1/b. \tag{4.231}
$$

Applying L'Hospital's rule in equation (4.230), it can be shown that

$$
r'(0) = \frac{1}{2}\left[\frac{d}{b} - \frac{a}{b}\right]. \tag{4.232}
$$

Equations (4.231) and (4.232) will be verified in a later section.

To obtain a numerical solution for $u(t, x)$ at any time t, equation (4.230) is integrated with initial condition (4.231) from $x = 0$ to $x = t$. At time $x = t$, equations (4.229) are adjoined and the entire system is integrated from $x = t$ to the interval length of interest. The application of the imbedding method to the buckling of columns is discussed in the following sections.

Application to Buckling Columns

Two examples will be given to illustrate the application of the imbedding method to the determination of the buckling loads of columns. The examples selected are the two examples given in the previous section.

Example 1. Consider a pin-ended column with an applied axial force p. The differential equation of the deflected beam is

$$\ddot{y}(t) + \lambda y(t) = 0, \qquad 0 \leq t \leq x, \tag{4.233}$$

where

$$\lambda = p/EI.$$

Here E is Young's modulus, I is the moment of inertia, t is the axial coordinate, and x is the length of the beam. The boundary conditions are

$$y(0) = 0, \qquad y(x) = 0. \tag{4.234}$$

The imbedding parameter utilized to derive the initial-value equations is the length of the beam, x. Rewriting equations (4.233) and (4.234) as functions of x gives

$$\begin{gathered} \ddot{y}(t, x) + \lambda y(t, x) = 0, \\ y(0, x) = 0, \qquad y(x, x) = 0. \end{gathered} \tag{4.235}$$

Differentiating with respect to x yields

$$\begin{gathered} \ddot{y}_x(t, x) + \lambda y_x(t, x) = 0, \\ y_x(0, x) = 0, \qquad \dot{y}(x, x) + y_x(x, x) = 0. \end{gathered} \tag{4.236}$$

Consider now the system

$$\begin{gathered} \ddot{m}(t, x) + \lambda m(t, x) = 0, \\ m(0, x) = 0, \qquad 1 + m(x, x) = 0. \end{gathered} \tag{4.237}$$

Defining

$$s(x) = \dot{m}(x, x) \tag{4.238}$$

it can be shown that

$$s'(x) = \lambda + s^2(x) \tag{4.239}$$

with initial condition

$$s(0) = -\infty. \tag{4.240}$$

In order to avoid the infinite initial condition, consider the system

$$\begin{gathered} \ddot{n}(t, x) + \lambda n(t, x) = 0, \\ n(0, x) = 0, \qquad x + n(x, x) = 0. \end{gathered} \tag{4.241}$$

These equations can be obtained simply by multiplying equations (4.237) by x and setting

$$n(t, x) = xm(t, x). \tag{4.242}$$

Equations (4.241) are similar in form to equations (4.224). Comparing equations (4.236) and (4.241), it is seen that

$$y_x(t, x) = \frac{y(x, x)}{x} n(t, x). \tag{4.243}$$

Differentiating equations (4.241) with respect to x gives

$$\begin{gathered} \ddot{n}_x(t, x) + \lambda n_x(t, x) = 0, \\ n_x(0, x) = 0, \qquad 1 + \dot{n}(x, x) + n_x(x, x) = 0. \end{gathered} \tag{4.244}$$

Comparing equations (4.241) and (4.244), it is seen that

$$n_x(t, x) = \frac{1 + \dot{n}(x, x)}{x} n(t, x). \tag{4.245}$$

Define

$$r(x) = \dot{n}(x, x). \tag{4.246}$$

Differentiating equation (4.246) with respect to x, we obtain

$$\begin{aligned} r'(x) &= \ddot{n}(x, x) + \dot{n}_x(x, x) \\ &= -\lambda n(x, x) + \frac{1 + \dot{n}(x, x)}{x} \dot{n}(x, x) \\ &= \lambda x + \frac{1 + r(x)}{x} r(x). \end{aligned} \tag{4.247}$$

From considerations similar to those utilized to derive equations (4.231) and (4.232), it is obvious that

$$r(0) = -1 \tag{4.248}$$

and

$$r'(0) = 0. \tag{4.249}$$

Define

$$z(x) = \dot{y}(x, x). \tag{4.250}$$

Differentiating equation (4.250) with respect to x gives

$$\begin{aligned} z'(x) &= \ddot{y}(x, x) + \dot{y}_x(x, x) \\ &= -\lambda y(x, x) + \frac{\dot{y}(x, x)}{x} \dot{n}(x, x) \\ &= \frac{1}{x} z(x) r(x). \end{aligned} \tag{4.251}$$

The initial condition for equation (4.251) is

$$z(0) = 0. \tag{4.252}$$

A numerical solution can be obtained for $y(t, x)$ at any time t, as discussed in the preceding paragraphs. However, since only the critical length of the column is desired, only the equation for $r(x)$ need be integrated. The integration stops when $r(x)$ becomes excessively large, indicating that a critical length has been reached.

The critical length can be determined with greater accuracy if the reciprocal of $r(x)$ is integrated, instead of $r(x)$, in the vicinity where $r(x)$ approaches infinity. Define

$$s(x) = 1/r(x). \tag{4.253}$$

Differentiating equation (4.253) gives

$$\begin{aligned} s'(x) &= -\frac{1}{r^2(x)} r'(x) \\ &= -\left[\lambda x s(x) + \frac{s(x) + 1}{x}\right]. \end{aligned} \tag{4.254}$$

Example 2. Consider the stability of a cantilever column supporting its own weight. The differential equation of the deflected beam is

$$\ddot{w}(t) + \lambda t w(t) = 0, \tag{4.255}$$

where

$$w(t) = \dot{y}(t), \qquad \lambda = q/EI$$

and q is the weight per unit length. The boundary conditions are

$$\dot{w}(0) = 0, \qquad w(x) = 0. \tag{4.256}$$

Following the same procedure as in Example 1, we can rewrite equations (4.255) and (4.256) as follows:

$$\begin{gathered}\ddot{w}(t, x) + \lambda t w(t, x) = 0, \\ \dot{w}(0, x) = 0, \qquad w(x, x) = 0.\end{gathered} \tag{4.257}$$

Differentiating with respect to x yields

$$\begin{gathered}\ddot{w}_x(t, x) + \lambda t w_x(t, x) = 0, \\ \dot{w}_x(0, x) = 0, \qquad \dot{w}(x, x) + w_x(x, x) = 0.\end{gathered} \tag{4.258}$$

Consider the system

$$\begin{gathered}\ddot{m}(t, x) + \lambda t m(t, x) = 0, \\ \dot{m}(0, x) = 0, \qquad 1 + m(x, x) = 0.\end{gathered} \tag{4.259}$$

Comparing equations (4.258) and (4.259), it is seen that

$$w_x(t, x) = \dot{w}(x, x) m(t, x). \tag{4.260}$$

Differentiating equations (4.259) with respect to x gives

$$\begin{gathered}\ddot{m}_x(t, x) + \lambda t m_x(t, x) = 0, \\ \dot{m}_x(0, x) = 0, \qquad \dot{m}(x, x) + m_x(x, x) = 0.\end{gathered} \tag{4.261}$$

Comparing equations (4.259) and (4.261), it is seen that

$$m_x(t, x) = \dot{m}(x, x) m(t, x). \tag{4.262}$$

Define

$$s(x) = \dot{m}(x, x) \tag{4.263}$$

and

$$z(x) = \dot{w}(x, x). \tag{4.264}$$

Then differentiating equations (4.263) and (4.264) with respect to x, we obtain

$$s'(x) = \dot{m}(x, x) + \dot{m}_x(x, x) = \lambda x + s^2(x) \tag{4.265}$$

and

$$z'(x) = \dot{w}(x, x) + \dot{w}_x(x, x) = z(x)s(x). \tag{4.266}$$

The initial conditions for equations (4.265) and (4.266) are easily obtained from equations (4.259) and (4.257), respectively:

$$s(0) = 0, \tag{4.267}$$

$$z(0) = 0. \tag{4.268}$$

To obtain the critical length, equation (4.265) with initial condition (4.267) is integrated until $s(x)$ becomes excessively large, indicating that the critical length has been reached.

Again, the critical length of the column can be obtained with greater accuracy if the reciprocal of $s(x)$ is integrated in the vicinity where $r(x)$ approaches infinity. Define

$$r(x) = 1/s(x). \tag{4.269}$$

Differentiating equation (4.269) yields

$$r'(x) = -[1 + \lambda x r^2(x)]. \tag{4.270}$$

Numerical Results and Discussion

Numerical results were obtained for the examples. The eigenvalues, λ, were selected such that the critical length is unity. References 17 and 18 show that these eigenvalues are π^2 and 7.84 for Examples 1 and 2, respectively. The differential equations were integrated using a fourth-order Runge–Kutta method with grid interval $\Delta x = 0.01$.

For Example 1, integration begins with $r(x)$ and then shifts to $s(x)$, in order to obtain greater accuracy as $r(x)$ approaches infinity. Equation (4.247) with initial condition (4.248) is integrated from $x = 0$ to $x = x_1$, where $r(x_1) = 2$. At $x = x_1$, integration is shifted to equation (4.254) with initial condition

$$s(x_1) = 1/r(x_1). \tag{4.271}$$

Table 4.11. Integral of Equation (4.254) for $x \geq 0.9$[a]

x	$s(x)$	$s'(x)$
0.9	0.114918	−1.3561
0.91	0.101625	−1.30333
0.92	8.88359 E−2	−1.25518
0.93	7.65071 E−2	−1.21126
0.94	6.45976 E−2	−1.17129
0.95	5.30697 E−2	−1.1349
0.96	4.18883 E−2	−1.10193
0.97	3.10206 E−2	−1.07212
0.98	2.04359 E−2	−1.0453
0.99	1.01052 E−2	−1.02131
1	8.41319 E−7	−1
1.01	−9.90342 E−3	−0.981272
1.02	−1.96329 E−2	−0.965025
1.03	−2.92119 E−2	−0.951188

[a] Numbers following E denote powers of ten.

Integration proceeds until $s(x)$ crosses zero. The computation time required is approximately three seconds. Table 4.11 shows the integral of $s'(x)$ for x greater than 0.9. Also shown is $s'(x)$. The sign of $s(x)$ changes from plus to minus between $x = 1$ and $x = 1.01$, indicating that $s(x)$ crosses through zero and $r(x)$ approaches infinity between these two values of x. Utilizing linear interpolation between $x = 1$ and $x = 1.01$, we obtain

$$s(x) = ax + b, \qquad 1.0 \leq x \leq 1.01. \tag{4.272}$$

Then we have

$$a = s'(1) = -1, \tag{4.273}$$

$$b = 1.00000084. \tag{4.274}$$

Thus we obtain

$$s(1.00000084) = 0. \tag{4.275}$$

That is, zero crossing occurs at $x = 1.00000084$. Since the critical length for $\lambda = \pi^2$ is unity, the error in the critical length is less than 1/10,000 of one percent.

For Example 2, equation (4.265) with initial condition (4.267) is integrated from $x = 0$ to $x = x_1$, where $s(x_1) = 1$. At $x = x_1$, integration is shifted to equation (4.270) with initial condition

$$r(x_1) = 1/s(x_1). \tag{4.276}$$

Zero crossing of $r(x)$ occurs between $x = 0.99$ and $x = 1.0$. The critical length of the column is equal to unity to within approximately three significant digits for $\lambda = 7.84$. Utilizing linear interpolation between $x = 0.99$ and $x = 1.0$, the critical length is obtained to an accuracy greater than three digits.

Verification of Initial Conditions

Verification of the initial conditions of $r(x)$ and $r'(x)$ given by equations (4.231) and (4.232) is obtained by expanding $u(t, x)$ and $v(t, x)$ into Maclaurin series expansions about the origin. Only the first two terms of the expansions are required, as shown by the following derivations. The series expansions for small x are

$$u(t, x) = u_1(x)t + u_2(x)t^2 + \cdots, \tag{4.277}$$

$$v(t, x) = v_0(x) + v_1(x)t + v_2(x)t^2 + \cdots. \tag{4.278}$$

Setting $x = t$ in equation (4.277) and utilizing the terminal boundary condition in equations (4.223)

$$u(x, x) = u_1(x)x + u_2(x)x^2 + \cdots = x. \tag{4.279}$$

Then dividing equation (4.279) by x and setting $x = 0$, we have

$$u_1(0) = 1. \tag{4.280}$$

Differentiating equations (4.277) and (4.278) with respect to t and then substituting equations (4.223), (4.277), and (4.278) yields

$$\begin{aligned} \dot{u}(t, x) &= u_1(x) + 2u_2(x)t + \cdots \\ &= a[u_1(x)t + u_2(x)t^2 + \cdots] + b[v_0(x) + v_1(x)t + \cdots], \end{aligned} \tag{4.281}$$

$$\begin{aligned} \dot{v}(t, x) &= v_1(x) + 2v_2(x)t + \cdots \\ &= c[u_1(x)t + u_2(x)t^2 + \cdots] + d[v_0(x) + v_1(x)t + \cdots]. \end{aligned} \tag{4.282}$$

Comparing the constant terms in equations (4.281) and (4.282) we obtain

$$\begin{aligned} u_1(x) &= bv_0(x), \\ v_1(x) &= dv_0(x). \end{aligned} \tag{4.283}$$

Setting $x = 0$ and substituting equation (4.280) gives

$$\begin{aligned} v_0(0) &= 1/b, \\ v_1(0) &= d/b. \end{aligned} \tag{4.284}$$

Differentiating equation (4.278) with respect to x and substituting equations (4.229) and (4.278) yields

$$\begin{aligned} v_x(t, x) &= v_0'(x) + v_1'(x)t + \cdots \\ &= \left[\frac{1 - ax - br(x)}{x}\right][v_0(x) + v_1(x)t + \cdots]. \end{aligned} \tag{4.285}$$

Comparing the constant terms in equation (4.285) we obtain

$$v_0'(x) = \left[\frac{1 - ax - br(x)}{x}\right]v_0(x). \tag{4.286}$$

Applying L'Hospital's rule in equation (4.286) yields

$$v_0'(0) = \frac{-a - br'(0)}{b}. \tag{4.287}$$

Now we have

$$\begin{aligned} r(x) &= v(x, x) \\ &= v_0(x) + v_1(x)x + v_2(x)x^2 + \cdots \end{aligned} \tag{4.288}$$

$$r'(x) = v_0'(x) + v_1(x) + v_1'(x)x + \cdots. \tag{4.289}$$

Setting $x = 0$ in equations (4.288) and (4.289) and substituting equations (4.284) and (4.287) yields

$$r(0) = v_0(0) = 1/b, \tag{4.290}$$

$$\begin{aligned} r'(0) &= v_0'(0) + v_1(0) \\ &= \frac{1}{2}\left(\frac{d}{b} - \frac{a}{b}\right). \end{aligned} \tag{4.291}$$

4.3.3. An Imbedding Method for a Nonlinear Two-Point Boundary-Value Problem

An imbedding method for a nonlinear two-point boundary-value problem is derived (Reference 24). The method makes use of Green's function to express the solution in terms of differential–integral equations.

The numerical solution is obtained using the method of lines. Numerical results are given and compared with the analytical solution.

Introduction

Consider the nonlinear two-point boundary-value problem

$$\ddot{u}(t) = \lambda e^{u(t)}, \qquad 0 \leq t \leq 1, \tag{4.292}$$

$$u(0) = 0, \quad u(1) = 0, \qquad 0 \leq \lambda \leq 1. \tag{4.293}$$

Problems of this type appear, for example, in one-dimensional magneto-hydrodynamics (Reference 1). The solution of the nonlinear two-point boundary-value problem is desired as a function of the parameter λ.

Using λ as an imbedding parameter (References 22, 25–27), differentiate $u = u(t, \lambda)$ with respect to λ:

$$\ddot{u}_\lambda(t, \lambda) = \lambda e^{u(t,\lambda)} u_\lambda(t, \lambda) + e^{u(t,\lambda)}, \tag{4.294}$$

$$u_\lambda(0, \lambda) = 0, \qquad u(1, \lambda) = 0, \tag{4.295}$$

where u_λ is the partial derivative of u with respect to λ. Equation (4.294) is a linear two-point boundary-value problem with variable coefficient, λe^u, and forcing function e^u.

Introduce the auxiliary function

$$\ddot{w}(t, \lambda) = \lambda e^{u(t,\lambda)} w(t, \lambda) + \phi(t, \lambda), \tag{4.296}$$

$$w(0, \lambda) = 0, \qquad w(1, \lambda) = 0 \tag{4.297}$$

with homogeneous boundary conditions and inhomogeneous forcing function, $\phi(t, \lambda)$. In order to evaluate this linear two-point boundary-value problem, use is made of Green's function.

Green's Function

The solution of equations (4.296) and (4.297) is given by

$$w(t, \lambda) = \int_0^1 G(t, y, \lambda)\phi(y, \lambda)\, dy, \tag{4.298}$$

where the kernel, $G(t, y, \lambda)$ is the Green's function of the problem. The solution is a linear function of the forcing term, $\phi(t, \lambda)$.

The method to be used for evaluating equation (4.298) is to express $w_\lambda(t, \lambda)$ in terms of Green's function in order to obtain an expression for $G_\lambda(t, y, \lambda)$. Differentiating equations (4.298), (4.296), and (4.297) with respect to λ gives

$$w_\lambda(t, \lambda) = \int_0^1 [G_\lambda(t, y, \lambda)\phi(y, \lambda) + G(t, y, \lambda)\phi_\lambda(y, \lambda)]\, dy, \tag{4.299}$$

$$\begin{aligned} \dot{w}_\lambda &= \lambda e^{u(t,\lambda)} w_\lambda + w e^u(\lambda u_\lambda + 1) + \phi_\lambda(t, \lambda), \\ &= \lambda e^{u(t,\lambda)} w_\lambda + F(t, \lambda), \end{aligned} \tag{4.300}$$

$$w_\lambda(0, \lambda) = 0, \qquad w_\lambda(1, \lambda) = 0, \tag{4.301}$$

where $F(t, \lambda)$ is the forcing function:

$$F(t, \lambda) = w(t, \lambda) e^{u(t,\lambda)}[\lambda u_\lambda(t, \lambda) + 1] + \phi_\lambda(t, \lambda). \tag{4.302}$$

The solution of equations (4.300) and (4.301) is given by

$$w_\lambda(t, \lambda) = \int_0^1 G(t, y, \lambda) F(y, \lambda)\, dy. \tag{4.303}$$

Substituting the expression for $F(t, \lambda)$ in equation (4.302) into equation (4.303) gives

$$w_\lambda(t, \lambda) = \int_0^1 G(t, y, \lambda)\{w(y, \lambda) e^{u(y,\lambda)}[\lambda u_\lambda(y, \lambda) + 1] + \phi_\lambda(y, \lambda)\}\, dy, \tag{4.304}$$

where t is replaced by y in $F(t, \lambda)$. Substituting the expression for $w(t, \lambda)$ in equation (4.298) into equation (4.304) yields

$$\begin{aligned} w_\lambda(t, \lambda) = \int_0^1 G(t, y, \lambda) \bigg\{ &\int_0^1 G(y, y', \lambda)\phi(y', \lambda)\, dy' \\ &\times e^{u(y,\lambda)}[\lambda u_\lambda(y, \lambda) + 1] + \phi_\lambda(y, \lambda) \bigg\}\, dy. \end{aligned} \tag{4.305}$$

Comparing equations (4.299) and (4.305), it is obvious that

$$\begin{aligned} &\int_0^1 G_\lambda(t, y, \lambda)\phi(y, \lambda)\, dy \\ &\quad = \int_0^1 G(t, y, \lambda) \bigg\{ \int_0^1 G(y, y', \lambda)\phi(y', \lambda)\, dy' e^{u(y,\lambda)}[\lambda u_\lambda(y, \lambda) + 1] \bigg\}\, dy \end{aligned} \tag{4.306}$$

since the last terms, $G(t, y, \lambda)\phi_\lambda(y, \lambda)$, in both equations (4.299) and (4.305) are identical.

Interchanging y and y' on the right side of equation (4.306) and rearranging yields

$$\int_0^1 G_\lambda(t, y, \lambda)\phi(y, \lambda)\, dy$$
$$= \int_0^1 G(t, y', \lambda)\left\{\int_0^1 G(y', y, \lambda)\phi(y, \lambda)\, dy e^{u(y',\lambda)}[\lambda u_\lambda(y', \lambda) + 1]\right\} dy'$$
$$= \int_0^1 \int_0^1 G(t, y', \lambda)\left\{e^{u(y',\lambda)}[\lambda u_\lambda(y', \lambda) + 1]\right\}G(y', y, \lambda)\, dy'\phi(y, \lambda)\, dy. \tag{4.307}$$

The coefficients of $\phi(y, \lambda)$ in equation (4.307) must be equal. Thus,

$$G_\lambda(t, y, \lambda) = \int_0^1 G(t, y', \lambda)\left\{e^{u(y',\lambda)}[\lambda u_\lambda(y', \lambda) + 1]\right\}G(y', y, \lambda)\, dy'. \tag{4.308}$$

Define

$$M(t, \lambda) = \int_0^1 G(t, y', \lambda)e^{u(y',\lambda)}\, dy', \tag{4.309}$$

$$u_\lambda(t, \lambda) = M(t, \lambda). \tag{4.310}$$

Then equation (4.308) becomes

$$G_\lambda(t, y, \lambda) = \int_0^1 G(t, y', \lambda)\left\{e^{u(y',\lambda)}[\lambda M(y', \lambda) + 1]\right\}G(y', y, \lambda)\, dy', \tag{4.311}$$

where $0 \le t, y \le 1$, and $0 \le \lambda \le 1$.

Equations (4.309) to (4.311) are the desired equations for solution. The differential–integral equations require an initial condition for $G(t, y, \lambda)$ at $\lambda = 0$.

Initial Conditions

In order to determine $G(t, y, 0)$, let $\lambda = 0$ in equations (4.296), (4.297), and (4.298). Then

$$\ddot{w}(t, 0) = \phi(t, 0), \tag{4.312}$$

$$w(0, 0) = 0, \qquad w(1, 0) = 1, \tag{4.313}$$

$$w(t, 0) = \int_0^1 G(t, y, 0)\phi(y, 0)\, dy. \tag{4.314}$$

The solution of differential equation (4.312) is given by

$$w(t, 0) = A + Bt + \int_0^t (t - y)\phi(y, 0)\, dy. \tag{4.315}$$

Using boundary conditions (4.313) to find A and B gives

$$A = 0, \tag{4.316}$$

$$B = \int_0^1 (y - 1)\phi(y, 0)\, dy. \tag{4.317}$$

Substituting the expressions for A and B into equation (4.315) yields

$$w(t, 0) = t\int_0^1 (y - 1)\phi(y, 0)\, dy + \int_0^t (t - y)\phi(y, 0)\, dy. \tag{4.318}$$

Comparing equations (4.314) and (4.318), it is obvious that the initial value of Green's function is given by

$$G(t, y, 0) = \begin{cases} y(t - 1), & 0 \le y \le t \\ t(y - 1), & t < y \le 1. \end{cases} \tag{4.319}$$

This function is continuous, but its derivative is discontinuous.

Method of Computation

The numerical solution was obtained using the method of lines (Reference 28). The quadrature formula used for approximating the integrals in equations (4.309) and (4.311) is the trapezoidal rule given by

$$\int_0^1 f(y)\, dy \cong \left[f(y_0) + f(y_n) + 2\sum_{m=1}^{n-1} f(y_m)\right]\frac{\Delta}{2}, \tag{4.320}$$

where the interval (0, 1) is divided into n equal parts of width $\Delta = 1/n$.

Introducing the nomenclature

$$M_i(\lambda) = M(i\Delta, \lambda) \tag{4.321}$$

$$u_i(\lambda) = u(i\Delta, \lambda) \tag{4.322}$$

$$G_{ij}(\lambda) = G(i\Delta, j\Delta, \lambda), \qquad 0 \le \lambda \le 1, \tag{4.323}$$

where

$$t = i\Delta, \tag{4.324}$$

$$y = j\Delta, \qquad i, j = 0, 1, 2, \ldots, n, \tag{4.325}$$

equations (4.309)–(4.311) become

$$M_i(\lambda) = \left[G_{i0}(\lambda)e^{u_0(\lambda)} + G_{in}(\lambda)e^{u_n(\lambda)} + 2\sum_{m=1}^{n-1} G_{im}(\lambda)e^{u_m(\lambda)} \right] \frac{\Delta}{2}, \tag{4.326}$$

$$\frac{du_i(\lambda)}{d\lambda} = M_i(\lambda), \tag{4.327}$$

$$\frac{dG_{ij}(\lambda)}{d\lambda} = \Big\{ G_{i0}(\lambda)e^{u_0(\lambda)}[\lambda M_0(\lambda)+1]G_{0j}(\lambda) + G_{in}(\lambda)e^{u_n(\lambda)}[\lambda M_n(\lambda)+1]G_{nj}(\lambda) + 2\sum_{m=1}^{n-1} G_{im}(\lambda)e^{u_m(\lambda)}[\lambda M_m(\lambda)+1]G_{mj}(\lambda) \Big\} \frac{\Delta}{2}. \tag{4.328}$$

This represents a system of $(n+1)+(n+1)^2$ ordinary differential equations which must be integrated. The initial conditions are

$$u_i(0) = 0, \qquad i = 0, 1, 2, \ldots, n, \tag{4.329}$$

$$G_{ij}(0) = \begin{cases} j\Delta(i\Delta - 1), & j \le i \\ i\Delta(j\Delta - 1), & i < j, \end{cases} \tag{4.330}$$

$$M_i(0) = \left[G_{i0}(0)e^{u_0(0)} + G_{in}(0)e^{u_n(0)} + 2\sum_{m=1}^{n-1} G_{im}(0)e^{u_m(0)} \right] \frac{\Delta}{2}, \tag{4.331}$$

where $i, j = 0, 1, 2, \ldots, n$.

Numerical Results

Numerical results were obtained using the trapezoidal rule with $n = 10$, i.e., 10 subintervals. Differential equations (4.327) and (4.328) were integrated using a fourth-order Runge–Kutta method with grid intervals, $\Delta\lambda = 0.1$.

A family of curves of $u(t, \lambda)$ as a function of t were obtained with λ as a parameter. Table 4.12 shows $u(t, \lambda)$ as a function of t for $\lambda = 1$. For $n = 10$, $u_i(\lambda)$ is available at increments of t equal to $1/n = 0.1$. Also shown in Table 4.12 is the analytical solution. The analytical solution (Reference 1) for $\lambda = 1$ is given by

$$u(t) = -\ln 2 + 2\ln\left\{ c \sec\left[\frac{c(t-1/2)}{2} \right] \right\},$$

where c is the root of the equation

$$\sqrt{2} = c\sec(c/4).$$

The value of c to eight digits is given by $c = 1.3360557$.

Table 4.12. Comparison of Imbedding and Analytical Solutions for $u(0, \lambda) = 0$, $\ddot{u}(t, \lambda) = e^{u(t,\lambda)}$; $u(1, \lambda) = 0$, $\lambda = 1$. Imbedding Method Uses Trapezoidal Rule with $n = 10$ and $\Delta\lambda = 0.1$

t	$u(t, \lambda)$	
	Imbedding method	Analytical solution
0	0	0
0.1	−0.0414044	−0.0414357
0.2	−0.0732143	−0.0732684
0.3	−0.0957303	−0.0957999
0.4	−0.109159	−0.109238
0.5	−0.113622	−0.113704
0.6	−0.109159	−0.109238
0.7	−0.0957303	−0.0957999
0.8	−0.0732143	−0.0732684
0.9	−0.0414044	−0.0414357
1	0	0

Table 4.12 shows that the imbedding method gives the correct solution to at least three digits. The maximum error is less than 0.0756%. The imbedding method requires the simultaneous integration of 132 differential equations. Computation time was less than 47 sec with no attempt made to optimize the computer program.

Numerical results were also obtained using the trapezoidal rule with $n = 5$ and integration grid intervals, $\Delta\lambda = 0.1$. The imbedding method gives the correct solution to at least two digits in this case, with maximum error less than 0.293%. The number of differential equations integrated simultaneously is 42. Computation time was less than 8 sec.

4.3.4. Post-Buckling Beam Configurations via an Imbedding Method

An imbedding method for the post-critical equilibrium forms of a beam built in at one end and free at the other is derived (Reference 29). The method makes use of Green's function to express the solution in terms of differential–integral equations. Since Green's function becomes extremely large in the vicinity of the critical force, the function must be initialized in the post-critical region. The appropriate initial conditions are derived and numerical results are obtained. The numerical results are compared with an analytical solution.

Introduction

Consider the post-critical deformation of a compressed vertical beam clamped at the lower end and with the upper end free (References 17, 18, 30, 31). A force p is applied at the free end, maintaining its direction along the original uncompressed axial position. Since the deflections cannot be considered small compared to the length of the beam, the exact expression for the curvature of the elastic curve must be used. The curvature, K, is given by

$$K = M/EI, \tag{4.332}$$

where E is the modulus of elasticity, I is the moment of inertia, and M is the bending moment. The differential equation of the elastic curve can be shown from (4.332) to be given by

$$\frac{d^2\theta}{ds^2} = -k \sin \theta, \tag{4.333}$$

with boundary conditions for a unit length arc

$$\theta(0) = 0, \qquad \dot{\theta}(1) = 0, \tag{4.334}$$

where $\theta(s)$ is the angle of inclination with the vertical, s is the arc length, $k = p/EI$, p is the applied force, and $\dot{\theta}(s) = d\theta/ds$. The critical force at which lateral buckling occurs is easily derived from the linear differential equation and boundary conditions for small deflections:

$$\frac{d^2\theta}{ds^2} = -k\theta, \tag{4.335}$$

$$\theta(0) = 0, \qquad \dot{\theta}(1) = 0, \tag{4.336}$$

and is given by

$$p_{\text{crit}} = \pi^2 EI/4 \tag{4.337}$$

or

$$k_{\text{crit}} = \pi^2/4. \tag{4.338}$$

An imbedding method for the post-critical equilibrium forms of the beam is derived. The method makes use of Green's function. Since Green's function becomes extremely large in the vicinity of the critical force, the function must be initialized in the post-critical region. This is done by evaluating $\theta(s)$ for an arbitrary value of k, i.e., $k = k_1 > k_{\text{crit}}$, in the post-critical region via quasilinearization. The imbedding equations are then integrated from $k = 0$ to the desired maximum $k = k_{\max}$, with only minor modifications in the imbedding equations at $k = k_1$. For $k \geq k_1$, a family of curves of $\theta(s)$ as a function of s with k as a parameter is obtained via the imbedding method.

The results show that problems of this type, which previously were solved using the elliptic integrals, can be effectively handled by computers via imbedding methods. Additional work in this area is required, however.

Imbedding Method

The imbedding equations (References 22, 25) are derived as follows. The nonlinear two-point boundary-value equations (4.333) and (4.334) are rewritten in the form

$$\ddot{u}(t) = -k \sin u, \tag{4.339}$$

$$u(0) = 0, \qquad \dot{u}(1) = 0, \tag{4.340}$$

where

$$u(t) = \theta(s), \tag{4.341}$$

$$\ddot{u}(t) = \frac{d^2\theta}{ds^2}, \tag{4.342}$$

$$t = s, \tag{4.343}$$

$$k = p/EI. \tag{4.344}$$

Using k as a parameter, differentiate equations (4.339) and (4.340) with respect to k:

$$\ddot{u}_k(t, k) = -k(\cos u)u_k - \sin u, \tag{4.345}$$

$$u_k(0, k) = 0, \qquad \dot{u}_k(1, k) = 0. \tag{4.346}$$

Introduce the auxiliary function

$$\ddot{w}(t, k) = -k(\cos u)w + \phi(t, k), \tag{4.347}$$

$$w(0, k) = 0, \qquad \dot{w}(1, k) = 0, \tag{4.348}$$

where $\phi(t, y)$ is the inhomogeneous forcing function

$$\phi(t, k) = -\sin u. \tag{4.349}$$

The solution of equations (4.347) and (4.348) is given by

$$w(t, k) = \int_0^1 G(t, y, k)\phi(y, k)\, dy, \tag{4.350}$$

where the kernel, $G(t, y, k)$ is the Green's function of the problem. The solution is a linear function of the forcing term, $\phi(t, k)$.

Differentiating equations (4.350), (4.347), and (4.348) with respect to k gives

$$w_k(t, k) = \int_0^1 [G_k(t, y, k)\phi(y, k) + G(t, y, k)\phi_k(y, k)]\, dy, \quad (4.351)$$

$$\begin{aligned} \ddot{w}_k &= -k(\cos u)w_k + k(\sin u)u_k w - (\cos u)w + \phi_k(t, k) \\ &= -k(\cos u)w_k + F(t, k), \end{aligned} \quad (4.352)$$

$$w_k(0, k) = 0, \qquad \dot{w}_k(1, k) = 0, \quad (4.353)$$

where $F(t, k)$ is the forcing function

$$F(t, k) = wn(t, k) + \phi_k(t, k) \quad (4.354)$$

and

$$n(t, k) = k(\sin u)u_k - \cos u. \quad (4.355)$$

The solution of equations (4.352) and (4.353) is given by

$$w_k(t, y) = \int_0^1 G(t, y, k)F(y, k)\, dy. \quad (4.356)$$

Substituting the expression for $F(t, k)$ in equation (4.354) into equation (4.356) yields

$$w_k(t, k) = \int_0^1 G(t, y, k)[w(y, k)n(y, k) + \phi_k(y, k)]\, dy, \quad (4.357)$$

where t is replaced by y in $F(t, k)$. Substituting the expression for $w(t, k)$ in equation (4.350) into equation (4.357) gives

$$w_k(t, k) = \int_0^1 G(t, y, k)\left[\int_0^1 G(y, y', k)\phi(y', k)\, dy' n(y, k) + \phi_k(y, k)\right] dy. \quad (4.358)$$

Comparing equations (4.351) and (4.358), it is obvious that

$$\begin{aligned} &\int_0^1 G_k(t, y, k)\phi(y, k)\, dy \\ &\quad = \int_0^1 G(t, y, k)\left[\int_0^1 G(y, y', k)\phi(y', k)\, dy' n(y, k)\right] dy \end{aligned} \quad (4.359)$$

since the last terms, $G(t, y, k)\phi_k(y, k)$, in both equations (4.351) and (4.358) are identical.

Interchanging y and y' on the right side of equation (4.359) and rearranging, we obtain

$$\begin{aligned}\int_0^1 G_k(t, y, k)\phi(y, k)\, dy &= \int_0^1 G(t, y', k)\left[\int_0^1 G(y', y, k)\phi(y, k)\, dy n(y', k)\right] dy' \\ &= \int_0^1 \int_0^1 G(t, y', k)n(y', k)G(y', y, k)\, dy'\phi(y, k)\, dy. \end{aligned} \tag{4.360}$$

The coefficients of $\phi(y, k)$ in equation (4.360) must be equal. Thus

$$G_k(t, y, k) = \int_0^1 G(t, y', k)n(y', k)G(y', y, k)\, dy'. \tag{4.361}$$

By definition, set equation (4.350) equal to $M(t, k)$. Then using equations (4.350), (4.361), and (4.355) we have

$$M(t, k) = -\int_0^1 G(t, y, k) \sin u(y, k)\, dy, \tag{4.362}$$

$$u_k(t, k) = M(t, k), \tag{4.363}$$

$$G_k(t, y, k) = \int_0^1 G(t, y', k)\, n(y', k)G(y', y, k)\, dy', \tag{4.364}$$

where

$$n(y', k) = k \sin u(y', k)M(y', k) - \cos u(y', k). \tag{4.365}$$

Equations (4.362) to (4.365) are the desired imbedding equations. These differential–integral equations require an initial condition at $k = 0$.

Initial Conditions

In order to determine $G(t, y, 0)$, let $k = 0$ in equations (4.347), (4.348), and (4.350):

$$\ddot{w}(t, 0) = \phi(t, 0), \tag{4.366}$$

$$w(0, 0) = 0, \qquad \dot{w}(1, 0) = 0, \tag{4.367}$$

$$w(t, 0) = \int_0^1 G(t, y, 0)\phi(y, 0)\, dy. \tag{4.368}$$

The solution of differential equation (4.366) is given by

$$w(t, 0) = A + Bt + \int_0^t (t - y)\phi(y, 0)\, dy. \tag{4.369}$$

Using boundary conditions (4.367) to find A and B it can be shown that

$$A = 0, \tag{4.370}$$

$$B = -\int_0^1 \phi(y, 0)\, dy. \tag{4.371}$$

Substituting the expressions for A and B into equation (4.369) yields

$$w(t, 0) = -t \int_0^1 \phi(y, 0)\, dy + \int_0^t (t - y)\phi(y, 0)\, dy. \tag{4.372}$$

Comparing equations (4.368) and (4.372), it is obvious that the initial value of Green's function is given by

$$G(t, y, 0) = \begin{cases} -y, & 0 \leq y \leq t \\ -t, & t < y \leq 1. \end{cases} \tag{4.373}$$

The initial condition for $u(t, k)$ at $k = 0$ is given by

$$u(t, 0) = 0. \tag{4.374}$$

If the numerical solution were obtained for imbedding equations (4.362) to (4.365) with initial conditions (4.373) and (4.374), Green's function would become extremely large in the vicinity of the critical value of k

$$k_{\text{crit}} = \pi^2/4 = 2.4674. \tag{4.375}$$

The computer would generate an overflow error message and the computation would come to a halt. Numerical results in the post-critical region, where $k > k_{\text{crit}}$, would not be obtained.

To overcome this obstacle, the solution of the nonlinear two-point boundary-value equations (4.339) and (4.340) is obtained at some convenient point in the post-critical region, for example at $k = 3$, using the method of quasilinearization. Green's function is obtained by integrating equation (4.364) from $k = 0$ to $k = 3$ with

$$n(y', k) = n(y', 3) = -\cos u(y', 3), \tag{4.376}$$

$$u(t, k) = u(t, 3) \tag{4.377}$$

and with initial condition (4.373). This gives Green's function at $k = 3$. Then adjoining equations (4.362) and (4.363) and with $n(y', k)$ given by equation (4.365), continue integrating from $k = 3$ to the desired maximum $k = k_{\max}$. This will give a complete set of solutions at each increment of Δk in the post-critical region between $k = 3$ and $k = k_{\max}$.

For convenience, equations (4.362)–(4.365) can be integrated directly from $k = 0$ to $k = k_{\max}$, with $M(t, k)$ set equal to zero at each increment of Δk from $k = 0$ to $k = 3$. This will yield Green's function for $u(t, 3)$ at $k = 3$. For $k > 3$, $M(t, k)$ is no longer set equal to zero, thus allowing $u(t, k)$ to vary as a function of k. The derivation of the equations for the computation of Green's function at $k = 3$ is given in the next section.

Computation of Green's Function at $k = 3$.

Starting with equations (4.345) and (4.346) introduce the auxiliary function

$$\begin{aligned} \ddot{w}(t, q) &= -q(\cos u)w - \sin u \\ &= qn(t)w + \phi(t) \end{aligned} \tag{4.378}$$

$$w(0, q) = 0, \qquad \dot{w}(1, q) = 0, \tag{4.379}$$

where k has been replaced by q. The functions $\phi(t)$ and $n(t)$ are independent of q and are given for $k = 3$ by

$$\phi(t) = -\sin u(t, 3), \tag{4.380}$$

$$n(t) = -\cos u(t, 3). \tag{4.381}$$

The solution of equations (4.378) and (4.379) is given by

$$w(t, q) = \int_0^1 H(t, y, q)\phi(y)\, dy, \tag{4.382}$$

where $H(t, y, q)$ is Green's function. Differentiating equations (4.382), (4.378), and (4.379) with respect to q yields

$$w_q(t, q) = \int_0^1 H_q(t, y, q)\phi(y)\, dy, \tag{4.383}$$

$$\ddot{w}_q(t, q) = qn(t)w_q + n(t)w, \tag{4.384}$$

$$w_q(0, q) = 0, \qquad \dot{w}_q(1, q) = 0. \tag{4.385}$$

The solution of equations (4.384) and (4.385) is given by

$$w_q(t, q) = \int_0^1 H(t, y, q)n(y)w(y, q)\,dy. \tag{4.386}$$

Substituting the expression for $w(t, q)$ in equation (4.382) into equation (4.386) gives

$$w_q(t, q) = \int_0^1 H(t, y, q)n(y) \int_0^1 H(y, y', q)\phi(y')\,dy'dy. \tag{4.387}$$

Interchanging y and y' in equation (4.387) and rearranging yields

$$w_q(t, q) = \int_0^1 \int_0^1 H(t, y', q)n(y')H(y', y, q)\,dy'\phi(y)\,dy. \tag{4.388}$$

Comparing equations (4.383) and (4.388) it is obvious that

$$H_q(t, y, q) = \int_0^1 H(t, y', q)n(y')H(y', y, q)\,dy', \tag{4.389}$$

where

$$n(y') = -\cos u(y', 3). \tag{4.390}$$

The initial conditions for Green's function can be shown to be given by

$$H(t, y, 0) = \begin{cases} -y, & 0 \le y \le t \\ -t, & t < y \le 1. \end{cases} \tag{4.391}$$

The numerical values of $u(t, 3)$ are derived in the next section. Comparing equations (4.364) and (4.373) with equations (4.389) and (4.391), it is seen that the equations have the same form. The only differences are in $n(y', k)$ and $u(t, k)$. Thus integrating equation (4.389) from $k = 0$ to $k = 3$ with initial condition (4.391) and $u(t, k) = u(t, 3)$ gives Green's function, $G(t, y, k)$, at $k = 3$. This is equivalent to integrating equation (4.364) with $M(t, k) = 0$ and $u(t, k) = u(t, 3)$.

Quasilinearization

The numerical solution of $u(t, k)$ at $k = 3$ is obtained by the method of quasilinearization. Let $x_1 = u(t)$ in equations (4.339) and (4.340). Then we have

$$\dot{x}_1 = x_2, \qquad x_1(0) = 0, \tag{4.392}$$

$$\dot{x}_2 = -k \sin x_1, \qquad x_2(1) = 0. \tag{4.393}$$

Let $x_1{}^0$ and $x_2{}^0$ be the current approximations to the functions $x_1(t)$ and $x_2(t)$. The new approximations are obtained by considering the linearized equations

$$\dot{x}_1 = x_2{}^0 + (x_2 - x_2{}^0), \tag{4.394}$$

$$\dot{x}_2 = -k \sin x_1{}^0 - k \cos x_1{}^0(x_1 - x_1{}^0). \tag{4.395}$$

Equations (4.394) and (4.395) can be expressed in the form

$$\begin{bmatrix} \dot{x}_1 \\ \dot{x}_2 \end{bmatrix} = \begin{bmatrix} 0 & 1 \\ -k \cos x_1{}^0 & 0 \end{bmatrix} \begin{bmatrix} x_1 \\ x_2 \end{bmatrix} + \begin{bmatrix} 0 \\ -k \sin x_1{}^0 + (k \cos x_1{}^0) x_1{}^0 \end{bmatrix} \tag{4.396}$$

or using matrix notation, the $i + 1$ approximation is

$$\dot{\mathbf{x}}_{i+1} = F_i(t)\mathbf{x}_{i+1} + \mathbf{y}_i(t). \tag{4.397}$$

The numerical solution to linear equation (4.397) is obtained using the method of complementary functions. The solution is given by the linear combination

$$\mathbf{x}_{i+1}(t) = \mathbf{p}_{i+1}(t) + H_{i+1}(t)\mathbf{c}, \tag{4.398}$$

where $\mathbf{p}_{i+1}$ and H_{i+1} are the solutions of the initial-value equations

$$\dot{\mathbf{p}}_{i+1}(t) = F_i(t)\mathbf{p}_{i+1}(t) + \mathbf{y}_i(t), \qquad \mathbf{p}_{i+1}(0) = 0, \tag{4.399}$$

$$\dot{H}_{i+1}(t) = F_i(t)H_{i+1}(t), \qquad H_{i+1}(0) = I \tag{4.400}$$

and the vector $\mathbf{c}$ is obtained by substituting the boundary conditions into equation (4.398) and solving for $\mathbf{c}$.

To obtain a numerical solution for $k = 3$, an initial guess is made for the vector $\mathbf{x}(t)$, for example

$$x_1(t) = 1.5 \sin(1.57t), \tag{4.401}$$

$$x_2(t) = 1.5 \cos(1.57t). \tag{4.402}$$

Equations (4.399) and (4.400) are integrated from $t = 0$ to $t = 1$. The values of $\mathbf{p}(t)$ and $H(t)$ at the boundaries are utilized to evaluate the vector $\mathbf{c}$, which is equal to the full set of initial conditions of the first approximation. Equation (4.397) is then integrated to obtain the first approximation. The first approximation is stored and the above sequence is repeated to obtain the second approximation, etc.

A fourth-order Runge–Kutta method was utilized for the integration with grid intervals equal to 1/100. Convergence was obtained in approxi-

Table 4.13. Angle of Inclination as a Function of Arc Length for $k = 3$; Quasilinearization Method

s	$\theta(s)$ (rad)
0	0
0.125	0.246932
0.25	0.482505
0.375	0.696494
0.5	0.8806
0.625	1.02873
0.75	1.13685
0.875	1.20252
1.0	1.22454

mately four iterations. Table 4.13 shows the numerical results in the post-buckling region for $k = 3$ where

$$u(t, 3) = x_1(t) = \theta(s). \tag{4.403}$$

It should be noted that if the initial guess for vector $\mathbf{x}(t)$ were $\mathbf{x}(t) = 0$, the solution would remain at zero.

Method of Computation

The numerical solution of imbedding equations (4.362)–(4.365) was obtained using the method of lines (Reference 28). The quadrature formula used for approximating the integrals in equations (4.362) and (4.364) is the trapezoidal rule given by

$$\int_0^1 f(y)\,dy \cong \left[f(y_0) + f(y_n) + 2 \sum_{m=1}^{n-1} f(y_m)\right] \frac{\Delta}{2}, \tag{4.404}$$

where the interval (0, 1) is divided into n equal parts of width $\Delta = 1/n$.

Introducing the nomenclature

$$M_i(k) = M(i\Delta, k), \tag{4.405}$$

$$u_i(k) = u(i\Delta, k), \tag{4.406}$$

$$G_{ij}(k) = G(i\Delta, j\Delta, k), \tag{4.407}$$

$$n_j(k) = n(j\Delta, k), \tag{4.408}$$

where

$$t = i\Delta, \tag{4.409}$$

$$y = j\Delta, \qquad i, j = 0, 1, 2, \ldots, n, \tag{4.410}$$

equations (4.362)–(4.365) become

$$M_i(k) = -\Big[G_{i0}(k) \sin u_0(k) + G_{in}(k) \sin u_n(k) + 2 \sum_{m=1}^{n-1} G_{im}(k) \sin u_m(k) \Big] \frac{\Delta}{2}, \tag{4.411}$$

$$\frac{du_i(k)}{dk} = M_i(k), \tag{4.412}$$

$$\frac{dG_{ij}(k)}{dk} = \Big[G_{i0}(k) n_0(k) G_{0j}(k) + G_{in}(k) n_n(k) G_{nj}(k) + 2 \sum_{m=1}^{n-1} G_{im}(k) n_m(k) G_{mj}(k) \Big] \frac{\Delta}{2}, \tag{4.413}$$

where

$$n_j(k) = k \sin u_j(k) M_j(k) - \cos u_j(k). \tag{4.414}$$

This represents a system of $(n + 1) + (n + 1)^2$ ordinary differential equations which must be integrated. The initial conditions are

$$G_{ij}(0) = \begin{cases} -j\Delta, & j \leq i \\ -i\Delta, & i < j. \end{cases} \tag{4.415}$$

Equations (4.412) and (4.413) are integrated from $k = 0$ to $k = 3$ with

$$u_i(k) = u_i(3) = \theta_i \tag{4.416}$$

given in Table 4.13 and $M_i(k)$ set equal to zero at each increment of Δk. This yields Green's function for $u_i(k)$ at $k = 3$. For $k > 3$, $M_i(k)$ is determined by equation (4.411), thus allowing $u_i(k)$ to vary as a function of k.

Elliptic Integral Solution

An analytical solution for the post-critical equilibrium forms of the beam can be obtained in terms of the elliptic integral. Vol'mir (Reference 17) shows that the elliptic integral is related to k by the equation

$$K(m) = \int_0^{\pi/2} (1 - m \sin^2 \alpha)^{-1/2} \, d\alpha = \sqrt{k}\, l, \tag{4.417}$$

where

$$l = \text{length (equal to 1)}, \quad (4.418)$$

$$m = \sin^2 \psi. \quad (4.419)$$

To obtain the deflection of the beam, y, at the free end [not related to the y in Green's function, $G(t, y, k)$], the elliptic integral was approximated by the polynomial (Reference 32)

$$K(m) = \sqrt{k}\, l = (a_0 + a_1 m_1 + \cdots + a_4 m_1^4) + (b_0 + b_1 m_1 + \cdots + b_4 m_1^4) \ln(1/m_1), \quad (4.420)$$

where

$$m_1 = 1 - m. \quad (4.421)$$

For a given value of k, the value of m was found from equation (4.420) using a simple computer subroutine. The deflection of the beam is then (Reference 17)

$$y = 2(m/k)^{1/2}. \quad (4.422)$$

The angle of inclination at the free end, $\theta(1)$, can be obtained for a given value of k from the equation

$$m = \sin^2 (\theta/2), \quad (4.423)$$

$$\theta(1) = 2 \sin^{-1} \sqrt{m}. \quad (4.424)$$

The deflection and the angle of inclination were computed as described above to an accuracy of approximately five digits.

Numerical Results

Numerical results were obtained using imbedding equations (4.411)–(4.414) with $n = 8$ for the trapezoidal rule. Differential equations (4.412) and (4.413) were integrated using a fourth-order Runge–Kutta method with grid intervals, $\Delta k = 0.1$.

A family of curves of $u(t, k)$ as a function of t was obtained with k as a parameter. In the nomenclature of the beam, these curves correspond to $\theta(s, k)$ as a function of s with k as a parameter. Table 4.14 shows the angle of inclination at the free end as a function of k. Shown are the solutions obtained via the imbedding and the elliptic integral methods. The imbedding method is correct to approximately three digits. Assuming the elliptic integral method gives the true solution, the maximum error of the imbedding solution is less than 0.089% and occurs at $k = 3.5$.

Table 4.14. Angle of Inclination at the Free End as a Function of k

k	$\theta(1)$ (rad)	
	Imbedding method	Elliptic integral method
3.0	1.22454	1.22452
3.5	1.61095	1.60952
4.0	1.86411	1.86264
4.5	2.04895	2.04768
5.0	2.19166	2.19066
5.5	2.30582	2.30510
6.0	2.39944	2.39899

The deflection of the beam at the free end is obtained by integrating

$$\frac{dy}{ds} = \sin\theta \tag{4.425}$$

to obtain

$$y = \int_0^1 \sin\theta(s, k)\, ds. \tag{4.426}$$

Using the imbedding method to obtain $\theta(s)$ for a given k, the deflection y was obtained using Simpson's rule. Table 4.15 shows the deflection as a function of k obtained via the imbedding and elliptic integral methods. The maximum error of the imbedding method solution is less than 0.069% and occurs at $k = 3.5$. Vol'mir shows the curve of the above deflection as

Table 4.15. Deflection as a Function of k

k	Deflection y	
	Imbedding method	Elliptic integral method
3.0	0.663656	0.663629
3.5	0.770952	0.770421
4.0	0.802915	0.802407
4.5	0.805731	0.805265
5.0	0.795664	0.795217
5.5	0.779743	0.779293
6.0	0.761327	0.760857

Table 4.16. Angle of Inclination as a Function of Arc Length for $k = 6$

s	$\theta(s)$ (rad)	
	Imbedding method	Quasilinearization method
0	0	0
0.125	0.565361	0.561929
0.25	1.0806	1.07542
0.375	1.51329	1.50826
0.5	1.85258	1.84874
0.625	2.10198	2.09953
0.75	2.2707	2.26935
0.875	2.36779	2.36714
1.0	2.39944	2.39901

a function of k. The imbedding method yields the numerical values for this type of curve routinely. The number of differential equations integrated for the imbedding method was 90.

Since $\theta(s)$ is available as a function of s at every increment of Δk between $k = 3$ and $k = 6$, it is of interest to compare $\theta(s)$ at $k = 6$ obtained via the imbedding method with $\theta(s)$ obtained via quasilinearization as shown in Table 4.16. The maximum error of the imbedding method solution is less than 0.61% and occurs at $s = 0.125$. At $s = 1$, the maximum error is less than 0.018%.

4.3.5. A Sequential Method for Nonlinear Filtering

Exact equations are presented for sequentially updating the optimal solution for a discrete time analog of the basic Sridhar nonlinear filtering problem as the process length increases and new observations are obtained (Reference 33). A tabular method is described for numerically implementing the sequential filtering equations. The accuracy and efficiency of the tabular method are illustrated by means of several numerical examples.

Introduction

In a series of studies (References 34–36) undertaken in the 1960s, R. Sridhar and associates developed a theory of filtering for nonlinear continuous time processes. The basic problem under consideration was the estimation of state variables $x(t)$ generated by a noisy nonlinear dynamical system

$$\dot{x}(t) = F(x(t)) + \varepsilon(t), \qquad t \in [0, T]$$

with observations $y(t)$ obtained in the form

$$y(t) = x(t) + \eta(t), \qquad t \in [0, T],$$

where little if any prior statistical knowledge was available concerning the modeling and observational error terms $\varepsilon(t)$ and $\eta(t)$. Various problems in rigid body dynamics having this format were then being investigated.

Sridhar emphasized that the objective of the estimation should be to recover the actual state values, not to obtain an ensemble averaging over possible state values. Since neither the dynamical equations nor the observations were exact, one should form a cost of estimation criterion function $J(x)$ which pays attention to both dynamical and observational potential sources of error. For example, one might consider a weighted sum of squared errors

$$J^*(x) \equiv \int_0^T [y(t) - x(t)]^2 \, dt + \int_0^T k(t)[\dot{x}(t) - F(x(t))]^2 \, dt$$

for some suitable weight function $k(t)$. Experiments with D. Detchmendy demonstrated the basic feasibility of this idea for problems in rigid body dynamics.

Originally the calculus of variations was used to obtain minimal cost solutions $\hat{x}$. Emphasis soon shifted to sequential solutions, and a merging was made with ideas developed in the theory of invariant imbedding for obtaining solutions of nonlinear two-point boundary-value problems as functions of interval length and parameters appearing in the boundary conditions. The resulting partial differential equations had to be solved on line to obtain updated current state estimates. Since these equations were typically intractable for practical implementation, Sridhar proposed a quadratic approximation technique for solving these equations. Later, M. Sugisaka and S. Sagara (Reference 37) demonstrated that significant improvement could be obtained by using higher-order terms in the approximation.

More recently (Reference 38), using imbedding methods, an exact procedure was obtained for sequentially updating the solution of the nonlinear two-point boundary-value problem associated with the discrete time analog of the basic Sridhar filtering problem with cost functional $J^*(x)$. The imbedding was based on two physically meaningful parameters, namely, the duration T of the dynamical process and the value of y_T of the final observation. The numerical instability problems which can arise when imbedding is based on artificially introduced parameters, as in the original Sridhar approach, were thus avoided. More importantly, this choice of imbedding parameters allowed the exact derivation of the sequential filtering equations, without any need for approximations. Although no statistical assumptions were used for the modeling and observational error terms, the sequential filtering equations obtained in the linear case were shown to be analogous in form to the standard Kalman filter state estimation equations. In contrast, the sequential state filtering equations obtained in the nonlinear case were shown *not* to be analogous in form to the standard extended Kalman filter state estimation equations.

In this section we focus on the numerical implementation of the sequential filtering equations derived in Reference 38. We first formulate the discrete time nonlinear filtering problem, and derive the associated nonlinear two-point boundary-value problem. The sequential filtering equations for generating solutions to the nonlinear two-point boundary-value problem are stated and validated. We then describe the tabular method for numerically implementing the sequential filtering equations, which appears to have certain advantages over previous imbedding approaches.

Specifically, in using the Sridhar imbedding approach, one first introduces a basic quadratic approximation

$$x(T, \mu) \cong \alpha(T) + \beta(T)\mu + \delta(T)\mu^2$$

for the optimal estimator of the current state $x(T)$ as a function of the current time T and some parameter μ appearing in the boundary conditions. The coefficients $\alpha(T)$, $\beta(T)$, and $\delta(T)$ are then sequentially updated as new observations are obtained. Although experiments have shown this approximation to be satisfactory for certain rigid body problems, there is always the possibility of significant error accumulation if the actual optimal state estimator is not quadratic. The same difficulty arises for higher-order approximations.

In contrast, apart from round-off error, the tabular method sequentially generates the *exact* optimal filtered and smoothed state estimates

$$x(0, \beta_T, T), x(1, \beta_T, T), \ldots, x(T, \beta_T, T)$$

for $x_0, \ldots, x_T$ as a function of the process duration T and the final observation β_T, over a grid of possible values for β_T. When an actual observation y_T for the current time T is realized, state estimates $\hat{x}(t \mid T) = x(t, y_T, T)$

are obtained by interpolation from this table of grid values. However, the error introduced into these latter estimates by interpolation does not affect the table, and hence is not cumulative.

The table itself is updated in an exceedingly straightforward manner, necessitating the introduction of only one new column of values at each time T. It is important to note that each of the table entries $x(t, \beta_T, T)$ is meaningful and potentially useful as a sensitivity check, being the optimal smoothed estimate for the state x_t at time t based on the observations $y_0, \ldots, y_{T-1}, \beta_T$. In contrast, the Sridhar state estimates $x(T, \mu)$ are obtained by an imbedding on an artificially introduced parameter μ, and they have no physical interpretation when $\mu \neq 0$.

The tabular method is illustrated by means of several numerical examples. The tabular solutions are compared with the solutions obtained using a Newton–Raphson procedure as a check on their accuracy and efficiency.

The Nonlinear Filtering Problem

Consider a nonlinear one-dimensional dynamical system described by

$$x_{t+1} = F(x_t) + \varepsilon_t, \qquad t = 0, \ldots, T-1. \tag{4.427}$$

where ε_t represents an unknown modeling error. The problem under consideration is the estimation of the state variables x_t on the basis of observations y_t obtained in the form

$$y_t = x_t + \eta_t, \qquad t = 0, \ldots, T, \tag{4.428}$$

where η_t represents an unknown observational error.

No statistical assumptions will be used for the error terms ε_t and η_t. Rather, we consider the problem of minimizing the least-squares criterion function

$$W(x, \varepsilon) \equiv \left[\sum_{t=0}^{T} (x_t - y_t)^2 + k \sum_{t=0}^{T-1} \varepsilon_t^2\right] \tag{4.429a}$$

with respect to $x_0, \ldots, x_T, \varepsilon_0, \ldots, \varepsilon_{T-1}$, subject to the restrictions

$$\mathbf{0} = V(x, \varepsilon) \equiv \begin{pmatrix} x_1 - F(x_0) - \varepsilon_0 \\ \vdots \\ x_T - F(x_{T-1}) - \varepsilon_{T-1} \end{pmatrix}, \tag{4.429b}$$

where k is a fixed positive scalar weight. Thus, state and error estimates are to be obtained at each time T by minimizing a sum of weighted squared residual errors.

An Associated Two-Point Boundary-Value Problem

There are several ways to derive a two-point boundary-value problem representation for the necessary conditions which must be satisfied by any optimal solution for problem (4.429). The representation established below proved to be particularly useful in the subsequent derivation of a sequential estimation procedure, for, as will be seen, it allows one to convert the two-point boundary-value problem into an initial-value problem via an imbedding on the two physically meaningful parameters T and y_T.

Specifically, defining the Lagrangian function by $L(x, \varepsilon, \mu) \equiv W(x, \varepsilon) + \mu V(x, \varepsilon)$, where $\mu \equiv (\mu_1, \ldots, \mu_T)$ is a vector of Lagrange multipliers, the Euler–Lagrange first-order conditions $\mathbf{0} = \partial L/\partial(x, \varepsilon, \mu)$ can be expressed in the form

$$0 = 2[x_t - y_t] + \mu_t - F'(x_t)\mu_{t+1}, \qquad t = 0, \ldots, T-1, \tag{4.430a}$$

$$0 = \mu_0, \tag{4.430b}$$

$$0 = x_T + \tfrac{1}{2}\mu_T - y_T, \tag{4.430c}$$

$$0 = 2k\varepsilon_t - \mu_{t+1}, \qquad t = 0, \ldots, T-1, \tag{4.430d}$$

$$\mathbf{0} = V(x, \varepsilon). \tag{4.430e}$$

Straightforward substitution then leads to the equivalent two-point boundary-value representation for problem 4.430:

$$0 = 2[x_t - y_t] + \mu_t - F'(x_t)\mu_{t+1}, \qquad t = 0, \ldots, T-1 \tag{4.431a}$$

$$0 = x_{t+1} - F(x_t) - \frac{1}{2k}\mu_{t+1}, \qquad t = 0, \ldots, T-1, \tag{4.431b}$$

$$0 = \mu_0, \tag{4.431c}$$

$$y_T = x_T + \tfrac{1}{2}\mu_T. \tag{4.431d}$$

Since the rank of $\partial V/\partial(x, \varepsilon)$ is T, independently of the trajectory point (x, ε) at which it is evaluated, equations (4.431) are necessary for a trajectory of values x_t and $\varepsilon_t \equiv \mu_{t+1}/2k$ to solve the original cost minimization problem (4.429). In subsequent sections it is assumed that equations (4.431) have a unique solution $(x_t, \mu_t)_{t=0}^{T}$ for each $T > 0$ and $y_T \in (-\infty, \infty)$. For $T = 0$, the unique solution is defined to be $x_0 = y_0$ and $\mu_0 = 0$. Thus, equations (4.431) are sufficient as well as necessary for a solution to (4.429), assuming a solution to (4.429) exists.

Finally, it is also assumed in subsequent sections that for each $T \geqq 0$ and $y_T \in (-\infty, \infty)$ the solution to equations (4.431) has a unique continuation over the interval $[T, T+1]$.

The Sequential Filtering Equations

The filtering and smoothing equations presented below provide an exact sequential procedure for updating the solution values $(x_t, \mu_t)_{t=0}^T$ for the nonlinear two-point boundary problem (4.431) when the duration of the process is increased from T to $T+1$ and an additional observation y_{T+1} is obtained.

Let $x(t, \beta_T, T)$ and $\mu(t, \beta_T, T)$ denote the tth period solution values for problem (4.431) when the duration of the process is T, $T \geqq t \geqq 0$, and the final observation takes on the value β_T, $-\infty < \beta_T < \infty$. Define

$$\varrho_T(\beta_T) \equiv x(T, \beta_T, T), \qquad T \geqq 0, \qquad -\infty < \beta_T < \infty, \tag{4.432a}$$

$$\gamma_T(\beta_T) \equiv \mu(T, \beta_T, T), \qquad T \geqq 0, \qquad -\infty < \beta_T < \infty, \tag{4.432b}$$

$$\hat{x}(t \mid T) \equiv x(t, y_T, T), \qquad T \geqq t \geqq 0, \tag{4.432c}$$

$$\hat{\mu}(t \mid T) \equiv \mu(t, y_T, T), \qquad T \geqq t \geqq 0. \tag{4.432d}$$

Note that (4.432c) and (4.432d) are the solution values for the nonlinear two-point boundary-value problem (4.431) with duration T and observations $y_0, \ldots, y_T$, hence also optimal solution values for the original least-squares estimation problem (4.429) under the uniqueness and existence assumptions stated above.

Let t now denote a fixed time point, $t \geqq 0$, and let the running variable T denote current time. The following procedure sequentially updates both the optimal state and error estimates $\hat{x}(T \mid T)$ and $\hat{\mu}(T \mid T)$ for the current time T, and the optimal state and error smoothed estimates $\hat{x}(t \mid T)$ and $\hat{\mu}(t \mid T)$ for time t, $T \geqq t$. The procedure can easily be modified to permit the simultaneous updating of smoothed estimates for arbitrarily many fixed time points t.

At time $T = 0$ the initial observation y_0 is known. For each β_0, $-\infty < \beta_0 < \infty$, determine in order

$$\varrho_0(\beta_0) = \beta_0, \tag{4.433a}$$

$$\gamma_0(\beta_0) = 0, \tag{4.433b}$$

$$x(0, \beta_0, 0) = \varrho_0(\beta_0), \tag{4.433c}$$

$$\mu(0, \beta_0, 0) = \gamma_0(\beta_0). \tag{4.433d}$$

The optimal state and error estimates for time $T = 0$ are

$$\hat{x}(0 \mid 0) = x(0, y_0, 0), \tag{4.434a}$$

$$\hat{\mu}(0 \mid 0) = \mu(0, y_0, 0). \tag{4.434b}$$

At time $T + 1$, $0 \leqq T < t$, if such a T exists, the observations $y_0, \ldots, y_T$ and functions $\varrho_T(\cdot)$ and $\gamma_T(\cdot)$ are known. For each β_T, $-\infty < \beta_T < \infty$, determine in order

$$\beta_{T+1} = F(\varrho_T(\beta_T)) + \left(\frac{k+1}{k}\right)\left[\frac{\beta_T - y_T}{F'(\varrho_T(\beta_T))}\right], \tag{4.435a}$$

$$\varrho_{T+1}(\beta_{T+1}) = \left(\frac{k}{k+1}\right)F(\varrho_T(\beta_T)) + \left(\frac{1}{k+1}\right)\beta_{T+1}, \tag{4.435b}$$

$$\gamma_{T+1}(\beta_{T+1}) = 2[\beta_{T+1} - \varrho_{T+1}(\beta_{T+1})], \tag{4.435c}$$

$$x(T+1, \beta_{T+1}, T+1) = \varrho_{T+1}(\beta_{T+1}), \tag{4.435d}$$

$$\mu(T+1, \beta_{T+1}, T+1) = \gamma_{T+1}(\beta_{T+1}). \tag{4.435e}$$

Obtain an additional observation y_{T+1}. The optimal updated state and error estimates for time $T + 1$ are

$$\hat{x}(T+1 \mid T+1) = x(T+1, y_{T+1}, T+1), \tag{4.436a}$$

$$\hat{\mu}(T+1 \mid T+1) = \mu(T+1, y_{T+1}, T+1). \tag{4.436b}$$

At time $T + 1$, $0 \leqq T = t$, the functions $\varrho_t(\cdot)$ and $\gamma_t(\cdot)$ are known. For each β_t, $-\infty < \beta_t < \infty$, set

$$x(t, \beta_t, t) = \varrho_t(\beta_t), \tag{4.437a}$$

$$\mu(t, \beta_t, t) = \gamma_t(\beta_t). \tag{4.437b}$$

At time $T + 1$, $T \geqq t$, the observations $y_0, \ldots, y_T$ and functions $\varrho_T(\cdot)$, $\gamma_T(\cdot)$, $x(t, \cdot, T)$, and $\mu(t, \cdot, T)$ are known. For each β_T, $-\infty < \beta_T < \infty$, determine in order

$$\beta_{T+1} = F(\varrho_T(\beta_T)) + \left(\frac{k+1}{k}\right)\left[\frac{\beta_T - y_T}{F'(\varrho_T(\beta_T))}\right], \tag{4.438a}$$

$$\varrho_{T+1}(\beta_{T+1}) = \left(\frac{k}{k+1}\right)F(\varrho_T(\beta_T)) + \left(\frac{1}{k+1}\right)\beta_{T+1}, \tag{4.438b}$$

$$\gamma_{T+1}(\beta_{T+1}) = 2[\beta_{T+1} - \varrho_{T+1}(\beta_{T+1})], \tag{4.438c}$$

$$x(t, \beta_{T+1}, T+1) = x(t, \beta_T, T), \tag{4.438d}$$

$$\mu(t, \beta_{T+1}, T+1) = \mu(t, \beta_T, T), \tag{4.438e}$$

$$x(T+1, \beta_{T+1}, T+1) = \varrho_{T+1}(\beta_{T+1}), \tag{4.438f}$$

$$\mu(T+1, \beta_{T+1}, T+1) = \gamma_{T+1}(\beta_{T+1}). \tag{4.438g}$$

Obtain an additional observation y_{T+1}. The optimal updated state and error estimates for time t and $T+1$ for the process of duration $T+1$ are

$$\hat{x}(t \mid T+1) = x(t, y_{T+1}, T+1), \tag{4.439a}$$

$$\hat{\mu}(t \mid T+1) = \mu(t, y_{T+1}, T+1), \tag{4.439b}$$

$$\hat{x}(T+1 \mid T+1) = x(T+1, y_{T+1}, T+1), \tag{4.439c}$$

$$\hat{\mu}(T+1 \mid T+1) = \mu(T+1, y_{T+1}, T+1). \tag{4.439d}$$

The uniqueness properties assumed for the solution of problem (4.431) for each $T \geqq 0$ and $y_T \in (-\infty, \infty)$ guarantee that the map $\beta_T \mapsto \beta_{T+1}$ defined by (4.438a) is one-to-one and onto. Thus, $\varrho_{T+1}(\cdot)$, $\gamma_{T+1}(\cdot)$, $x(t, \cdot, T+1)$ and $\mu(t, \cdot, T+1)$ are well-defined functions of β_{T+1}, $-\infty < \beta_{T+1} < \infty$, as required for the feasibility of (4.439). Interpretations for the basic recurrence relations (4.438a) and (4.438b) are provided in Reference 38.

Validation for the Sequential Filtering Equations

Consider the two-point boundary-value problem (4.431) for arbitrary duration time $T \geqq 1$ and arbitrary inhomogeneous value $\beta_T \in (-\infty, \infty)$ for the terminal boundary condition,

$$0 = 2[x_t - y_t] + \mu_t - F'(x_t)\mu_{t+1}, \qquad t = 0, \ldots, T-1, \tag{4.440a}$$

$$0 = x_{t+1} - F(x_t) - \frac{1}{2k}\mu_{t+1}, \qquad t = 0, \ldots, T-1, \tag{4.440b}$$

$$0 = \mu_0, \tag{4.440c}$$

$$\beta_T = x_T + \tfrac{1}{2}\mu_T. \tag{4.440d}$$

It will be shown that the values $x(t, \beta_T, T)$ and $\mu(t, \beta_T, T)$ generated by the sequential estimation equations developed above are solution values for problem (4.440).

The proof will proceed by induction on the process duration length T. The obvious way to carry out the induction proof is to begin with a proof for $T = 1$, then establish a proof for duration length $T+1$ assuming a proof has previously been established for duration length T, $T \geqq 1$. However, a more straightforward and revealing induction proof is obtained by proceeding equation by equation.

Consider equation (4.440a) for arbitrarily selected t, $t \geqq 0$. Let $T = t + 1$. This is the first value of T for which equation (4.440a) must be validated. It must be shown that

$$0 = 2[x(t, \beta_{t+1}, t+1) - y_t] + \mu(t, \beta_{t+1}, t+1) \\ - F'(x(t, \beta_{t+1}, t+1))\mu(t+1, \beta_{t+1}, t+1) \tag{4.441}$$

for arbitrary β_{t+1} in R. By (4.437)–(4.438), there exists some β_t in R such that

$$\begin{aligned} 2[y_t - x(t, \beta_{t+1}, t+1)] &+ F'(x(t, \beta_{t+1}, t+1))\mu(t+1, \beta_{t+1}, t+1) \\ &= 2[y_t - x(t, \beta_t, t)] + F'(x(t, \beta_t, t))\mu(t+1, \beta_{t+1}, t+1) \\ &= 2[y_t - \varrho_t(\beta_t)] + F'(\varrho_t(\beta_t))\gamma_{t+1}(\beta_{t+1}) \\ &= 2[y_t - \varrho_t(\beta_t)] + 2F'(\varrho_t(\beta_t))[\beta_{t+1} - \varrho_{t+1}(\beta_{t+1})] \\ &= 2[y_t - \varrho_t(\beta_t)] + 2[\beta_t - y_t] \\ &= 2[\beta_t - \varrho_t(\beta_t)], \end{aligned} \tag{4.442}$$

and, by (4.433)–(4.438),

$$\mu(t, \beta_{t+1}, t+1) = \mu(t, \beta_t, t) = \gamma_t(\beta_t) = 2[\beta_t - \varrho_t(\beta_t)]. \tag{4.443}$$

Combining (4.442) and (4.443), (4.441) is established.

Now suppose it has been established that

$$0 = 2[x(t, \beta_T, T) - y_t] + \mu(t, \beta_T, T) \\ - F'(x(t, \beta_T, T))\mu(t+1, \beta_T, T) \tag{4.444}$$

for some value $T \geqq t+1$ and arbitrary β_T in R. By (4.438), for any β_{T+1} in R there exists some β_T in R such that

$$x(t, \beta_{T+1}, T+1) = x(t, \beta_T, T), \tag{4.445a}$$

$$\mu(t, \beta_{T+1}, T+1) = \mu(t, \beta_T, T), \tag{4.445b}$$

$$\mu(t+1, \beta_{T+1}, T+1) = \mu(t+1, \beta_T, T). \tag{4.445c}$$

Combining (4.445) with the induction hypothesis (4.444), we obtain

$$0 = 2[x(t, \beta_{T+1}, T+1) - y_t] + \mu(t, \beta_{T+1}, T+1) \\ - F'(x(t, \beta_{T+1}, T+1))\mu(t+1, \beta_{T+1}, T+1) \tag{4.446}$$

for arbitrary β_{T+1} in R. It follows, by induction, that equation (4.440a) is satisfied for arbitrary $t \geqq 0$, $T \geqq t+1$, and β_T in R by the values $x(t, \beta_T, T)$, $\mu(t, \beta_T, T)$, and $\mu(t+1, \beta_T, T)$ generated by the sequential filtering equations. Validation of equation (4.440b) proceeds in an entirely analogous manner.

Consider the boundary condition (4.440c). It must be shown that

$$0 = \mu(0, \beta_T, T) \tag{4.447}$$

for arbitrary $T \geqq 1$ and β_T in R. By (4.433), $\mu(0, \beta_0, 0) = 0$ for arbitrary β_0 in R. Suppose (4.447) holds for arbitrary β_T in R for some $T \geqq 0$. Letting $t = 0$, it follows by (4.438) that for arbitrary β_{T+1} in R there exists β_T in R such that

$$\mu(0, \beta_{T+1}, T+1) = \mu(0, \beta_T, T). \tag{4.448}$$

Combining (4.448) with the induction hypothesis, we obtain

$$0 = \mu(0, \beta_{T+1}, T+1) \tag{4.449}$$

for arbitrary β_{T+1} in R. It follows by induction that equation (4.440c) is satisfied for arbitrary $T \geqq 1$ and β_T in R by the values $\mu(0, \beta_T, T)$ generated by the sequential filtering equations.

Finally, consider the terminal boundary condition (4.440d). It must be shown that

$$\beta_T = x(T, \beta_T, T) + \tfrac{1}{2}\mu(T, \beta_T, T) \tag{4.450}$$

for arbitrary $T \geqq 1$ and β_T in R. By (4.433), (4.50) holds for arbitrary β_0 in R when $T = 0$. Suppose (4.450) holds for arbitrary β_T in R for some $T \geqq 0$. Letting $t = 0$, it follows by (4.438) that for arbitrary β_{T+1} in R,

$$\begin{aligned} x(T+1, \beta_{T+1}, T+1) &+ \tfrac{1}{2}\mu(T+1, \beta_{T+1}, T+1) \\ &= \varrho_{T+1}(\beta_{T+1}) + \tfrac{1}{2}\gamma_{T+1}(\beta_{T+1}) \\ &= \varrho_{T+1}(\beta_{T+1}) + [\beta_{T+1} - \varrho_{T+1}(\beta_{T+1})] = \beta_{T+1}. \end{aligned} \tag{4.451}$$

It follows by induction that equation (4.440d) is satisfied for arbitrary $T \geqq 1$ and β_T in R by the values $x(T, \beta_T, T)$ and $\mu(T, \beta_T, T)$ generated by the sequential filtering equations.

The validation of the sequential filtering equations is thus complete.

Numerical Implementation

The numerical implementation of the sequential filtering equations proceeds by what may appropriately be termed the tabular method. Consider, first, the basic filtering equations for the state estimates $\hat{x}(T+1 \mid T+1), T \geqq 0$,

$$\beta_{T+1} = F(\varrho_T(\beta_T)) + \left(\frac{k+1}{k}\right)\left[\frac{\beta_T - y_T}{F'(\varrho_T(\beta_T))}\right], \tag{4.452a}$$

$$\varrho_{T+1}(\beta_{T+1}) = \left(\frac{k}{k+1}\right)F(\varrho_T(\beta_T)) + \left(\frac{1}{k+1}\right)\beta_{T+1}, \quad -\infty < \beta_T < \infty, \tag{4.452b}$$

$$\hat{x}(T+1 \mid T+1) = \varrho_{T+1}(y_{T+1}). \tag{4.452c}$$

The process (4.452) is initialized by setting

$$\varrho_0(\beta_0) = \beta_0, \; -\infty < \beta_0 < \infty, \tag{4.452d}$$

$$\hat{x}(0 \mid 0) = \varrho_0(y_0). \tag{4.452e}$$

At time $T = 0$ the function $\varrho_0(\cdot)$ is stored in tabular form,

$$\begin{matrix} 0 & \beta_0 \\ \begin{bmatrix} r_{00} \\ \vdots \\ r_{0m} \end{bmatrix} & \begin{bmatrix} b_{00} \\ \vdots \\ b_{0m} \end{bmatrix} \end{matrix}, \tag{4.453}$$

for a suitably fine grid of values $b_{00}, \ldots, b_{0m}$ for β_0, with $b_{0l} \geqq b_{0,l+1}$, $0 \leqq l \leqq m-1$. An initial observation y_0 is obtained, and the initial state estimate $\hat{x}(0 \mid 0)$ is set equal to y_0.

Table (4.453) is then directly updated via (4.452a) and (4.452b). Specifically, for each value b_{0j} for β_0 one calculates corresponding values

$$b_{1j} \equiv F(r_{0j}) + \left(\frac{k+1}{k}\right)\left[\frac{b_{0j} - y_0}{F'(r_{0j})}\right], \tag{4.454a}$$

$$r_{1j} \equiv \left(\frac{k}{k+1}\right)F(r_{0j}) + \left(\frac{1}{k+1}\right)b_{1j}; \tag{4.454b}$$

for β_1 and $\varrho_1(\beta_1)$. Table (4.453) is then replaced by

$$\begin{matrix} 1 & \beta_1 \\ \begin{bmatrix} r_{10} \\ \vdots \\ r_{1m} \end{bmatrix} & \begin{bmatrix} b_{10} \\ \vdots \\ b_{1m} \end{bmatrix} \end{matrix}. \tag{4.454}$$

An observation y_1 is obtained which satisfies

$$b_{1,j-1} \geqq y_1 \geqq b_{1j} \tag{4.455}$$

for some j. If, for example, linear interpolation is to be used, an approximation for $\hat{x}(1 \mid 1) = \varrho_1(y_1)$ is then obtained by setting

$$\hat{x}(1 \mid 1) = \left[\frac{r_{1j-1} - r_{1j}}{b_{1j-1} - b_{1j}}\right][y_1 - b_{1j}] + r_{1j}. \tag{4.456}$$

Schematically, one simply moves across table (4.454) from right to left at the appropriate β_1 level:

$$\begin{array}{ccccc} 1 & & \beta_1 & & \\ \begin{bmatrix} r_{10} \\ \vdots \\ r_{1,j-1} \\ r_{1j} \\ \vdots \\ r_{1m} \end{bmatrix} & \longleftarrow & \begin{bmatrix} b_{10} \\ \vdots \\ b_{1,j-1} \\ b_{1j} \\ \vdots \\ b_{1m} \end{bmatrix} & \longleftarrow & y_1 \end{array} . \tag{4.457}$$

For arbitrary time $T \geqq 0$ the values

$$\begin{array}{cc} T & \beta_T \\ \begin{bmatrix} r_{T0} \\ \vdots \\ r_{Tm} \end{bmatrix} & \begin{bmatrix} b_{T0} \\ \vdots \\ b_{Tm} \end{bmatrix} \end{array} \tag{4.458}$$

are in storage, and a new observation y_{T+1} is obtained. The equations for generating $\hat{x}(T+1 \mid T+1)$ then take the form

$$b_{T+1,l} = F(r_{Tl}) + \left(\frac{k+1}{k}\right)\left[\frac{b_{Tl} - y_T}{F'(r_{Tl})}\right], \tag{4.459a}$$

$$r_{T+1,l} = \left(\frac{k}{k+1}\right)F(r_{Tl}) + \left(\frac{1}{k+1}\right)b_{T+1,l}, \qquad l = 0, \ldots, m, \tag{4.459b}$$

$$\hat{x}(T+1 \mid T+1) = \left(\frac{r_{T+1,j-1} - r_{T+1,j}}{b_{T+1,j-1} - b_{T+1,j}}\right)[y_{T+1} - b_{T+1,j}] + r_{T+1,j}, \tag{4.459c}$$

for $b_{T+1,j-1} \geqq y_{T+1} \geqq b_{T+1,j}$, and the new table

$$\overset{T+1}{\begin{bmatrix} r_{T+1,0} \\ \vdots \\ r_{T+1,m} \end{bmatrix}} \quad \overset{\beta_{T+1}}{\begin{bmatrix} b_{T+1,0} \\ \vdots \\ b_{T+1,m} \end{bmatrix}} \tag{4.460}$$

is stored in place of (4.458).

To obtain the complete trajectory of estimates $\hat{x}(t \mid T+1)$, $0 \leqq t \leqq T+1$, for time $T+1$, the only modification is that the intermediate tables labeled $0, 1, \ldots, T$ must be retained. The estimates $\hat{x}(t \mid T+1)$ are found by interpolating across all of the tables simultaneously, i.e.,

$$\hat{x}(t \mid T+1) = \left(\frac{r_{t,j-1} - r_{tj}}{b_{T+1,j-1} - b_{T+1,j}}\right)[y_{T+1} - b_{T+1,j}] + r_{tj}, \quad 0 \leqq t \leqq T+1, \tag{4.461}$$

for $b_{T+1,j-1} \geqq y_{T+1} \geqq b_{T+1,j}$. Schematically, we have

$$\overset{0}{\begin{bmatrix} r_{00} \\ \vdots \\ r_{0,j-1} \\ r_{0j} \\ \vdots \\ r_{0m} \end{bmatrix}} \leftarrow \overset{1}{\begin{bmatrix} r_{10} \\ \vdots \\ r_{1,j-1} \\ r_{1j} \\ \vdots \\ r_{1m} \end{bmatrix}} \cdots \leftarrow \overset{T}{\begin{bmatrix} r_{T0} \\ \vdots \\ r_{T,j-1} \\ r_{Tj} \\ \vdots \\ r_{Tm} \end{bmatrix}} \leftarrow \overset{T+1}{\begin{bmatrix} r_{T+1,0} \\ \vdots \\ r_{T+1,j-1} \\ r_{T+1,j} \\ \vdots \\ r_{T+1,m} \end{bmatrix}} \leftarrow \overset{\beta_{T+1}}{\begin{bmatrix} b_{T+1,0} \\ \vdots \\ \beta_{T+1,j-1} \\ \beta_{T+1,j} \\ \vdots \\ b_{T+1,m} \end{bmatrix}} \leftarrow y_{T+1}. \tag{4.462}$$

It is important to note that the solution trajectory (4.461) is exact, apart from round-off errors, if the observation y_{T+1} at time $T+1$ coincides with a value $b_{T+1,j}$ in the β_{T+1} table (4.460). Errors are introduced by interpolation into the state estimates $\hat{x}(t \mid T+1)$. However, these interpolation errors do not affect the tables labeled $0, \ldots, T+1$, β_{T+1}, hence they are not cumulative.

An Illustrative Numerical Example

The numerical sequential filtering procedure outlined above will be used to generate state estimates $\hat{x}(t \mid T)$, $0 \leqq t \leqq T$, for an illustrative nonlinear filtering problem. The resulting estimates will be compared with the state estimates obtained for the same nonlinear filtering problem using a Newton–Raphson method.

Consider the nonlinear one-dimensional dynamic system described by

$$x_{t+1} = ax_t + bx_t^2, \qquad t = 0, \ldots, T-1, \tag{4.463a}$$

$$x_0 = c, \tag{4.463b}$$

with measurements

$$y_t = x_t + \eta_t, \qquad t = 0, \ldots, T, \tag{4.463c}$$

and coefficient values

$$a = 0.99, \qquad b = 0.2, \qquad c = 1.0. \tag{4.463d}$$

Two different specifications were tested for the measurement noise η_t, namely, a constant bias specification

$$\eta_t = 0.05, \qquad t = 0, \ldots, T, \tag{4.464}$$

and a zero-mean Gaussian white noise specification

$$\eta_t \sim N(0, 0.02), \qquad t = 0, \ldots, T, \tag{4.465a}$$

$$E\eta_t\eta_s = 0 \qquad \text{for } t \neq s. \tag{4.465b}$$

System (4.463) was first simulated for $\eta_t \equiv 0.05$, and the resulting values $\{x_0^*, y_0^*, \ldots, x_T^*, y_T^*\}$ were recorded. Using the observation values $\{y_0^*, \ldots, y_T^*\}$, state estimates $\hat{x}(t \mid T)$, $0 \leqq t \leqq T$, were then determined for system (4.463) using the numerical sequential filtering procedure outlined above with k arbitrarily set equal to 1.0. The initial table selected for $\varrho_0(\beta_0) \equiv \beta_0$ and β_0 was

$$\begin{array}{cc} 0 & \beta_0 \\ \begin{bmatrix} 1.050 \\ 1.049 \\ \vdots \\ 1.031 \end{bmatrix} & \begin{bmatrix} 1.050 \\ 1.049 \\ \vdots \\ 1.031 \end{bmatrix} \end{array}. \tag{4.466}$$

Both linear and quadratic interpolation were used to generate the state estimates $\hat{x}(t \mid T)$. For linear interpolation, equation (4.461) was used. For quadratic interpolation, the second-degree Lagrange polynomials were used. For example, suppose $b_{T+1,i-1} > b_{T+1,i} > b_{T+1,i+1}$ are three neighboring values in the table for β_{T+1}, and the observation y_{T+1} for period $T+1$ lies between $b_{T+1,i-1}$ and $b_{T+1,i}$. The needed Lagrange polynomials are then

$$L_0 \equiv \frac{(y_{T+1}^* - b_{T+1,i})(y_{T+1}^* - b_{T+1,i+1})}{(b_{T+1,i-1} - b_{T+1,i})(b_{T+1,i-1} - b_{T+1,i+1})}, \tag{4.467a}$$

$$L_1 \equiv \frac{(y_{T+1}^* - b_{T+1,i-1})(y_{T+1}^* - b_{T+1,i+1})}{(b_{T+1,i} - b_{T+1,i-1})(b_{T+1,i} - b_{T+1,i+1})}, \tag{4.467b}$$

$$L_2 \equiv \frac{(y_{T+1}^* - b_{T+1,i-1})(y_{T+1}^* - b_{T+1,i})}{(b_{T+1,i+1} - b_{T+1,i-1})(b_{T+1,i+1} - b_{T+1,i})}. \tag{4.467c}$$

The smoothed state estimate $\hat{x}(t \mid T)$ for x_t based on the observations y_0^*, $\ldots$, y_T^* is then given in approximate form by

$$\hat{x}(t \mid T) = r_{t,i-1}L_0 + r_{ti}L_1 + r_{t,i+1}L_2, \tag{4.468}$$

$0 \leqq t \leqq T$.

Similar tests were conducted for system (4.463) using various sample runs for η_t specified as in (4.465).

State estimates were also generated for system (4.463) using a Newton–Raphson method. Specifically, a Newton–Raphson method was first used to generate approximate solution values for the two-point nonlinear boundary-value problem with $\eta_t \equiv 0.05$, i.e.,

$$0 = 2[x_t - y_t^*] + \mu_t - F'(x_t)\mu_{t+1}, \qquad t = 0, \ldots, T-1, \tag{4.469a}$$

$$0 = x_{t+1} - F(x_t) - \frac{1}{2k}\mu_{t+1}, \qquad t = 0, \ldots, T-1, \tag{4.469b}$$

$$0 = \mu_0, \tag{4.469c}$$

$$y_T^* = x_T + \tfrac{1}{2}\mu_T, \tag{4.469d}$$

where

$$F(x) \equiv ax + bx^2, \tag{4.469e}$$

$$F'(x) \equiv a + 2bx, \tag{4.469f}$$

and the observation values $y_0^*, \ldots, y_T^*$ are the same as were used to generate the sequential filter state estimates $\hat{x}(t \mid T)$, $0 \leqq t \leqq T$. The weight factor k was again set equal to 1.0.

Specifically, let $x(t, c)$ and $\mu(t, c)$, $0 \leqq t \leqq T$, denote the solution values for system (4.469) with initial condition $x_0 = c$. Expanding the boundary condition (4.469d) in a Taylor series around the nth iteration approximation c^n for c, and retaining only the linear terms,

$$y_T^* = x(T, c^n) + \tfrac{1}{2}\mu(T, c^n) + x_c(T, c^n)(c^{n+1} - c^n) + \tfrac{1}{2}\mu_c(T, c^n)(c^{n+1} - c^n), \tag{4.470}$$

where $x_c(\cdot) \equiv dx(\cdot)/dc$ and $\mu_c(\cdot) \equiv d\mu(\cdot)/dc$. Thus

$$c^{n+1} = c^n - [x_c(T, c^n) + \tfrac{1}{2}\mu_c(T, c^n)]^{-1}[x(T, c^n) + \tfrac{1}{2}\mu(T, c^n) - y_T^*]. \tag{4.471}$$

Equations for $x_c(t + 1, c^n)$ and $\mu_c(t + 1, c^n)$, $0 \leqq t \leqq T - 1$, are obtained by differentiating equations (4.469a) and (4.469b) with respect to c, i.e.,

$$\mu_c(t + 1, c) = \frac{\partial [F'(x(t, c))]^{-1}}{\partial c} \{2[x(t, c) - y_t^*] + \mu(t, c)\} + F'(x(t, c))^{-1}[2x_c(t, c) + \mu_c(t, c)], \tag{4.472a}$$

$$x_c(t + 1, c) = \frac{\partial F(x(t, c))}{\partial c} + \tfrac{1}{2}\mu_c(t + 1, c), \tag{4.472b}$$

with initial conditions

$$\mu_c(0, c) = 0, \tag{4.472c}$$

$$x_c(0, c) = 0. \tag{4.472d}$$

To obtain approximate solution values $x(0, \tilde{c})$, $\mu(0, \tilde{c})$, $\ldots$, $x(T, \tilde{c})$, $\mu(T, \tilde{c})$ for (4.469), an initial approximation c^0 is first selected for the initial state c. The difference equations (4.469a)–(4.469c) and (4.472) are then integrated from time $t = 0$ to time $t = T - 1$ to obtain $x(T, c^0)$, $\mu(T, c^0)$, $x_c(T, c^0)$, and $\mu_c(T, c^0)$. A new approximation c^1 for c is then determined by equation (4.471), and the above procedure is repeated with c^1 in place of c^0. Iterations continue until the sequence $c^0, c^1, c^2, \ldots$ appears to have approximately converged to some limit $\tilde{c}$. For the example at hand,

$$\frac{\partial [F'(x(t, c))]^{-1}}{\partial c} = -\frac{2b}{[a + 2bx(t, c)]^2} x_c(t, c), \tag{4.473a}$$

$$\frac{\partial F(x(t, c))}{\partial c} = [a + 2bx(t, c)]x_c(t, c). \tag{4.473b}$$

A similar procedure was used to obtain Newton–Raphson state estimates $x(t, \tilde{c})$ for system (4.463) with η_t specified as in (4.465).

Table 4.17 gives the sequentially filtered state estimates $\hat{x}(T \mid T)$ and Newton–Raphson filtered state estimates $x(T, \tilde{c})$ for the terminal time T

Table 4.17. Filtered Terminal State Estimates for Measurement Noises η_t Set Identically Equal to 0.05

Terminal time T	Actual state $x_T{}^*$	Actual observation $y_T{}^*$	Sequential solution $\hat{x}(T \mid T)$		Newton–Raphson solution $x(T, \tilde{c})$
			Linear interpolation	Quadratic interpolation	
0	1.0	1.05	1.05	1.05	1.05
1	1.19	1.24	1.24502	1.24502	1.24502
2	1.46132	1.51132	1.51989	1.51989	1.51989
3	1.8738	1.9238	1.93496	1.93496	1.93496
4	2.55728	2.60728	2.62062	2.62062	2.62062
5	3.83965	3.88965	3.9048$\underline{5}$	3.90483	3.90483
6	6.74984	6.79984	6.81$\underline{603}$	6.8158$\underline{7}$	6.81588
7	15.7944	15.8444	15.85$\underline{9}$	15.8585	15.8585

varying from $T = 0$ to $T = 7$. The measurement noise terms η_t are set identically equal to 0.05. The sequential estimates $\hat{x}(T \mid T)$ are given for both linear and quadratic interpolation. The sequential solution digits which do not compare with the Newton–Raphson solution are shown underlined in the table. Approximately four-digit accuracy is obtained for linear interpolation and approximately six-digit accuracy is obtained for quadratic interpolation, assuming the Newton–Raphson method gives the correct solution. It should be noted that the Newton–Raphson method requires an iterative solution based on all observations each time a new observation is obtained, whereas the sequential filtering method only requires the successive updating of tables for $\varrho_T(\beta_T)$ and β_T and interpolation to determine the value of $\hat{x}(T \mid T)$, $T \geq 0$.

Table 4.18 gives the sequentially determined smoothed state estimates $\hat{x}(t \mid T)$ and Newton–Raphson smoothed state estimates $x(t, \tilde{c})$ for t varying from $t = 0$ to $t = T$, and the terminal time T set equal to 7. The measurement noise terms η_t were again set identically equal to 0.05. The sequential solution with quadratic interpolation agrees exactly with the Newton–Raphson solution to at least six digits.

Table 4.19 gives the sequentially determined smoothed state estimates $\hat{x}(t \mid T)$ and Newton–Raphson smoothed state estimates $x(t, \tilde{c})$, $0 \leqq t \leqq T$, with terminal time T set equal to 7 and measurement noise terms η_t generated by a zero mean Gaussian white noise process with standard deviation $\sigma = 0.02$. The initial grid interval for the β_0 table was increased from 0.001

Table 4.18. Smoothed State Estimates for Terminal Time $T = 7$ and Measurement Noises η_t Set Identically Equal to 0.05

Time t	Actual state x_t^*	Actual observation y_t^*	Sequential solution quadratic interpolation $\hat{x}(t \mid T)$	Newton–Raphson solution $x(t, \tilde{c})$
0	1.0	1.05	1.0332	1.0332
1	1.19	1.24	1.22439	1.22439
2	1.46132	1.51132	1.49332	1.49332
3	1.8738	1.9238	1.90131	1.90131
4	2.55728	2.60728	2.57926	2.57926
5	3.83965	3.88965	3.85725	3.85725
6	6.74984	6.79984	6.77101	6.77101
7	15.7944	15.8444	15.8585	15.8585

to 0.002 with 20 values in the table in all, starting from the upper bound 1.010. The sequential solution with quadratic interpolation agrees with the Newton–Raphson solution to at least five digits.

The accuracy of the sequential solution can be increased by decreasing the initial grid interval for the β_0 table and increasing the number of values

Table 4.19. Smoothed State Estimates for Terminal Time $T = 7$ and Measurement Noises η_t Generated by a Zero Mean Gaussian White Noise Process with Standard Deviation $\sigma = 0.02$

Time t	Actual state x_t^*	Actual observation y_t^*	Sequential solution quadratic interpolation $\hat{x}(t \mid T)$	Newton–Raphson solution $x(t, \tilde{c})$
0	1.0	0.98842	1.00054	1.00054
1	1.19	1.22034	1.19947	1.19947
2	1.46132	1.46408	1.46696	1.46696
3	1.8738	1.87889	1.87928	1.87928
4	2.55728	2.59243	2.56509	2.56509
5	3.83965	3.8313	3.84095	3.84095
6	6.74984	6.75597	6.75121	6.75122
7	15.7944	15.7959	15.7975	15.7977

in the table. The initial ranges of β_0 values were small, i.e., 1.031–1.050 for η_t identically equal to 0.05 and 0.972–1.010 for η_t generated by the Gaussian process. These ranges were found to be adequate, in the sense that the observation values $y_0^*, \ldots, y_T^*$ were well within the ranges of the updated β tables.

Discussion

Exact equations have been presented for sequentially updating the solution of a certain nonlinear two-point boundary-value problem associated with a discrete time analog of the basic Sridhar nonlinear filtering problem. A tabular method for implementing the sequential filtering equations has been described and illustrated.

The particular criterion function (4.429a) selected for the Sridhar problem can be generalized in a variety of ways, e.g., by the incorporation of statistical considerations, without affecting the basic tabular method. The underlying estimation problem is inherently nonlinear, independently of the selected criterion function.

More generally, the basic tabular method can be applied to the solution of nonlinear two-point boundary-value problems for difference equations, irrespective of their origin. The tabular method can be used to sequentially update the solutions of such problems as the process length increases.

Exercises

1. Verify the numerical results in Section 4.2.1 using the Newton–Raphson method.
2. Derive the two-point boundary-value equations for the example in Section 4.2.2 using Pontryagin's maximum principle.
3. Derive the Euler–Lagrange equations for the example given in Section 4.2.3.
4. Derive the quasilinearization equations for obtaining the solution of the two-point boundary-value equations in Exercise 3.
5. Derive the two-point boundary-value equations for minimizing the cost functional

$$J = \frac{1}{2}\int_0^T (x^2 + y^2)\,dt$$

subject to the nonlinear differential constraint and initial condition

$$\dot{x} = -x + ax^3 + y, \qquad x(0) = c.$$

Use Pontryagin's maximum principle.

6. Derive the quasilinearization equations for obtaining the solution of the two-point boundary-value equations in Exercise 5.

7. Derive the Newton–Raphson equations for obtaining the solution of the two-point boundary-value equations in Exercise 5.

8. Given the nonlinear two-point boundary-value problem

$$\ddot{\mu}(t) = e^{\mu}, \qquad \mu(0) = 0, \qquad \mu(1) = 0,$$

use quasilinearization to obtain the linearized problem and find the solution, $\mu_1(t)$. Assume $\mu_0(t) = 0$, $0 \leq t \leq 1$.

PART III

System Identification

5

Gauss–Newton Method for System Identification

The unknown system parameters are estimated using least-squares methods (References 1–16). When the measurements are linear in the parameters to be estimated, the least-squares estimates of constant unknown parameters can be obtained using a one-step procedure. No a priori estimates of the unknown parameters are required. For dynamic systems with measurements nonlinear in the parameters, such as those of interest in this text, iterative methods are required as well as initial estimates of the unknown parameters.

In this chapter, least-squares estimation, maximum likelihood estimation, and the Cramér–Rao lower bound are briefly discussed followed by the Gauss–Newton iterative method. Several examples are given with the equations used in the Gauss–Newton method derived.

5.1. Least-Squares Estimation

Least-squares methods are used for the estimation of the unknown parameters. The performance criterion for least-squares estimation is given by

$$J = \mathbf{e}^T\mathbf{e} = \sum_{i=1}^{L} \mathbf{e}_i^2, \tag{5.1}$$

where $\mathbf{e}_i$ is the m-dimensional column vector

$$\mathbf{e}_i = \mathbf{z}(t_i) - \mathbf{y}(t_i), \qquad i = 1, 2, \ldots, L, \tag{5.2}$$

$\mathbf{z}(t_i)$ is the vector of observations at time t_i, $\mathbf{y}(t_i)$ is the model response at time t_i, and L is the number of observation times.

Assume that $\mathbf{x}$ is a constant n-dimensional vector of parameters to be estimated with the observations linear in $\mathbf{x}$

$$\mathbf{z} = H\mathbf{x} + \mathbf{v}, \tag{5.3}$$

where H is an $mL \times n$ matrix and $\mathbf{v}$ is the measurement noise. It is desired to find the estimate of $\mathbf{x}$ that minimizes the sum of the squares of the elements of $\mathbf{z} - H\mathbf{x}$. The performance criterion corresponding to equation (5.1) is

$$J = (\mathbf{z} - H\mathbf{x})^T(\mathbf{z} - H\mathbf{x}). \tag{5.4}$$

Setting the derivative of J with respect to $\mathbf{x}$ equal to zero* gives

$$\frac{\partial J}{\partial \mathbf{x}} = -2H^T(\mathbf{z} - H\mathbf{x}) = 0. \tag{5.5}$$

Then we have

$$H^T H\mathbf{x} = H^T\mathbf{z}. \tag{5.6}$$

Denoting the solution of equation (5.6) by $\hat{\mathbf{x}}$, the least-squares estimate is

$$\hat{\mathbf{x}} = (H^T H)^{-1} H^T \mathbf{z}. \tag{5.7}$$

The following examples illustrate the use of this equation.

5.1.1. Scalar Least-Squares Estimation

Assume L measurements of a scalar x are given. Then H is the $L \times 1$ unit vector

$$H = (1, 1, \ldots, 1)^T, \tag{5.8}$$

$$H^T H = \sum_{i=1}^{L} (1)_i^2 = L, \tag{5.9}$$

$$H^T \mathbf{z} = \sum_{i=1}^{L} z_i. \tag{5.10}$$

* The derivative is obtained using the equation $(\partial \mathbf{y}^T\mathbf{y}/\partial \mathbf{x}) = 2(\partial \mathbf{y}^T/\partial \mathbf{x})\mathbf{y}$, where $\mathbf{y} = \mathbf{y}(x)$.

The least-squares estimate is obtained using equation (5.7):

$$\hat{x} = (H^T H)^{-1} H^T \mathbf{z} = \frac{1}{L} \sum_{i=1}^{L} z_i. \tag{5.11}$$

The estimate is simply the average of the observations.

5.1.2. Linear Least-Squares Estimation

Assume $x(t)$ is a linear function of time

$$x(t) = x_0 + \dot{x}t, \tag{5.12}$$

where x_0 and $\dot{x}$ are the parameters to be estimated. In matrix notation, the model response at time t_i is

$$x(t_i) = H_i \mathbf{x} = [1 \quad t_i] \begin{bmatrix} x_0 \\ \dot{x} \end{bmatrix}. \tag{5.13}$$

The m-dimensional observation vector is the scalar

$$z(t_i) = H_i \mathbf{x} + v(t_i). \tag{5.14}$$

The mL vector of observations is

$$\begin{bmatrix} z(t_1) \\ z(t_2) \\ \vdots \\ z(t_L) \end{bmatrix} = \begin{bmatrix} 1 & t_1 \\ 1 & t_2 \\ \vdots & \vdots \\ 1 & t_L \end{bmatrix} \begin{bmatrix} x_0 \\ \dot{x} \end{bmatrix} + \begin{bmatrix} v(t_1) \\ v(t_2) \\ \vdots \\ v(t_L) \end{bmatrix} \tag{5.15}$$

or

$$\mathbf{z} = H\mathbf{x} + \mathbf{v}. \tag{5.16}$$

The least-squares estimate is the two-dimensional vector

$$\hat{\mathbf{x}} = \begin{bmatrix} \hat{x}_0 \\ \hat{\dot{x}} \end{bmatrix} = (H^T H)^{-1} H^T \mathbf{z}. \tag{5.17}$$

5.2. Maximum Likelihood Estimation

Consider once again the system given by equation (5.3)

$$\mathbf{z} = H\mathbf{x} + \mathbf{v}, \tag{5.18}$$

where $\mathbf{x}$ is a constant $n \times 1$ vector of parameters to be estimated, H is an $mL \times n$ matrix, $\mathbf{z}$ is the $mL \times 1$ vector of observations, and $\mathbf{v}$ is the $mL \times 1$ measurement noise. Assume now that $\mathbf{v}$ is a zero mean Gaussian noise process with covariance matrix R,

$$E[\mathbf{v}] = 0, \qquad E[\mathbf{v} \quad \mathbf{v}^T] = R, \tag{5.19}$$

where $E[\cdot]$ denotes the expected value. The conditional probability density of $\mathbf{z}$ given $\mathbf{x}$ is defined as the likelihood function (References 1–5)

$$p(\mathbf{z} \mid \mathbf{x}) = \frac{1}{(2\pi)^{mL/2} | R |^{1/2}} \exp\left\{-\left[\frac{1}{2}(\mathbf{z} - H\mathbf{x})^T R^{-1} (\mathbf{z} - H\mathbf{x})\right]\right\}. \tag{5.20}$$

The logarithm of the probability density function is

$$\ln p(\mathbf{z} \mid \mathbf{x}) = -\tfrac{1}{2} \ln[(2\pi)^{mL} | R |] - \tfrac{1}{2}[(\mathbf{z} - H\mathbf{x})^T R^{-1} (\mathbf{z} - H\mathbf{x})]. \tag{5.21}$$

After the observation $\mathbf{z}$ has been received, the parameter $\mathbf{x}$ is chosen such that $\mathbf{z}$ has a high probability of occurrence.

The estimate $\hat{\mathbf{x}}$ of $\mathbf{x}$ that maximizes equation (5.21) is

$$\frac{\partial}{\partial \hat{\mathbf{x}}} [\ln p(\mathbf{z} \mid \mathbf{x})] = 0. \tag{5.22}$$

Substituting the right-hand side of equation (5.21) into (5.22) gives

$$\frac{\partial}{\partial \hat{\mathbf{x}}} [(\mathbf{z} - H\mathbf{x})^T R^{-1} (\mathbf{z} - H\mathbf{x})] = 0 \tag{5.23}$$

or

$$H^T R^{-1} H \hat{\mathbf{x}} - H^T R^{-1} \mathbf{z} = 0. \tag{5.24}$$

If $H^T R^{-1} H$ is nonsingular then the maximum likelihood estimate is

$$\hat{\mathbf{x}} = (H^T R^{-1} H)^{-1} H^T R^{-1} \mathbf{z}. \tag{5.25}$$

If all the components of the observation vector have the same noise variance, σ^2, and the noise is white, then

$$R^{-1} = (1/\sigma^2) I \tag{5.26}$$

and equation (5.25) becomes

$$\hat{\mathbf{x}} = (H^T H)^{-1} H^T \mathbf{z}. \tag{5.27}$$

Equation (5.27) is the same as the least-squares estimate given by equation (5.7). Thus, for the above conditions, the least-squares estimates are the same as the maximum likelihood estimates.

If R^{-1} is considered to be any $mL \times mL$ symmetric positive definite weighting matrix, then equation (5.25) gives the weighted least-squares estimate. Furthermore, if the covariance of the noise is unknown, then it is appropriate to assume $R = \sigma^2 I$, in which case (5.25) and (5.27) are identical.

5.3. Cramér–Rao Lower Bound

The accuracy of the parameter estimation obtainable from maximum likelihood estimation is given by the Cramér–Rao inequality (References 1–5). For any unbiased estimator $\hat{\mathbf{x}}$ of $\mathbf{x}$, the covariance matrix

$$\mathrm{cov}(\mathbf{x}) \geq M^{-1}, \tag{5.28}$$

where M is the Fisher information matrix

$$M = E\left[\frac{\partial \ln p(\mathbf{z} \mid \mathbf{x})}{\partial \mathbf{x}}\right]\left[\frac{\partial \ln p(\mathbf{z} \mid \mathbf{x})}{\partial \mathbf{x}}\right]^T. \tag{5.29}$$

The inequality is easily proved for the scalar parameter case. Since $\hat{x}$ is unbiased, we have

$$E[\hat{x} - x] = \int_{-\infty}^{\infty} (\hat{x} - x) p(\mathbf{z} \mid x)\, d\mathbf{z} = 0. \tag{5.30}$$

Differentiating with respect to x gives

$$\frac{d}{dx} \int_{-\infty}^{\infty} (\hat{x} - x) p(\mathbf{z} \mid x)\, d\mathbf{z} = -\int_{-\infty}^{\infty} p(\mathbf{z} \mid x)\, d\mathbf{z} + \int_{-\infty}^{\infty} (\hat{x} - x) \frac{\partial p(\mathbf{z} \mid x)}{\partial x}\, d\mathbf{z}. \tag{5.31}$$

The first integral on the right-hand side is equal to 1. Thus

$$\int_{-\infty}^{\infty} (\hat{x} - x) \frac{\partial p(\mathbf{z} \mid x)}{\partial x}\, d\mathbf{z} = 1. \tag{5.32}$$

Since

$$\frac{\partial \ln p(\mathbf{z} \mid x)}{\partial x} = \frac{1}{p(\mathbf{z} \mid x)} \frac{\partial p(\mathbf{z} \mid x)}{\partial x}, \tag{5.33}$$

then combining equations (5.32) and (5.33) gives

$$\int_{-\infty}^{\infty} (\hat{x} - x) \frac{\partial \ln p(\mathbf{z} \mid x)}{\partial x} p(\mathbf{z} \mid x)\, d\mathbf{z} = 1. \tag{5.34}$$

Equation (5.34) can be written

$$\int_{-\infty}^{\infty} (\hat{x} - x)[p(\mathbf{z} \mid x)]^{1/2} \frac{\partial \ln p(\mathbf{z} \mid x)}{\partial x} [p(\mathbf{z} \mid x)]^{1/2}\, d\mathbf{z} = 1. \tag{5.35}$$

The Schwarz inequality states that if

$$\int_{-\infty}^{\infty} a(\mathbf{z})b(\mathbf{z})\, d\mathbf{z} = 1 \tag{5.36}$$

then

$$\left[\int_{-\infty}^{\infty} a^2(\mathbf{z})\, d\mathbf{z}\right]\left[\int_{-\infty}^{\infty} b^2(\mathbf{z})\, d\mathbf{z}\right] \geq 1. \tag{5.37}$$

Applying the Schwarz inequality to equation (5.35) gives

$$\left[\int_{-\infty}^{\infty} (\hat{x} - x)^2 p(\mathbf{z} \mid x)\, d\mathbf{z}\right]\left[\int_{-\infty}^{\infty} \left(\frac{\partial \ln p(\mathbf{z} \mid x)}{\partial x}\right)^2 p(\mathbf{z} \mid x)\, d\mathbf{z}\right] \geq 1 \tag{5.38}$$

or

$$E[\hat{x} - x]^2 \geq 1 \Big/ E\left[\frac{\partial \ln p(\mathbf{z} \mid x)}{\partial x}\right]^2. \tag{5.39}$$

An unbiased estimator is called efficient if its covariance is equal to the Cramér–Rao lower bound. An estimator is called consistent if $\hat{x}$ converges in probability to the true value x as the number of observations increases.

5.4. Gauss–Newton Method

For parameter estimation with the observations nonlinear in the parameters, the Gauss–Newton method is used (References 5, 6). Linearity in the parameters is obtained using the Taylor series expansion of the state vector $\mathbf{x}$:

$$\mathbf{x}(\mathbf{p} + \Delta\mathbf{p}) \doteq \mathbf{x}(\mathbf{p}) + \frac{\partial \mathbf{x}}{\partial \mathbf{p}^T} \Delta\mathbf{p} = \mathbf{x}(\mathbf{p}) + X\Delta\mathbf{p}, \tag{5.40}$$

where $\mathbf{p}$ is the N vector of unknown parameters and X is the $(nL \times N)$-

parameter influence matrix. The elements of X are given by

$$[X]_{ij} = \frac{\partial x_i}{\partial p_j}. \tag{5.41}$$

The $k+1$ estimate of the N vector of unknown parameters, p_{k+1}, is obtained using the relation

$$\mathbf{p}_{k+1} = \mathbf{p}_k + \Delta\mathbf{p}_k, \tag{5.42}$$

where $\mathbf{p}_k$ is the estimate of the unknown parameters at iteration k and $\Delta\mathbf{p}_k$ is the least-squares estimate of the change in the parameters from iteration k to iteration $k+1$.

Consider the observation error

$$\mathbf{z} - \mathbf{x}(\mathbf{p} + \Delta\mathbf{p}) = \mathbf{z} - \mathbf{x}(\mathbf{p}) - X\Delta\mathbf{p} = \mathbf{e} - X\Delta\mathbf{p}. \tag{5.43}$$

The performance criterion is

$$J = (\mathbf{e} - X\Delta\mathbf{p})^T(\mathbf{e} - X\Delta\mathbf{p}). \tag{5.44}$$

Differentiating with respect to $\Delta\mathbf{p}$, the least-squares estimate is given by

$$\Delta\mathbf{p}_k = (X_k^T X_k)^{-1} X_k^T \mathbf{e}. \tag{5.45}$$

The Gauss–Newton method uses iteration equation (5.42) where $\Delta\mathbf{p}_k$ is the least-squares estimate given by equation (5.45). The method requires the solution of nN auxiliary equations obtained by taking the partial derivatives of the system differential equations with respect to the unknown parameters.

In many cases, not all the components of the vector $\mathbf{x}$ can be measured. For example, assume that equation (5.2) is given by (Reference 7)

$$\mathbf{e}_i = \mathbf{z}(t_i) - H_i\mathbf{x}(t_i), \qquad i = 1, 2, \ldots, L, \tag{5.46}$$

where H_i is the $m \times n$ measurement matrix. Equation (5.43) then becomes

$$\mathbf{z} - H\mathbf{x}(\mathbf{p} + \Delta\mathbf{p}) = \mathbf{z} - H\mathbf{x} - HX\Delta\mathbf{p} = \mathbf{e} - HX\Delta\mathbf{p}, \tag{5.47}$$

where H is an $mL \times nL$ matrix, X is an $nL \times N$ matrix, and $\mathbf{x}$ is an $nL \times 1$ vector. The least-squares estimate of $\Delta\mathbf{p}$ is then given by

$$\Delta\mathbf{p}_k = (X_k^T H^T H X_k)^{-1} X_k^T H^T \mathbf{e}. \tag{5.48}$$

When H_i is the $n \times n$ identity matrix, I, equation (5.48) reduces to equation (5.45).

5.5. Examples of the Gauss–Newton Method

The following examples illustrate the use of the Gauss–Newton method.

5.5.1. First-Order System with Single Unknown Parameter

Given the scalar system equation,

$$\dot{x} = -ax + u, \qquad x(0) = c, \tag{5.49}$$

where a is an unknown parameter, c is a known initial condition, and $u(t)$ is a known input function, derive the equations for the Gauss–Newton iteration.

The measurements are given by

$$z(t_i) = x(t_i) + v(t_i), \qquad i = 1, 2, \ldots, L, \tag{5.50}$$

where

$$x(t_i) = H_i x(t_i), \qquad H_i = I = (1) \tag{5.51}$$

and $v(t_i)$ is the measurement noise. The iteration equation for the least-squares estimate is

$$p_{k+1} = p_k + \Delta p_k, \tag{5.52}$$

where

$$\Delta p = (X_k{}^T X_k)^{-1} X_k{}^T \mathbf{e} \tag{5.53}$$

$$\mathbf{e} = \begin{bmatrix} z(t_1) - x(t_1, p) \\ z(t_2) - x(t_2, p) \\ \vdots \\ z(t_L) - x(t_L, p) \end{bmatrix}, \qquad L \text{ vector} \tag{5.54}$$

$$X_k = \begin{bmatrix} x_a(t_1) \\ x_a(t_2) \\ \vdots \\ x_a(t_L) \end{bmatrix}, \qquad L \times 1 \text{ parameter influence matrix,} \tag{5.55}$$

$p_k = a_k$ is an estimate of the unknown parameter a at iteration k, L is the number of observations, $z(t_i)$ is the scalar observation at time t_i, $x(t_i, p)$

is the scalar state, and

$$x_a(t_i) = \frac{\partial x(t_i)}{\partial a}. \tag{5.56}$$

The elements of the parameter influence matrix are obtained from the auxiliary equation derived by differentiating equation (5.49) with respect to a:

$$\dot{x}_a = -ax_a - x, \qquad x_a(0) = 0. \tag{5.57}$$

Using an initial approximation for the unknown parameter a, equations (5.49) and (5.57) are integrated from time $t = 0$ to $t = t_L$. A new estimate of a is computed using equations (5.52) and (5.53) and the cycle is repeated, etc., until convergence is obtained.

5.5.2. First-Order System with Unknown Initial Condition and Single Unknown Parameter

The scalar system equation is

$$\dot{x} = -ax + u, \qquad x(0) = c, \tag{5.58}$$

where a is an unknown parameter, c is an unknown initial condition, and $u(t)$ is a known input function. The equations for the measurements and the Gauss–Newton iteration are the same as for the previous example except that p is now a vector and the parameter influence matrix an $L \times 2$ matrix:

$$\mathbf{p} = \begin{bmatrix} c \\ a \end{bmatrix}, \tag{5.59}$$

$$X_k = \begin{bmatrix} x_c(t_1) & x_a(t_1) \\ x_c(t_2) & x_a(t_2) \\ \vdots & \\ x_c(t_L) & x_a(t_L) \end{bmatrix}, \qquad L \times 2 \text{ matrix}, \tag{5.60}$$

where

$$x_c(t_i) = \frac{\partial x(t_i)}{\partial c}, \qquad x_a(t_i) = \frac{\partial x(t_i)}{\partial a}. \tag{5.61}$$

The elements of the parameter influence matrix are obtained from the auxiliary equations derived by differentiating equation (5.58) with respect to c and a, respectively:

$$\dot{x}_c = -ax_c, \qquad x_c(0) = 1, \tag{5.62}$$

$$\dot{x}_a = -ax_a - x, \qquad x_a(0) = 0. \tag{5.63}$$

The method of solution is then the same as above except that an initial approximation for both c and a is required and equations (5.58), (5.62), and (5.63) must be integrated for each cycle.

5.5.3. Second-Order System with Two Unknown Parameters and Vector Measurement

The system equations are given by

$$\dot{x}_1 = x_2, \qquad x_1(0) = c_1, \tag{5.64}$$

$$\dot{x}_2 = -ax_2 - bx_1 + u, \qquad x_2(0) = c_2, \tag{5.65}$$

where a, b are the unknown parameters, c_1, c_2 are known initial conditions, and $u(t)$ is the known input function. Both components, x_1 and x_2, of the system equations are assumed to be measured

$$\begin{bmatrix} z_1(t_i) \\ z_2(t_i) \end{bmatrix} = \begin{bmatrix} x_1(t_i) \\ x_2(t_i) \end{bmatrix} + \begin{bmatrix} v_1(t_i) \\ v_2(t_i) \end{bmatrix} \tag{5.66}$$

or

$$\mathbf{z}(t_i) = \mathbf{x}(t_i) + \mathbf{v}(t_i), \tag{5.67}$$

where

$$\mathbf{x}(t_i) = H_i\mathbf{x}(t_i), \qquad H_i = \begin{bmatrix} 1 & 0 \\ 0 & 1 \end{bmatrix} = I \tag{5.68}$$

and $\mathbf{v}(t_i)$ is the vector measurement noise. The iteration equation is

$$\mathbf{p}_{k+1} = \mathbf{p}_k + \Delta\mathbf{p}_k, \tag{5.69}$$

where

$$\Delta\mathbf{p}_k = (X_k{}^T X_k)^{-1} X_k{}^T \mathbf{e}, \tag{5.70}$$

$$X_k = \begin{bmatrix} x_{1a}(t_1) & x_{1b}(t_1) \\ x_{2a}(t_1) & x_{2b}(t_1) \\ \vdots & \\ x_{1a}(t_L) & x_{1b}(t_L) \\ x_{2a}(t_L) & x_{2b}(t_L) \end{bmatrix}, \qquad 2L \times 2 \text{ matrix}, \tag{5.71}$$

$$\mathbf{p} = \begin{bmatrix} a \\ b \end{bmatrix}. \tag{5.72}$$

The auxiliary equations are derived by differentiating equations (5.64) and (5.65) with respect to a and b:

$$\dot{x}_{1a} = x_{2a}, \qquad x_{1a}(0) = 0, \tag{5.73}$$

$$\dot{x}_{2a} = -ax_{2a} - x_2 - bx_{1a}, \qquad x_{2a}(0) = 0, \tag{5.74}$$

$$\dot{x}_{1b} = x_{2b}, \qquad x_{1b}(0) = 0, \tag{5.75}$$

$$\dot{x}_{2b} = -ax_{2b} - bx_{1b} - x_1, \qquad x_{2b}(0) = 0. \tag{5.76}$$

5.5.4. Second-Order System with Two Unknown Parameters and Scalar Measurement

The system equations are the same as for the previous example, equations (5.64) and (5.65) with unknown parameters a and b. In this case, however, only the component x_1 is measured.

$$z(t_i) = x_1(t_i) + v(t_i), \tag{5.77}$$

where

$$x_1(t_i) = H_i\mathbf{x}(t_i), \qquad H_i = [1 \quad 0]. \tag{5.78}$$

The iteration equation is

$$\mathbf{p}_{k+1} = \mathbf{p}_k + \mathbf{\Delta p}_k, \tag{5.79}$$

where the least-squares estimate is given by equation (5.48)

$$\mathbf{\Delta p}_k = (X_k^T H^T H X_k)^{-1} X_k^T H^T \mathbf{e}, \tag{5.80}$$

$$HX_k = \begin{bmatrix} x_{1a}(t_1) & x_{1b}(t_1) \\ x_{1a}(t_2) & x_{1b}(t_2) \\ \vdots & \\ x_{1a}(t_L) & x_{1b}(t_L) \end{bmatrix}, \qquad L \times 2 \text{ matrix}. \tag{5.81}$$

The auxiliary equations are the same as for the previous example.

Exercises

1. Find the least-squares estimates of the constant parameters x_0, $\dot{x}$, and $\ddot{x}$ assuming that $x(t)$ is the quadratic function of time

$$x(t) = x_0 + \dot{x}t + \tfrac{1}{2}\ddot{x}t^2$$

with the scalar observations

$$z(t_i) = x(t_i) + v(t_i), \qquad i = 1, \ldots, L.$$

2. Find the maximum likelihood estimates in Exercise 1 if both the position $x(t)$ and the velocity $\dot{x}(t)$ are measured

$$z_1(t_i) = x(t_i) + v_1(t_i),$$
$$z_2(t_i) = \dot{x}(t_i) + v_2(t_i)$$

and the noise is a zero mean Gaussian white-noise process

$$E[\mathbf{v}_i] = 0, \qquad E[\mathbf{v}_i \ \mathbf{v}_i^T] = R_i = \begin{bmatrix} \sigma_1^2 & 0 \\ 0 & \sigma_2^2 \end{bmatrix},$$

where

$$\mathbf{v}_i = \begin{bmatrix} v_1(t_i) \\ v_2(t_i) \end{bmatrix}, \qquad i = 1, \ \ldots, \ L.$$

3. Given the system, $\dot{x} = a$; $x(0) = 1$, with unknown parameter a, known initial condition $x(0)$, and measurements, $z(1) = 1.2$, $z(2) = 1.4$. Find the analytical solution for $x_a(t)$ and show that the parameter influence matrix of the Gauss–Newton method is

$$X_k = \begin{bmatrix} x_a(1) \\ x_a(2) \end{bmatrix} = \begin{bmatrix} 1 \\ 2 \end{bmatrix}.$$

4. In Exercise 3, integrate $\dot{x}(t)$ to evaluate

$$e_i = z(t_i) - x(t_i).$$

Assume the initial approximation of a is $a_0 = 0.25$. Then estimate a using the Gauss–Newton method, equations (5.52) and (5.53).

Answer: $a_1 = 0.2$.

5. Given the system $\dot{x} = a$, with known parameter $a = 0.2$, unknown initial condition $x(0) = c$, and noisy measurements, $z(0) = 0.55$, $z(1) = 0.65$. Find analytical solutions for $x(t)$ and $x_c(t)$. Assume the initial approximation of the initial condition c is $c_0 = 1$. Estimate the initial condition using the Gauss–Newton method.

Answer: $c_1 = 0.5$.

6. Given the system $\dot{x} = -ax$, $x(0) = 1$, with unknown parameter a, known initial condition $x(0)$, and measurement $z(1) = 0.904837$. Find analytical solutions for $x(t)$ and $x_a(t)$. Assume the initial approximation of a is $a_0 = 0.15$. Obtain the first estimate of a using the Gauss–Newton method.

Answer: $x_a(t) = -te^{-a_0 t}$, $a_1 = 0.098729$.

7. Obtain the first estimate of a in Exercise 6 if $a_0 = 0.1$ instead of 0.15.

   Answer: $a_1 = 0.1$.

8. Obtain the first estimate of a in Exercise 6 if $\dot{x} = -ax + y$; $x(0) = 1$; y is a unit step function, $y(t) = 1$; and $z(1) = 1.856463$. Assume $a_0 = 0.15$.

   Answer:

$$x(t) = e^{-a_0 t} + \frac{1}{a_0}(1 - e^{-a_0 t}),$$

$$x_a(t) = -te^{-a_0 t} - \frac{1}{a_0^2}[1 - (1 + a_0 t)\, e^{-a_0 t}], \quad a = 0.098880.$$

6

Quasilinearization Method for System Identification

The method of quasilinearization was introduced in Chapter 4 as a successive approximation method for finding the solution of nonlinear two-point boundary problems. In this chapter quasilinearization is used for system identification (References 1–9) using the measurements to formulate the problem as a multipoint boundary-value problem. The least-squares criterion is used to estimate the unknown initial conditions and/or unknown parameters.

The general formulation of the quasilinearization method for system identification is given in Section 6.1 (Reference 6) followed by detailed examples for first- and second-order systems in Section 6.2. The examples are derived for general nonlinear dynamical systems but are applied directly to the same examples as in Chapter 5. Numerical results are given in Chapter 7.

6.1. System Identification via Quasilinearization

Consider the nonlinear dynamical system

$$\dot{\mathbf{x}} = \mathbf{f}(\mathbf{x}, \boldsymbol{\alpha}), \qquad x_i(0) = c_i, \qquad i = 1, 2, \ldots, R, \tag{6.1}$$

where $\mathbf{x}(t)$ is a n-dimensional state vector, $\boldsymbol{\alpha}$ is a N-dimensional vector of unknown parameters, and c_i are known initial conditions. The first R components of $\mathbf{x}(0)$ are specified. It is desired to estimate the $n - R$ remaining initial conditions and the N components of the constant unknown parameter vector $\boldsymbol{\alpha}$. The problem of estimating the unknown parameters is first transformed into one in which the initial conditions are estimated for a system of differential equations (Reference 6). Consider the vector $\boldsymbol{\alpha}$ to be a function of time that satisfies the differential equation

$$\dot{\boldsymbol{\alpha}} = 0. \tag{6.2}$$

Then redefine the vector $\mathbf{x}(t)$ as an $(n + N)$-dimensional vector with initial conditions such that

$$\mathbf{x}(t) = \begin{bmatrix} x_1 \\ x_2 \\ \vdots \\ x_R \\ \vdots \\ x_n \\ \alpha_1 \\ \alpha_2 \\ \vdots \\ \alpha_N \end{bmatrix}, \qquad \mathbf{x}(0) = \begin{bmatrix} c_1 \\ c_2 \\ \vdots \\ c_R \\ c_{R+1} \\ c_{R+2} \\ \vdots \\ c_{n+N} \end{bmatrix}. \tag{6.3}$$

The vector, $\mathbf{x}(t)$, satisfies the system of nonlinear differential equations

$$\dot{\mathbf{x}} = \mathbf{f}(\mathbf{x}), \qquad \mathbf{x}(0) = \mathbf{c}. \tag{6.4}$$

The first R components of $\mathbf{x}(0)$ are specified and the remaining $n + N - R$ components are free to be chosen such as to minimize the sum S:

$$S = \sum_{i=1}^{L} [\boldsymbol{\beta}^T \mathbf{x}(t_i) - b_i]^2 \tag{6.5}$$

where $\boldsymbol{\beta}$ is a constant weighting vector and b_i is the observed state of the system at time t_i. Differential equation (6.4) is linearized around the kth approximation by expanding the function f in a Taylor series and retaining only the linear terms. The $k + 1$ approximation is

$$\dot{\mathbf{x}}_{k+1} = \mathbf{f}(\mathbf{x}_k) + J(\mathbf{x}_k)(\mathbf{x}_{k+1} - \mathbf{x}_k), \tag{6.6}$$

where $J(x)$ is the $(n + N)\times(n + N)$ Jacobian matrix with elements

$$J_{ij} = \frac{\partial f_i}{\partial x_j}. \qquad (6.7)$$

The solution, $\mathbf{x}_{k+1}$, of the system of linear differential equations (6.6) is the sum of a particular solution, $\mathbf{p}(t)$, and a linear combination of $n + N - R$ independent solutions of the homogeneous equations, $\mathbf{h}_j(t)$:

$$\mathbf{x}_{k+1} = \mathbf{p}(t) + \sum_{j=R+1}^{n+N} c_j\mathbf{h}_j(t), \qquad (6.8)$$

where $c_{R+1}, c_{R+2}, \ldots, c_{n+N}$ are constants to be determined. The function $\mathbf{p}(t)$ satisfies the equation

$$\dot{\mathbf{p}} = f(\mathbf{x}_k) + J(\mathbf{x}_k)(\mathbf{p} - \mathbf{x}_k) \qquad (6.9)$$

subject to the initial conditions

$$p_i(0) = \begin{cases} c_i, & i = 1, 2, \ldots, R \\ 0, & i = R + 1, R + 2, \ldots, n + N. \end{cases} \qquad (6.10)$$

The functions $h_j(t)$ are solutions of the homogeneous equations

$$\dot{\mathbf{h}}_j = J(\mathbf{x}_k)\mathbf{h}_j, \qquad j = R + 1, R + 2, \ldots, n + N, \qquad (6.11)$$

where $\mathbf{h}_j(t)$ is an $(n + N)$-dimensional vector with initial conditions

$$\mathbf{h}_j(0) = \delta_{ij}, \qquad j = R + 1, R + 2, \ldots, n + N. \qquad (6.12)$$

The unknown constants, $c_{R+1}, c_{R+2}, \ldots, c_{n+N}$ are calculated as solutions of the linear algebraic equations subject to multipoint boundary conditions

$$\frac{\partial S}{\partial c_j} = 0, \qquad j = R + 1, R + 2, \ldots, n + N, \qquad (6.13)$$

where for $\mathbf{x}(t_i)$ is substituted

$$\mathbf{x}(t_i) = \mathbf{p}(t_i) + \sum_{j=R+1}^{n+N} c_j\mathbf{h}_j(t_i), \qquad i = 1, 2, \ldots, L. \qquad (6.14)$$

The known initial conditions plus solution of the linear algebraic equation yield the next approximation of the initial state vector,

$$\mathbf{x}_{k+1}(0) = \begin{bmatrix} c_1 \\ c_2 \\ \vdots \\ c_R \\ c_{R+1} \\ c_{R+2} \\ \vdots \\ c_{n+N} \end{bmatrix} = \begin{bmatrix} x_1(0) \\ x_2(0) \\ \vdots \\ x_R(0) \\ \vdots \\ x_n(0) \\ \alpha_1 \\ \alpha_2 \\ \vdots \\ \alpha_N \end{bmatrix} \tag{6.15}$$

where $c_1, c_2, \ldots, c_R$ are the known initial conditions and $c_{R+1}, c_{R+2}, \ldots, c_{n+N}$ are the $k + 1$ approximations for the remaining $n - R$ initial conditions and the N unknown parameters.

To obtain a numerical solution, an initial approximation of the missing initial conditions is selected and equation (6.4) is integrated to obtain the $k = 0$ vector $\mathbf{x}(t)$. The initial-value equations for p and h are integrated and may be stored as a function of time from $t = 0$ to $t = t_L$. As each boundary condition is passed at times t_i, $i = 1, 2, \ldots, L$, the sums formed in equations (6.13) are updated. These equations can then be solved for the c_j's, $j = R + 1, R + 2, \ldots, n + N$, when $t = t_L$ is reached. The $k = 1$ approximation for $\mathbf{x}(t)$ may then be obtained by the integration of the linear equations with the initial conditions just found, or by the linear combination of $p(t)$ and $h(t)$ if they have been stored. The process is then repeated to obtain the next approximation, etc., until no further change in the initial conditions is obtained.

6.2. Examples of the Quasilinearization Method

The following examples illustrate the use of the quasilinearization method.

6.2.1. First-Order System with Single Unknown Parameter

Given the scalar nonlinear dynamical system

$$\dot{x} = f(x, u, a, t), \qquad x(0) = c, \tag{6.16}$$

the unknown system parameter a is to be estimated by minimizing the sum S:

$$S = \sum_{i=1}^{L} [x(t_i, a) - b_i]^2, \tag{6.17}$$

where $x(t)$ is a scalar state variable, $u(t)$ is a known input function, a is an unknown system parameter, c is a known initial condition, and b_i is the observed value of $x(t_i, a)$ at time t_i, $i = 1, 2, \ldots, L$.

Though the formal quasilinearization equations derived in the previous section can be utilized directly, it is informative to rederive the equations for the scalar case.

The method of quasilinearization is utilized for minimizing the sum S given by equation (6.17). Differential equation (6.16) is first linearized around the kth approximation by expanding the function f in a power series and retaining only the linear terms. The $k + 1$ approximation is

$$\dot{x}_{k+1} = f_k + (a_{k+1} - a_k)\frac{\partial f}{\partial a}\bigg|_k + (x_{k+1} - x_k)\frac{\partial f}{\partial x}\bigg|_k, \qquad x_{k+1}(0) = c. \tag{6.18}$$

The general solution of equation (6.18) is

$$x_{k+1}(t) = p(t) + a_{k+1}h(t). \tag{6.19}$$

The function $p(t)$ is the solution of the initial-value equation

$$\dot{p}(t) = f_k - a_k\frac{\partial f}{\partial a}\bigg|_k + (p - x_k)\frac{\partial f}{\partial x}\bigg|_k, \qquad p(0) = c \tag{6.20}$$

and the function $h(t)$ is the solution of the initial-value equation

$$\dot{h}(t) = \frac{\partial f}{\partial a}\bigg|_k + \frac{\partial f}{\partial x}\bigg|_k h(t), \qquad h(0) = 0. \tag{6.21}$$

Substituting equation (6.19) into (6.17) yields

$$S = \sum_{i=1}^{N} [p(t_i) + a_{k+1}h(t_i) - b_i]^2. \tag{6.22}$$

The condition for minimum S is given by

$$\frac{\partial S}{\partial a_{k+1}} = 2\sum_{i=1}^{N} h(t_i)[p(t_i) + a_{k+1}h(t_i) - b_i] = 0 \tag{6.23}$$

solving for a_{k+1} gives

$$a_{k+1} = \frac{\sum_{i=1}^{N} [b_i - p(t_i)]h(t_i)}{\sum_{i=1}^{N} h^2(t_i)}. \tag{6.24}$$

In a typical cycle, the two initial-value equations for p and h are integrated and stored as a function of time from $t = 0$ to $t = t_L$. As each point t_i, $i = 1, 2, \ldots, L$ is passed, the sums in (6.24) are updated. Equation (6.24) is solved for the constant a_{k+1} when $t = t_L$ is reached. Equation (6.19) is then evaluated for $x_{k+1}(t)$ and the latter is stored as a function of time. Utilizing the new value of a and the stored values of $x(t)$, the process is then repeated to obtain the next approximation.

Assume as an example that equation (6.16) is given by

$$\dot{x} = - ax + u, \qquad x(0) = c. \tag{6.25}$$

The partial derivatives in equation (6.18) are then

$$\left.\frac{\partial f}{\partial a}\right|_k = -x_k, \qquad \left.\frac{\partial f}{\partial x}\right|_k = -a_k. \tag{6.26}$$

6.2.2. First-Order System with Unknown Initial Condition and Single Unknown Parameter

The scalar system equation is

$$\dot{x} = f(x, u, a, t), \qquad x(0) = c, \tag{6.27}$$

where $x(t)$ is the scalar state variable, $u(t)$ is a known input function, a is an unknown system parameter, and c is an unknown initial condition. The unknown initial condition c and the unknown system parameter a are to be estimated by minimizing the sum S:

$$S = \sum_{i=1}^{L} [x(t_i, a) - b_i]^2, \tag{6.28}$$

where, as before, b_i is the observed value of $x(t_i, a)$ at time t_i, $i = 1, 2, \ldots, L$. Following the same procedure as for the last example, the linearized equation for the $k + 1$ approximation of x^{k+1} is

$$\dot{x}^{k+1} = f^k + (a^{k+1} - a^k)f_a^{\,k} + (x^{k+1} - x^k)f_x^{\,k}, \qquad \dot{x}^{k+1}(0) = c, \tag{6.29}$$

where the notation has been simplified by defining the partial derivatives as follows:

$$f_a^k = \left.\frac{\partial f}{\partial a}\right|_k, \qquad f_x^k = \left.\frac{\partial f}{\partial x}\right|_k. \tag{6.30}$$

The general solution of equation (6.29) is

$$x^{k+1}(t) = p(t) + c^{k+1}q_1(t) + a^{k+1}r_1(t). \tag{6.31}$$

The function $p(t)$ is the solution of the initial-value equation

$$\dot{p}(t) = f^k - a^k f_a^k + (p - x^k)f_x^k, \qquad p(0) = 0, \tag{6.32}$$

and the functions $q_1(t)$ and $r_1(t)$ are the solutions of the initial-value equations

$$\dot{q}_1(t) = f_x^k q_1(t), \qquad q_1(0) = 1, \tag{6.33}$$

$$\dot{r}_1(t) = f_a^k + f_x^k r_1(t), \qquad r_1(0) = 0. \tag{6.34}$$

Substituting equation (6.31) into (6.28) yields

$$S = \sum_{i=1}^{L} [p(t_i) + c^{k+1}q_1(t_i) + a^{k+1}r_1(t_i) - b_i]^2. \tag{6.35}$$

The conditions for minimum S are

$$\frac{\partial S}{\partial c^{k+1}} = 2\sum_{i=1}^{L} q_1(t_i)[p(t_i) + c^{k+1}q_1(t_i) + a^{k+1}r_1(t_i) - b_i] = 0, \tag{6.36}$$

$$\frac{\partial S}{\partial a^{k+1}} = 2\sum_{i=1}^{L} r_1(t_i)[p(t_i) + c^{k+1}q_1(t_i) + a^{k+1}r_1(t_i) - b_i] = 0. \tag{6.37}$$

These equations yield two linear algebraic equations for the unknowns, c^{k+1} and a^{k+1}. Multiplying and rearranging the terms, the equations become

$$\sum_{i=1}^{L} q_1^2(t_i)c^{k+1} + \sum_{i=1}^{L} q_1(t_i)r_1(t_i)a^{k+1} = \sum_{i=1}^{L} [b_i - p(t_i)]q_1(t_i), \tag{6.38}$$

$$\sum_{i=1}^{L} q_1(t_i)r_1(t_i)c^{k+1} + \sum_{i=1}^{L} r_1^2(t_i)a^{k+1} = \sum_{i=1}^{L} [b_i - p(t_i)]r_1(t_i). \tag{6.39}$$

This is of the form

$$m_{11}c^{k+1} + m_{12}a^{k+1} = g_1, \tag{6.40}$$

$$m_{21}c^{k+1} + m_{22}a^{k+1} = g_2, \tag{6.41}$$

from which c^{k+1} and a^{k+1} can be computed.

The quasilinearization equations can also be derived using the more formal approach involving the Jacobian matrix. Define the vector $\mathbf{x}(t)$:

$$\mathbf{x}(t) = \begin{bmatrix} x \\ a \end{bmatrix}. \tag{6.42}$$

The equation (6.6) becomes

$$\dot{\mathbf{x}}_{k+1} = \mathbf{f}(\mathbf{x}_k) + J(\mathbf{x}_k)(\mathbf{x}_{k+1} - \mathbf{x}_k) \tag{6.43}$$

or

$$\begin{bmatrix} \dot{x}^{k+1} \\ \dot{a}^{k+1} \end{bmatrix} = \begin{bmatrix} f(x^k) \\ 0 \end{bmatrix} + \begin{bmatrix} f_x^{\ k} & f_a^{\ k} \\ 0 & 0 \end{bmatrix} \begin{bmatrix} x^{k+1} - x^k \\ a^{k+1} - a^k \end{bmatrix}. \tag{6.44}$$

The solution is given by equation (6.8):

$$\begin{aligned} \mathbf{x}_{k+1} &= \mathbf{p}(t) + \sum_{j=1}^{2} c_j \mathbf{h}_j(t) \\ &= \mathbf{p}(t) + c_1 \mathbf{h}_1(t) + c_2 \mathbf{h}_2(t), \end{aligned} \tag{6.45}$$

where

$$c_1 = c^{k+1}, \qquad c_2 = a^{k+1}. \tag{6.46}$$

The summation is from $j = R + 1$ to $n + N$, where $R = 0$ is the number of components of $\mathbf{x}(0)$ specified, and $n + N = 2$ is the dimension of $\mathbf{x}$. Equations (6.9)–(6.12) then become

$$\dot{\mathbf{p}} = \begin{bmatrix} \dot{p}_1 \\ \dot{p}_2 \end{bmatrix} = \begin{bmatrix} f(x^k) \\ 0 \end{bmatrix} + \begin{bmatrix} f_x^{\ k} & f_a^{\ k} \\ 0 & 0 \end{bmatrix} \begin{bmatrix} p_1 - x^k \\ p_2 - a^k \end{bmatrix}, \qquad \begin{bmatrix} p_1(0) \\ p_2(0) \end{bmatrix} = \begin{bmatrix} 0 \\ 0 \end{bmatrix}, \tag{6.47}$$

$$\dot{\mathbf{h}}_1 = \begin{bmatrix} \dot{q}_1 \\ \dot{q}_2 \end{bmatrix} = \begin{bmatrix} f_x^{\ k} & f_a^{\ k} \\ 0 & 0 \end{bmatrix} \begin{bmatrix} q_1 \\ q_2 \end{bmatrix}, \qquad \begin{bmatrix} q_1(0) \\ q_2(0) \end{bmatrix} = \begin{bmatrix} 1 \\ 0 \end{bmatrix}, \tag{6.48}$$

$$\dot{\mathbf{h}}_2 = \begin{bmatrix} \dot{r}_1 \\ \dot{r}_2 \end{bmatrix} = \begin{bmatrix} f_x^{\ k} & f_a^{\ k} \\ 0 & 0 \end{bmatrix} \begin{bmatrix} r_1 \\ r_2 \end{bmatrix}, \qquad \begin{bmatrix} r_1(0) \\ r_2(0) \end{bmatrix} = \begin{bmatrix} 0 \\ 1 \end{bmatrix}. \tag{6.49}$$

It is obvious that

$$p_2(t) = 0, \qquad q_2(t) = 0, \qquad r_2(0) = 1, \tag{6.50}$$

since

$$\dot{p}_2(t) = 0, \qquad p_2(0) = 0, \tag{6.51}$$

$$\dot{q}_2(t) = 0, \qquad q_2(0) = 0, \tag{6.52}$$

$$\dot{r}_2(t) = 0, \qquad r_2(0) = 1. \tag{6.53}$$

Then it can be shown that equations (6.47)–(6.49) are identical to equations (6.32)–(6.34). The vector $\boldsymbol{\beta}$ in equation (6.5) is $\boldsymbol{\beta} = (1, 0)^T$.

For the example given by equation (6.25),

$$\dot{x} = -ax + u, \qquad x(0) = c. \tag{6.54}$$

The partial derivatives are, as previously,

$$f_a^k = -x^k, \qquad f_x^k = -a^k. \tag{6.55}$$

6.2.3. Second-Order System with Two Unknown Parameters and Vector Measurement

The system equations are

$$\dot{x}_1 = f_1(x_1, x_2, u, a, b, t), \qquad x_1(0) = c_1, \tag{6.56}$$

$$\dot{x}_2 = f_2(x_1, x_2, u, a, b, t), \qquad x_2(0) = c_2, \tag{6.57}$$

where a, b are the unknown parameters, c_1, c_2 are known initial conditions, and $u(t)$ is a known input function. The unknown system parameters, a and b, are to be estimated by minimizing the sum

$$S = \sum_{i=1}^{L} \{\lambda_1[x_1(t_i) - z_1(t_i)]^2 + \lambda_2[x_2(t_i) - z_2(t_i)]^2\}, \tag{6.58}$$

where $z_1(t_i)$, $z_2(t_i)$ are the observed values of $x_1(t_i)$, $x_2(t_i)$, respectively, at time t_i, $i = 1, 2, \ldots, L$, and λ_1, λ_2, are weighting factors. The linearized equations for the $k+1$ approximations of x_1^{k+1} and x_2^{k+1} are

$$\begin{aligned} \dot{x}_1^{k+1} = {} & f_1^k + (a^{k+1} - a^k)f_{1a}^k + (b^{k+1} - b^k)f_{1b}^k \\ & + (x_1^{k+1} - x_1^k)f_{1x_1}^k + (x_2^{k+1} - x_2)f_{1x_2}^k, \end{aligned} \tag{6.59}$$

$$\begin{aligned} \dot{x}_2^{k+1} = {} & f_2^k + (a^{k+1} - a^k)f_{2a}^k + (b^{k+1} - b^k)f_{2b}^k \\ & + (x_1^{k+1} - x_1^k)f_{2x_1}^k + (x_2^{k+1} - x_2^k)f_{2x_2}^k, \end{aligned} \tag{6.60}$$

where as previously the partial derivatives are defined by

$$\begin{aligned} f_{1a}^k &= \left.\frac{\partial f_1}{\partial a}\right|_k, \\ &\vdots \\ f_{1x_1}^k &= \left.\frac{\partial f_1}{\partial x_1}\right|_k \\ &\vdots \end{aligned} \tag{6.61}$$

The general solution of equations (6.59) and (6.60) is

$$x_1^{k+1}(t) = p_1(t) + a^{k+1}q_1(t) + b^{k+1}r_1(t), \tag{6.62}$$

$$x_2^{k+1}(t) = p_2(t) + a^{k+1}q_2(t) + b^{k+1}r_2(t). \tag{6.63}$$

The functions $p_1(t)$ and $p_2(t)$ are the solutions of the initial-value equations

$$\begin{aligned}\dot{p}_1(t) = f_1^k - a^k f_{1a}^k - b^k f_{1b}^k + (p_1 - x_1^k) f_{1x_1}^k \\ + (p_2 - x_2^k) f_{1x_2}^k, \qquad p_1(0) = c_1,\end{aligned} \tag{6.64}$$

$$\begin{aligned}\dot{p}_2(t) = f_2^k - a^k f_{2a}^k - b^k f_{2b}^k + (p_1 - x_1^k) f_{2x_1}^k \\ + (p_2 - x_2^k) f_{2x_2}^k, \qquad p_2(0) = c_2.\end{aligned} \tag{6.65}$$

The functions $q_1(t)$, $q_2(t)$, $r_1(t)$, and $r_2(t)$ are the solutions of the initial-value equations:

$$\dot{q}_1(t) = f_{1a}^k + f_{1x_1}^k q_1(t) + f_{1x_2}^k q_2(t), \qquad q_1(0) = 0, \tag{6.66}$$

$$\dot{q}_2(t) = f_{2a}^k + f_{2x_1}^k q_1(t) + f_{2x_2}^k q_2(t), \qquad q_2(0) = 0, \tag{6.67}$$

$$\dot{r}_1(t) = f_{1b}^k + f_{1x_1}^k r_1(t) + f_{1x_2}^k r_2(t), \qquad r_1(0) = 0, \tag{6.68}$$

$$\dot{r}_2(t) = f_{2b}^k + f_{2x_2}^k r_1(t) + f_{2x_2}^k r_2(t), \qquad r_2(0) = 0. \tag{6.69}$$

Assume that $\lambda_1 = \lambda_2 = 1$, then substituting equations (6.62) and (6.63) into (6.58) gives

$$\begin{aligned}S = \sum_{i=1}^{L} \{[p_1(t_i) + a^{k+1}q_1(t_i) + b^{k+1}r_1(t_i) - z_1(t_i)]^2 \\ + [p_2(t_i) + a^{k+1}q_2(t_i) + b^{k+1}r_2(t_i) - z_2(t_i)]^2\}.\end{aligned} \tag{6.70}$$

The conditions for minimum S are

$$\begin{aligned}\frac{\partial S}{\partial a^{k+1}} &= 2 \sum_{i=1}^{L} q_1(t_i)[p_1(t_i) + a^{k+1}q_1(t_i) + b^{k+1}r_1(t_i) - z_1(t_i)] \\ &\quad + 2 \sum_{i=1}^{L} q_2(t_i)[p_2(t_i) + a^{k+1}q_2(t_i) + b^{k+1}r_2(t_i) - z_2(t_i)] \\ &= 0,\end{aligned} \tag{6.71}$$

$$\begin{aligned}\frac{\partial S}{\partial b^{k+1}} &= 2 \sum_{i=1}^{L} r_1(t_i)[p_1(t_i) + a^{k+1}q_1(t_i) + b^{k+1}r_1(t_i) - z_1(t_i)] \\ &\quad + 2 \sum_{i=1}^{L} r_2(t_i)[p_2(t_i) + a^{k+1}q_2(t_i) + b^{k+1}r_2(t_i) - z_2(t_i)] \\ &= 0.\end{aligned} \tag{6.72}$$

These equations yield two linear algebraic equations for the unknown parameters a^{k+1} and b^{k+1} of the form

$$m_{11}a^{k+1} + m_{12}b^{k+1} = g_1, \tag{6.73}$$

$$m_{21}a^{k+1} + m_{22}b^{k+1} = g_2, \tag{6.74}$$

where

$$m_{11} = \sum_{i=1}^{L} [q_1^2(t_i) + q_2^2(t_i)], \tag{6.75}$$

$$m_{12} = \sum_{i=1}^{L} [q_1(t_i)r_1(t_i) + q_2(t_i)r_2(t_i)], \tag{6.76}$$

$$m_{21} = \sum_{i=1}^{L} [q_1(t_i)r_1(t_i) + q_2(t_i)r_2(t_i)], \tag{6.77}$$

$$m_{22} = \sum_{i=1}^{L} [r_1^2(t_i) + r_2^2(t_i)], \tag{6.78}$$

$$g_1 = \sum_{i=1}^{L} \{[z_1(t_i) - p_1(t_i)]q_1(t_i) + [z_2(t_i) - p_2(t_i)]q_2(t_i)\}, \tag{6.79}$$

$$g_2 = \sum_{i=1}^{L} \{[z_1(t_i) - p_1(t_i)]r_1(t_i) + [z_2(t_i) - p_2(t_i)]r_2(t_i)\}. \tag{6.80}$$

The quasilinearization equations can also be derived using the approach involving the Jacobian matrix. Define the vector $\mathbf{x}(t)$:

$$\mathbf{x}(t) = \begin{bmatrix} x_1 \\ x_2 \\ a \\ b \end{bmatrix}. \tag{6.81}$$

Then equation (6.6) becomes

$$\dot{\mathbf{x}}_{k+1} = \mathbf{f}(\mathbf{x}_k) + J(\mathbf{x}_k)(\mathbf{x}_{k+1} - \mathbf{x}_k) \tag{6.82}$$

or

$$\begin{bmatrix} \dot{x}_1^{k+1} \\ \dot{x}_2^{k+1} \\ \dot{a}^{k+1} \\ \dot{b}^{k+1} \end{bmatrix} = \begin{bmatrix} f_1^k \\ f_2^k \\ 0 \\ 0 \end{bmatrix} + \begin{bmatrix} f_{1x_1}^k & f_{1x_2}^k & f_{1a}^k & f_{1b}^k \\ f_{2x_1}^k & f_{2x_2}^k & f_{2a}^k & f_{2b}^k \\ 0 & 0 & 0 & 0 \\ 0 & 0 & 0 & 0 \end{bmatrix} \begin{bmatrix} x_1^{k+1} - x_1^k \\ x_2^{k+1} - x_2^k \\ a^{k+1} - a^k \\ b^{k+1} - b^k \end{bmatrix}. \tag{6.83}$$

The solution is given by equation (6.8):

$$\mathbf{x}^{k+1} = \mathbf{p}(t) + \sum_{j=3}^{4} c_j \mathbf{h}_j$$
$$= \mathbf{p}(t) + c_3 \mathbf{h}_3 + c_4 \mathbf{h}_4, \tag{6.84}$$

where

$$c_3 = a^{k+1}, \qquad c_4 = b^{k+1}. \tag{6.85}$$

The summation is from $j = R + 1$ to $n + N$, where $R = 2$ and $n + N = 4$. Equations (6.9)–(6.12) then become

$$\dot{\mathbf{p}} = \begin{bmatrix} \dot{p}_1 \\ \dot{p}_2 \\ \dot{p}_3 \\ \dot{p}_4 \end{bmatrix} = \begin{bmatrix} f_1^k \\ f_2^k \\ 0 \\ 0 \end{bmatrix} + \begin{bmatrix} f_{1x_1}^k & f_{1x_2}^k & f_{1a}^k & f_{1b}^k \\ f_{2x_1}^k & f_{2x_2}^k & f_{2a}^k & f_{2b}^k \\ 0 & 0 & 0 & 0 \\ 0 & 0 & 0 & 0 \end{bmatrix} \begin{bmatrix} p_1 - x_1^k \\ p_2 - x_2^k \\ p_3 - a^k \\ p_4 - b^k \end{bmatrix}, \quad \begin{bmatrix} p_1(0) \\ p_2(0) \\ p_3(0) \\ p_4(0) \end{bmatrix} = \begin{bmatrix} c_1 \\ c_2 \\ 0 \\ 0 \end{bmatrix}, \tag{6.86}$$

$$\dot{\mathbf{h}}_3 = \begin{bmatrix} \dot{q}_1 \\ \dot{q}_2 \\ \dot{q}_3 \\ \dot{q}_4 \end{bmatrix} = \begin{bmatrix} f_{1x_1}^k & f_{1x_2}^k & f_{1a}^k & f_{1b}^k \\ f_{2x_1}^k & f_{2x_2}^k & f_{2a}^k & f_{2b}^k \\ 0 & 0 & 0 & 0 \\ 0 & 0 & 0 & 0 \end{bmatrix} \begin{bmatrix} q_1 \\ q_2 \\ q_3 \\ q_4 \end{bmatrix}, \quad \begin{bmatrix} q_1(0) \\ q_2(0) \\ q_3(0) \\ q_4(0) \end{bmatrix} = \begin{bmatrix} 0 \\ 0 \\ 1 \\ 0 \end{bmatrix}, \tag{6.87}$$

$$\dot{\mathbf{h}}_4 = \begin{bmatrix} \dot{r}_1 \\ \dot{r}_2 \\ \dot{r}_3 \\ \dot{r}_4 \end{bmatrix} = \begin{bmatrix} f_{1x_1}^k & f_{1x_2}^k & f_{1a}^k & f_{1b}^k \\ f_{2x_1}^k & f_{2x_2}^k & f_{2a}^k & f_{2b}^k \\ 0 & 0 & 0 & 0 \\ 0 & 0 & 0 & 0 \end{bmatrix} \begin{bmatrix} r_1 \\ r_2 \\ r_3 \\ r_4 \end{bmatrix}, \quad \begin{bmatrix} r_1(0) \\ r_2(0) \\ r_3(0) \\ r_4(0) \end{bmatrix} = \begin{bmatrix} 0 \\ 0 \\ 0 \\ 1 \end{bmatrix}. \tag{6.88}$$

It is obvious that

$$p_3(t) = p_4(t) = 0, \tag{6.89}$$
$$q_3(t) = 1, \qquad q_4(t) = 0, \tag{6.90}$$
$$r_3(t) = 0, \qquad r_4(t) = 1, \tag{6.91}$$

since

$$\dot{p}_3(t) = 0, \qquad p_3(0) = 0, \qquad \dot{p}_4(t) = 0, \qquad p_4(0) = 0, \tag{6.92}$$
$$\dot{q}_3(t) = 0, \qquad q_3(0) = 1, \qquad \dot{q}_4(t) = 0, \qquad q_4(0) = 0, \tag{6.93}$$
$$\dot{r}_3(t) = 0, \qquad r_3(0) = 0, \qquad \dot{r}_4(t) = 0, \qquad r_4(0) = 1. \tag{6.94}$$

It can then be shown that equations (6.86)–(6.88) are identical to equations (6.64)–(6.69).

Assume as an example that equations (6.56) and (6.57) are given by

$$\dot{x}_1 = x_2, \qquad x_1(0) = c_1, \tag{6.95}$$

$$\dot{x}_2 = -ax_2 - bx_1 + u, \qquad x_2(0) = c_2. \tag{6.96}$$

The partial derivatives in equations (6.95) and (6.96) are then

$$f_{1a}^k = 0, \qquad f_{2a}^k = -x_2, \tag{6.97}$$

$$f_{1b}^k = 0, \qquad f_{2b}^k = -x_1, \tag{6.98}$$

$$f_{1x_1}^k = 0, \qquad f_{2x_1}^k = -b, \tag{6.99}$$

$$f_{1x_2}^k = 1, \qquad f_{2k_2}^k = -a. \tag{6.100}$$

Equation (6.59) is thus greatly simplified since three of its terms are zero.

6.2.4. Second-Order System with Two Unknown Parameters and Scalar Measurements

The system equations are the same as for the previous example, equations (6.56) and (6.57) with unknown parameters a and b. In this case however, only the component x_1 is observed. Thus we have

$$S = \sum_{i=1}^{L} [x_1(t_i) - z_1(t_i)]^2. \tag{6.101}$$

The linearized equations are the same as for the previous example. Substituting equation (6.62) into (6.101) gives

$$S = \sum_{i=1}^{L} [p_1(t_i) + a^{k+1}q_1(t_i) + b^{k+1}r_1(t_i) - z_1(t_i)]^2. \tag{6.102}$$

The conditions for minimum S are

$$\frac{\partial S}{\partial a^{k+1}} = 2 \sum_{i=1}^{L} q_1(t_i)[p_1(t_i) + a^{k+1}q_1(t_i) + b^{k+1}r_1(t_i) - z_1(t_i)] = 0, \tag{6.103}$$

$$\frac{\partial S}{\partial b^{k+1}} = 2 \sum_{i=1}^{L} r_1(t_i)[p_1(t_i) + a^{k+1}q_1(t_i) + b^{k+1}r_1(t_i) - z_1(t_i)] = 0. \tag{6.104}$$

These equations yield two linear algebraic equations for the unknown parameters a^{k+1} and b^{k+1} of the form

$$m_{11}a^{k+1} + m_{12}b^{k+1} = g_1, \tag{6.105}$$

$$m_{21}a^{k+1} + m_{22}b^{k+1} = g_2, \tag{6.106}$$

where

$$m_{11} = \sum_{i=1}^{L} q_1^2(t_i), \tag{6.107}$$

$$m_{12} = \sum_{i=1}^{L} q_1(t_i)r_1(t_i), \tag{6.108}$$

$$m_{21} = \sum_{i=1}^{L} q_1(t_i)r_1(t_i), \tag{6.109}$$

$$m_{22} = \sum_{i=1}^{L} r_1^2(t_i), \tag{6.110}$$

$$g_1 = \sum_{i=1}^{L} [z_1(t_i) - p_1(t_i)]q_1(t_i), \tag{6.111}$$

$$g_2 = \sum_{i=1}^{L} [z_1(t_i) - p_1(t_i)]r_1(t_i). \tag{6.112}$$

Exercises

1. Given the system $\dot{x} = a$, $x(0) = 1$, with unknown parameter a, known initial condition $x(0)$, and measurement $b(1) = 1.2$, show that the particular and homogeneous quasilinearization differential equations for system identification are given by

$$\dot{p} = 0, \qquad p(0) = 1,$$
$$\dot{h} = 1, \qquad h(0) = 0.$$

2. Find the analytical solutions for $p(t)$ and $h(t)$ in Exercise 1. Then estimate the value of a using equation (6.24).

   Answer: $a_1 = 0.2$.

3. Derive the quasilinearization equations and recursive relations for system identification of a first-order system with known parameters and unknown initial condition.

   Answer: $c_{k+1} = \sum_{i=1}^{N} [b_i - p(t_i)]q(t_i) \Big/ \sum_{i=1}^{N} q^2(t_i)$.

4. Solve Exercise 5 in Chapter 5 using the quasilinearization method for system identification.

   Answer: $c_1 = 0.5$.

5. Given the system $\dot{x} = -ax$, $x(0) = 1$, with unknown parameter a, known initial condition $x(0)$, and measurement $b(1) = 0.904837$, show that the particular and homogeneous quasilinearization differential equations for system identification are given by

$$\dot{p} = -a_k p + a_k x_k, \qquad p(0) = 1,$$
$$\dot{h} = -a_k h - x_k, \qquad h(0) = 0.$$

6. Find the analytical solutions for $p(t)$ and $h(t)$ in Exercise 5 assuming the initial approximation for the trajectory $x(t)$ is $x_0(t) = 1$. Then obtain the first estimate of the parameter a assuming the initial approximation is $a_0 = 0.15$.

   Answer: $p(t) = 1$, $h(t) = -(1/a_0)(1 - e^{-a_0 t})$, $a_1 = 0.102478$.

7. Obtain the first estimate of a in Exercise 6 if $a_0 = 0.1$ instead of 0.15.

   Answer: $a_1 = 0.1$.

8. Obtain the first estimate of a in Exercise 5 if $\dot{x} = -ax + y$, $x(0) = 1$; y is a unit step function, $y(t) = 1$; and $z(1) = 1.856463$. Use the initial approximations given in Exercise 6.

   Answer:

$$p(t) = 1 + (1/a_0)(1 - e^{-a_0 t}), \; h(t) = -(1/a_0)(1 - e^{-a_0 t}), \; a_1 = 0.077697.$$

9. Obtain the first estimate of a in Exercise 8 if $a_0 = 0.1$ instead of 0.15.

   Answer: $a_1 = 0.1$.

10. Why is the first estimate of a in Exercise 8 so poor?

    Answer: Because $x_0(t) = 1$ is a poor estimate.

7

Applications of System Identification

System identification problems occur in many diverse fields. In this chapter, two examples are given with numerical results. The first example is for blood glucose regulation parameter estimation. The blood glucose concentration increases when glucose is administered in mammals. This results in an increase in the plasma insulin concentration. The insulin accelerates the rate of disappearance of glucose from the plasma compartment and the blood sugar quickly returns to its normal value. A linear two-compartment model is used to model the process. The parameters of the model are estimated using the methods of Chapters 5 and 6.

The second example is concerned with the fitting of nonlinear models of drug metabolism to experimental data. Linear models have been shown to be inadequate to describe some metabolic processes. Quasilinearization is used to fit a nonlinear model.

7.1. Blood Glucose Regulation Parameter Estimation

A glucose tolerance test was performed on dogs by injecting glucose intravenously and measuring the plasma glucose and insulin concentrations as a function of time. Quasilinearization and the Gauss–Newton methods were utilized to fit the data to a minimal model and to estimate the parameters of the blood glucose regulation process. A relatively good fit was obtained for the simple model.

7.1.1. Introduction

The physiology of the blood glucose regulation process is extremely complex. As the model of the process increases in complexity, the estimation of the parameters becomes more difficult. In this section the parameters of a minimal model of the blood glucose regulation process are estimated.

The blood glucose concentration in normal mammals is finely regulated. The administration of glucose, orally or in a vein, results in a brisk, precipitous rise in the plasma insulin concentration due to a direct effect of the rising blood sugar upon the beta cells of the pancreas. The insulin, in turn, accelerates the rate of disappearance of glucose from the plasma compartment, and the blood sugar quickly returns to a normal value of 80–100 mg/100 ml.

In certain metabolic diseases, e.g., juvenile diabetes mellitus, the responsiveness of the beta cells of the pancreas may be severely diminished, resulting in a failure of the blood glucose to return quickly to its normal value. In an alternative form of the disease, i.e., maturity onset diabetes, the pancreatic sensitivity to glucose may be normal, or supranormal, but the sensitivity of glucose uptake by peripheral tissues to the elevated insulin concentration may be obtunded. The altered time course of the blood glucose concentration following a glucose load may be similar, but traditional clinical tests (such as the glucose tolerance test) lack the ability to discriminate effectively between various forms of diabetes.

It may be possible to use parameter identification and optimization techniques to design new diagnostic tests which will more effectively discriminate between normal individuals, and those who suffer more or less severely from various forms of diabetes. In principle, what is done is to utilize a model for the system which regulates the blood glucose concentration in humans. The system of interest (i.e., the subject) may be perturbed (glucose is administered) and the dynamic response of the system to the perturbation is used to estimate the model parameters. It is to be expected that certain of the parameter values estimated (for example, that which represents the sensitivity of the beta cells to glucose) will be a sensitive index of the severity of diabetes in a given individual, while other parameters (e.g., that representing the sensitivity of glucose uptake by peripheral tissue to insulin) will be an index of peripheral causes of systemic hyperglycemia.

Historical Background

In 1961, Bolie (Reference 1) utilized a linear two-compartment model to describe the regulation of the blood glucose concentration. Bolie extrapolated experimental data to estimate the parameters of the differential equations. The parameters are of interest because initial parameter estimates are required to obtain optimal inputs for blood glucose regulation parameter estimation as in Chapter 10. Previous studies have shown that optimal inputs can be utilized to increase the accuracy of estimated parameters. Thus optimal inputs not only yield a better description of the blood glucose response, but may also have possible clinical application. Estimates of the parameters using nonoptimal inputs, however, are not readily available in the literature in a form that can easily be used for obtaining optimal inputs.

Ackerman *et al.* (References 2, 3) used experimental data to estimate the damped natural frequency and the damping constant of the blood glucose response to orally administered glucose. Similar studies for infused glucose were made by Ceresa *et al.* (Reference 4) who also estimated the damped natural frequency and damping constant using glucose concentration measurements. More recently Segre *et al.* (Reference 5) estimated the parameters of the differential equations using measurements of both insulin and glucose. In this chapter the parameters are estimated using data obtained by Bergman *et al.* (Reference 6) from experiments on dogs.

Minimal Model

The minimal model is the linear two-compartment model introduced by Bolie (Reference 1) and is given by

$$\begin{aligned} \dot{x}_1 &= -ax_1 + bx_2, \qquad & x_1(0) &= c_1, \\ \dot{x}_2 &= -cx_1 - dx_2, \qquad & x_2(0) &= c_2, \end{aligned} \tag{7.1}$$

where x_1 is the deviation of the extracellular insulin concentration from the mean, μU/ml, and x_2 is the deviation of the extracellular glucose concentration from the mean, mg/100 ml. Assuming noisy measurements, the observations are given by

$$\begin{aligned} z_1(t_i) &= x_1(t_i) + \zeta_1(t_i), & \\ z_2(t_i) &= x_2(t_i) + \zeta_2(t_i), \qquad & i = 1, 2, \ldots, L, \end{aligned} \tag{7.2}$$

where $\zeta_1(t_i)$ and $\zeta_2(t_i)$ are the insulin and glucose measurement noise, respectively.

The parameter, b, is of particular interest because it determines the rate of insulin production. For diabetic dogs it is expected that this parameter will be considerably different from that for the average normal dog.

In this section, the results of physiological experiments to obtain the insulin and glucose responses of dogs is described. This is followed by a description of the computational methods utilized to estimate the parameters, a, b, c, and d, of the linear differential equation model. Numerical results are given, followed by a discussion of the results.

7.1.2. Physiological Experiments

A glucose tolerance test was performed on two different conscious, intact dogs. For data set number one, 3 g of glucose were injected intravenously into a 59-lb dog over a time interval of 4 min. Measurements of the plasma glucose and insulin concentrations were made every 10 min for a total time period of approximately 3 hr. The first measurements were made 10 min after the start of the glucose injection. This is assumed to provide sufficient time for the initial mixing to occur. The measurement

Table 7.1. Deviations of the Plasma Insulin and Glucose Concentrations from the Mean—Data Set Number 1[a]

Measurement number	Measurement time (min)	Insulin (μU/ml)	Glucose (mg/100 ml)
0	0	73	32
1	10	0	9
2	20	−6.7	6.5
3	30	−5.	5.5
4	40	−3.	5
5	50	(−2.5)	2.5
6	60	(−2)	0
7	70	6.8	1
8	80	−2.2	−4
9	90	1.3	−1.5
10	100	−7.5	−1
11	110	−7.7	0.5
12	120	−4.2	0
13	130	1.	1.5
14	140	−2.5	0.5
15	150	−2.	−1.
16	160	−4.7	−7
17	170	0.5	−2
18	180	−0.7	−3

[a] Three grams glucose injected at −10 to −6 min. Mean levels: insulin, 17.5 μU/ml; glucose, 92 mg/100 ml. Parentheses indicate estimated values. For relative least-squares estimation, zero observation values are replaced by 0.1.

deviations from the mean levels are given in Table 7.1. Measurements which have had to be estimated because of either poor or missing data are enclosed in parentheses. It should be noted that the insulin concentration has already reached its peak value at the time of the first measurement and is back to its mean level at the time of the second measurement. For these reasons, measurements for data set number two were made at more frequent time intervals.

For data set number two, 9 g of glucose were injected intravenously into a 66-lb dog over a time interval of 1 min. Measurements of the plasma glucose and insulin concentrations were made every 2.5 min for the first

Table 7.2. Deviations of the Plasma Insulin and Glucose Concentrations from the Mean—Data Set Number 2[a]

Measurement number	Measurement time (min)	Insulin (μU/ml)	Glucose (mg/100 ml)
0	0	34	143
1	2.5	166	118
2	5.0	123	100
3	7.5	149	75
4	10	89	72
5	12.5	202	61
6	15.	170	46
7	17.5	83	32
8	20.	(59)	(22.5)
9	22.5	35	13
10	25	(26)	(6)
11	27.5	17	−1
12	32.5	7	−1.5
13	37.5	0	−11
14	42.5	10	−7
15	47.5	0	−7
16	52.5	0	0
17	57.5	10	0
18	62.5	−3	1
19	67.5	−3	0
20	72.5	0	5
21	77.5	12	6
22	82.5	0	1
23	87.5	2	1

[a] Nine grams glucose injected at −2.5 to −1.5 min. Mean levels: insulin 23 μU/ml; glucose, 95 mg/100 ml. Parentheses indicate estimated values.

20 min and every 5 min thereafter for an additional 70 min. The first measurements were made 2.5 min after the start of the glucose injection. The measurement deviations from the mean levels are given in Table 7.2. The insulin response appears to be somewhat oscillatory for approximately the first 15 min. This is to be expected since it is well known that a step input of glucose causes a biphasic insulin response (References 7–11). The simple Bolie equations, however, do not model this initial response.

Glucose was measured by the glucose oxidase technique using a Beckman glucose analyzer. Insulin was determined by radioimmunoassay using the dextran charcoal separation technique.

7.1.3. Computational Methods

The unknown parameters were estimated using three different methods. In most cases the numerical results were the same regardless of the method used. In other cases, however, not all of the methods gave a convergent solution. The three methods used were

1. Gauss–Newton method;
2. quasilinearization;
3. modified quasilinearization.

A brief description of the use of each of these methods follows. See also Chapters 5 and 6.

Gauss–Newton Method

The performance criterion for least-squares estimation is given by

$$J = \mathbf{e}^T\mathbf{e} = \sum_{i=1}^{L} e_i^2, \tag{7.3}$$

where for equations (7.1), $\mathbf{e}_i$ is the two-dimensional column vector

$$\mathbf{e}_i(p) = \begin{bmatrix} e_1(t_i, \mathbf{p}) \\ e_2(t_i, \mathbf{p}) \end{bmatrix} = \begin{bmatrix} z_1(t_i) - x_1(t_i, \mathbf{p}) \\ z_2(t_i) - x_2(t_i, \mathbf{p}) \end{bmatrix}$$

$$= \mathbf{z}(t_i) - \mathbf{x}(t_i, \mathbf{p}), \quad i = 1, 2, \ldots, L. \tag{7.4}$$

Here L is the number of observation times, $\mathbf{z}(t_i)$ is a vector of observations at time t_i, $\mathbf{x}(t_i, \mathbf{p})$ is a state vector, $\mathbf{p}$ is a vector of unknown parameters, and

$$\mathbf{e} = \begin{bmatrix} \mathbf{z}(t_1) - \mathbf{x}(t_1, \mathbf{p}) \\ \mathbf{z}(t_2) - \mathbf{x}(t_2, \mathbf{p}) \\ \vdots \\ \mathbf{z}(t_L) - \mathbf{x}(t_L, \mathbf{p}) \end{bmatrix}, \qquad 2L \text{ vector.} \tag{7.5}$$

The Gauss–Newton method is based on the Taylor series expansion

$$\mathbf{x}(\mathbf{p} + \Delta\mathbf{p}) \doteq \mathbf{x}(\mathbf{p}) + \frac{\partial \mathbf{x}}{\partial \mathbf{p}} \Delta\mathbf{p} = \mathbf{x}(\mathbf{p}) + X\,\Delta\mathbf{p}, \tag{7.6}$$

where

$$X = \begin{bmatrix} \dfrac{\partial x_1(t_1)}{\partial a} & \dfrac{\partial x_1(t_1)}{\partial b} & \dfrac{\partial x_1(t_1)}{\partial c} & \dfrac{\partial x_1(t_1)}{\partial d} \\ \dfrac{\partial x_2(t_1)}{\partial a} & \dfrac{\partial x_2(t_1)}{\partial b} & \dfrac{\partial x_2(t_1)}{\partial c} & \dfrac{\partial x_2(t_1)}{\partial d} \\ \vdots & & & \\ \dfrac{\partial x_1(t_L)}{\partial a} & \cdots & & \\ \dfrac{\partial x_2(t_L)}{\partial a} & \cdots & & \end{bmatrix} \tag{7.7}$$

is a $(2L\times 4)$-parameter influence matrix. The iteration equation for the least-squares estimate is

$$\mathbf{p}_{j+1} = \mathbf{p}_j + \Delta\mathbf{p}_j, \tag{7.8}$$

where

$$\Delta\mathbf{p}_j = (X^T X)^{-1} X^T \mathbf{e} \tag{7.9}$$

and

$$\mathbf{p}_j = \begin{bmatrix} a \\ b \\ c \\ d \end{bmatrix} \tag{7.10}$$

is the estimate of the unknown parameters at iteration j. The method requires the solution of the eight auxiliary equations obtained by taking the partial derivatives of equations (7.1) with respect to the unknown parameters.

Defining

$$x_{1a} = \frac{\partial x_1}{\partial a},$$

$$x_{2a} = \frac{\partial x_2}{\partial a},$$

$$x_{1b} = \frac{\partial x_1}{\partial b},$$

$$\vdots$$

the auxiliary equations can be conveniently expressed in the form

$$\begin{aligned}\dot{P} &= \begin{bmatrix} -a & b \\ -c & -d \end{bmatrix} \begin{bmatrix} x_{1a} & x_{1b} & x_{1c} & x_{1d} \\ x_{2a} & x_{2b} & x_{2c} & x_{2d} \end{bmatrix} + \begin{bmatrix} -x_1 & x_2 & 0 & 0 \\ 0 & 0 & -x_1 & -x_2 \end{bmatrix} \\ &= FP + U \end{aligned} \tag{7.11}$$

with initial condition

$$P(0) = 0. \tag{7.12}$$

The sequence for obtaining the least-squares estimate is as follows. Using an initial approximation for the unknown parameters, integrate equations (7.1) and (7.11) from $t = 0$ to $t = T_f$. Compute a new set of parameters using equations (7.8) and (7.9) and repeat the cycle, starting with the integration of equations (7.1) and (7.11).

The performance criterion for weighted least-squares estimation is given by

$$J = \mathbf{e}^T W \mathbf{e}, \tag{7.13}$$

where W is the weighting matrix. Equation (7.9) then becomes

$$\mathbf{\Delta p}_j = (X^T W X)^{-1} X^T W \mathbf{e}. \tag{7.14}$$

If it is desired to minimize the sum of squares of relative deviations between the observational and the state variable values, then W becomes the $2L \times 2L$ diagonal matrix

$$W = \begin{bmatrix} \dfrac{\lambda_1}{z_1(t_1)^2} & & & & 0 \\ & \dfrac{\lambda_2}{z_2(t_1)^2} & & & \\ & & \ddots & & \\ & 0 & & \dfrac{\lambda_1}{z_1(t_L)^2} & \\ & & & & \dfrac{\lambda_2}{z_2(t_L)^2} \end{bmatrix}, \tag{7.15}$$

where λ_1 and λ_2 are constant weighting factors. If relative deviation weighting is not utilized, W becomes

$$W = \begin{bmatrix} \lambda_1 & & & & 0 \\ & \lambda_2 & & & \\ & & \ddots & & \\ & & & \lambda_1 & \\ 0 & & & & \lambda_2 \end{bmatrix}. \tag{7.16}$$

For simplicity, estimation using W given by equation (7.16) will be referred to as least-squares estimation, whereas estimation using W given by equation (7.15) will be referred to as relative least-squares estimation. Note that if $\lambda_1 = \lambda_2 = 1$, then W becomes the identity matrix and equation (7.14) becomes equal to equation (7.9).

Quasilinearization

Using the method of quasilinearization, equations (7.1) are written in the form

$$\begin{aligned} \dot{x}_1 &= f_1 = -ax_1 + bx_2, \qquad x_1(0) = c_1, \\ \dot{x}_2 &= f_2 = -cx_1 - dx_2, \qquad x_2(0) = c_2. \end{aligned} \tag{7.17}$$

The differential equations for x_1 and x_2 are linearized about the current approximations by expanding f_1 and f_2 in a power series and retaining only the linear terms. The new approximations for x_1 and x_2 are

$$\begin{aligned} \dot{x}_1 &= f_1^0 + (a - a^0)f_{1a}^0 + (b - b^0)f_{1b}^0 + (x_1 - x_1^0)f_{1x_1}^0 + (x_2 - x_2^0)f_{1x_2}^0, \\ \dot{x}_2 &= f_2^0 + (c - c^0)f_{2c}^0 + (d - d^0)f_{2d}^0 + (x_1 - x_1^0)f_{2x_1}^0 + (x_2 - x_2^0)f_{2x_2}^0, \end{aligned} \tag{7.18}$$

where the superscript zero indicates that the quantity is evaluated for the current approximation and the partial derivatives are defined by the notation

$$f_{1a}^0 = \left.\frac{\partial f_1}{\partial a}\right|^0$$

$$\vdots$$

$$f_{1x_1} = \left.\frac{\partial f_1}{\partial x_1}\right|^0$$

$$\vdots$$

The general solution for equations (7.18) is

$$\begin{aligned} x_1(t) &= p_1(t) + aq_1(t) + br_1(t) + cv_1(t) + dw_1(t), \\ x_2(t) &= p_2(t) + aq_2(t) + br_2(t) + cv_2(t) + dw_2(t). \end{aligned} \tag{7.19}$$

The functions p_1 and p_2 are solutions of the initial-value equations

$$\begin{aligned} \dot{p}_1 &= f_1^0 - a^0 f_{1a}^0 - b^0 f_{1b}^0 + (p_1 - x_1^0) f_{1x_1}^0 + (p_2 - x_2^0) f_{1x_2}^0, \\ \dot{p}_2 &= f_2^0 - c^0 f_{2c}^0 - d^0 f_{2d}^0 + (p_1 - x_1^0) f_{2x_1}^0 + (p_2 - x_2^0) f_{2x_2}^0 \end{aligned} \tag{7.20}$$

with initial conditions

$$\begin{aligned} p_1(0) &= c_1, \\ p_2(0) &= c_2. \end{aligned} \tag{7.21}$$

The functions q, r, v, and w are solutions of the initial-value equations

$$\begin{aligned} \dot{q}_1 &= f_{1a}^0 + f_{1x_1}^0 q_1 + f_{1x_2}^0 q_2, \\ \dot{q}_2 &= f_{2x_1}^0 q_1 + f_{2x_2} q_2, \\ \dot{r}_1 &= f_{1b}^0 + f_{1x_1}^0 r_1 + f_{1x_2}^0 r_2, \\ \dot{r}_2 &= f_{2x_1}^0 r_1 + f_{2x_2}^0 r_2, \\ \dot{v}_1 &= f_{1x_1}^0 v_1 + f_{1x_2}^0 v_2, \\ \dot{v}_2 &= f_{2c}^0 + f_{2x_1}^0 v_1 + f_{2x_2}^0 v_2, \\ \dot{w}_1 &= f_{1x_1}^0 w_1 + f_{1x_2}^0 w_2, \\ \dot{w}_2 &= f_{2d}^0 + f_{2x_1}^0 w_1 + f_{2x_2}^0 w_2 \end{aligned} \tag{7.22}$$

with all initial conditions equal to zero.

The weighted least-squares estimates of the parameters are obtained by minimizing the sum of the squares of the relative deviations between the state variable and observational values:

$$S = \sum_{i=1}^{L} \left\{ \lambda_1 \left[\frac{x_1(t_i) - z_1(t_i)}{z_1(t_i)} \right]^2 + \lambda_2 \left[\frac{x_2(t_i) - z_2(t_i)}{z_2(t_i)} \right]^2 \right\}. \tag{7.23}$$

Substituting equations (7.19) into equation (7.23) and setting the partial derivatives

$$\frac{\partial S}{\partial a}, \quad \frac{\partial S}{\partial b}, \quad \frac{\partial S}{\partial c}, \quad \frac{\partial S}{\partial d}$$

equal to zero, a set of linear algebraic equations for the unknown parameters is obtained. These equations can be expressed in the form

$$\begin{bmatrix} h_{11} & h_{12} & h_{13} & h_{14} \\ h_{21} & h_{22} & h_{23} & h_{24} \\ h_{31} & h_{32} & h_{33} & h_{34} \\ h_{41} & h_{42} & h_{43} & h_{44} \end{bmatrix} \begin{bmatrix} a \\ b \\ c \\ d \end{bmatrix} = \begin{bmatrix} g_1 \\ g_2 \\ g_3 \\ g_4 \end{bmatrix}, \tag{7.24}$$

where

$$h_{11} = \sum_{i=1}^{L} \left[\lambda_1 \frac{q_1(t_i) q_2(t_i)}{z_1(t_i)^2} + \lambda_2 \frac{q_2(t_i)^2}{z_2(t_i)^2} \right],$$

$$h_{12} = \sum_{i=1}^{L} \left[\lambda_1 \frac{q_1(t_i) r_1(t_i)}{z_1(t_i)^2} + \lambda_2 \frac{q_2(t_i) r_2(t_i)}{z_2(t_i)^2} \right],$$

$$h_{21} = \sum_{i=1}^{L} \left[\lambda_1 \frac{q_1(t_i) r_1(t_i)}{z_1(t_i)^2} + \lambda_2 \frac{q_2(t_i) r_2(t_i)}{z_2(t_i)^2} \right],$$

$$\vdots$$

and

$$g_1 = \sum_{i=1}^{L} \left\{ \frac{\lambda_1 q_1(t_i)}{z_1(t_i)^2} [z_1(t_i) - p_1(t_i)] + \frac{\lambda_2 q_2(t_i)}{z_2(t_i)^2} [z_2(t_i) - p_2(t_i)] \right\}$$

$$\vdots$$

In matrix notation equation (7.24) becomes

$$H\mathbf{p} = \mathbf{g}. \tag{7.25}$$

Solving for the unknown parameters, we obtain

$$\mathbf{p} = H^{-1}\mathbf{g}. \tag{7.26}$$

The sequence for the quasilinearization method is as follows. Using an initial approximation for the unknown parameters, obtain an initial approximation for $x_1(t)$ and $x_2(t)$. Integrate equations (7.20) and (7.22). Solve the linear algebraic equations (7.24) for the unknown parameters. Obtain the new approximation for $x_1(t)$ and $x_2(t)$ by the numerical integration of the linearized equations (7.18). Then repeat the cycle, starting with the integration of equations (7.20) and (7.22).

Modified Quasilinearization (References 12, 13)

The quasilinearization method described above has the disadvantage of requiring the storage of $x_1^0(t)$ and $x_2^0(t)$ in order to integrate the linearized equations (7.18). This storage is not required, however, if equations (7.17) are used directly to obtain the new approximations for $x_1(t)$ and $x_2(t)$, after the linear algebraic equations have been solved for the unknown parameters. Equations (7.17), (7.20), and (7.22) are integrated simultaneously. This method, however, may reduce the range of convergence of the algorithm and thus require very good initial estimates of the unknown parameters.

Initial Parameter Estimates

The above algorithms require initial estimates of the parameters. Based on the data given in Tables 7.1 and 7.2, the following values were crudely estimated:

$$
\begin{aligned}
a &= 0.1 \quad && 1/\text{min}, \\
b &= 0.2 \quad && 100\ \mu\text{U/mg/min}, \\
c &= 0.04 \quad && \text{mg}/100\ \mu\text{U/min}, \\
d &= 0.001 \quad && 1/\text{min}.
\end{aligned}
$$

These values were used as the initial parameter estimates for all cases.

The estimates were obtained as follows. Parameter a defines the dependence of the rate of decrease of insulin on the plasma concentration of the hormone. Insulin has a plasma half-time of approximately 10 min. Thus the value of a, which is the inverse of the insulin time constant, is approximately 0.1 min^{-1}.

Parameter b represents the sensitivity of insulin secretion by the pancreas to the increase in the plasma glucose concentration above the normal value. Assuming that an increase in the plasma glucose concentration of 100 mg/100 ml causes an increase of 200 μU/ml in the plasma insulin level within 10 min, then ignoring the decrease in the insulin rate we have

$$
\begin{aligned}
b &\doteq \frac{\Delta \dot{x}_1}{\Delta x_2} = \frac{200(\mu\text{U/ml})/(10\ \text{min})}{100(\text{mg}/100\ \text{ml})} \\
&\doteq 0.2\,\frac{(\mu\text{U/ml})/\text{min}}{(\text{mg}/100\ \text{ml})}.
\end{aligned}
$$

Parameter d represents the rate of glucose disappearance with no additional insulin secreted from the pancreas. Assuming that such a diabetic condition has a time constant greater than 360 min, it was assumed that the value of d is 0.001 min^{-1}.

Finally, the parameter c describes the effect of insulin to accelerate the glucose uptake. Assuming that in the presence of 100 μU/ml of insulin, 100 mg/100 ml glucose disappears in about 25 min, then ignoring the insulin-independent glucose disappearance, we have

$$c \doteq -\frac{\Delta \dot{x}_2}{\Delta x_1} = \frac{100(\text{mg}/100\ \text{ml})/(25\ \text{min})}{100(\mu\text{U}/\text{ml})}$$
$$\doteq 0.04\,\frac{(\text{mg}/100\ \text{ml})/\text{min}}{(\mu\text{U}/\text{ml})}.$$

7.1.4. Numerical Results

To obtain the least-squares estimation of the unknown parameters, the algorithms described above were utilized with a fourth-order Runge–Kutta method for the integration of the differential equations. In most cases, grid intervals of 0.5 min were found to be adequate.

For data set number one, measurements of the insulin and glucose concentrations were made every 10 min as described previously. The first measurements were made 10 min after the start of the glucose injection. These measurements provided the assumed initial conditions for the linear model, given by equations (7.1), at time $t = 0$. At this time, however, both the insulin and glucose concentrations have already reached their peak values and are rapidly decreasing toward their steady-state values.

The modified quasilinearization method was utilized first, using the initial parameter estimates given previously. A convergent set of parameter estimates could not be obtained, however. A convergent set of parameter estimates was obtained using the Gauss–Newton method, for a time interval of 50 min, as shown in Figure 7.1. The least-squares performance criterion with $\lambda_1 = \lambda_2 = 1$ in equation (7.16) was utilized. The final magnitude of the parameter b turned out incorrectly to be negative. The reason for this is easily explained since the magnitudes of the insulin observations do not increase before reaching a peak value. Thus the rate of change of the insulin concentration is negative. An attempt to remedy this situation by assuming the initial condition for insulin is zero, and estimating an unknown initial condition for glucose, failed to produce a convergent solution.

For data set number two, measurements of the insulin and glucose concentrations were made every 2.5 min for the first 20 min, and every 5 min thereafter for an additional 70 min. The first measurements were made 2.5 min after the start of the glucose injection and provided the assumed initial conditions for the linear model. With the smaller time between measurements, the biphasic insulin response was clearly evident.

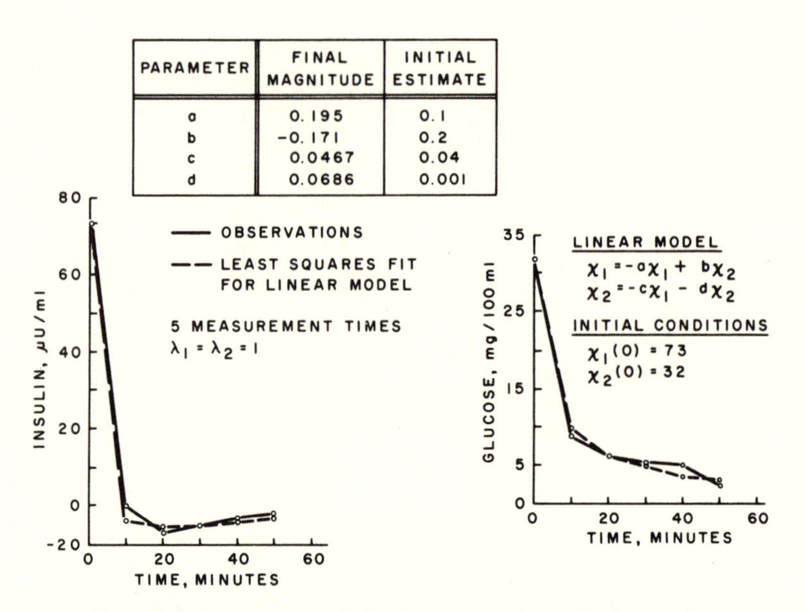

Figure 7.1. Least-squares fit for data set number 1 (Reference 6).

Using the Gauss–Newton method for a time interval of 25 min, a convergent set of parameter estimates was obtained. The least-squares performance criterion with $\lambda_1 = \lambda_2 = 1$ in equation (7.16) was utilized. As anticipated, the parameter b turned out to be positive in this case. Figure 7.2 gives the values of the final convergent set of parameter estimates, based on a 25-min interval, and shows the effect of integrating the linear model (7.1) with these parameters, out to a time interval of 62.5 min. The curves show that while the simple linear model cannot follow the biphasic insulin response, the curve fit is otherwise excellent.

Figure 7.3 shows the effects of using different performance criteria. Since the glucose measurements are more accurate than the insulin measurements, weighting factors of $\lambda_1 = 1/3$ for insulin and $\lambda_2 = 1$ for glucose were utilized. Curves 1 and 2 in Figure 7.3 show that this weighting factors had negligible effect on the curves, but did cause the parameter b to increase from 0.341 for $\lambda_1 = 1$ to 0.362 for $\lambda_1 = 1/3$.

Using the relative least-squares performance criterion, with weighting matrix defined by equation (7.15), a more pronounced effect was obtained for the estimated insulin response, as shown by curves 3 and 4 in Figure 7.3. Again, the effect of using $\lambda_1 = 1/3$ instead of 1 did not appear to make much difference in the shape of the curves. It did, however, make a difference in the parameter estimates, the magnitude of b decreasing from 0.308 for

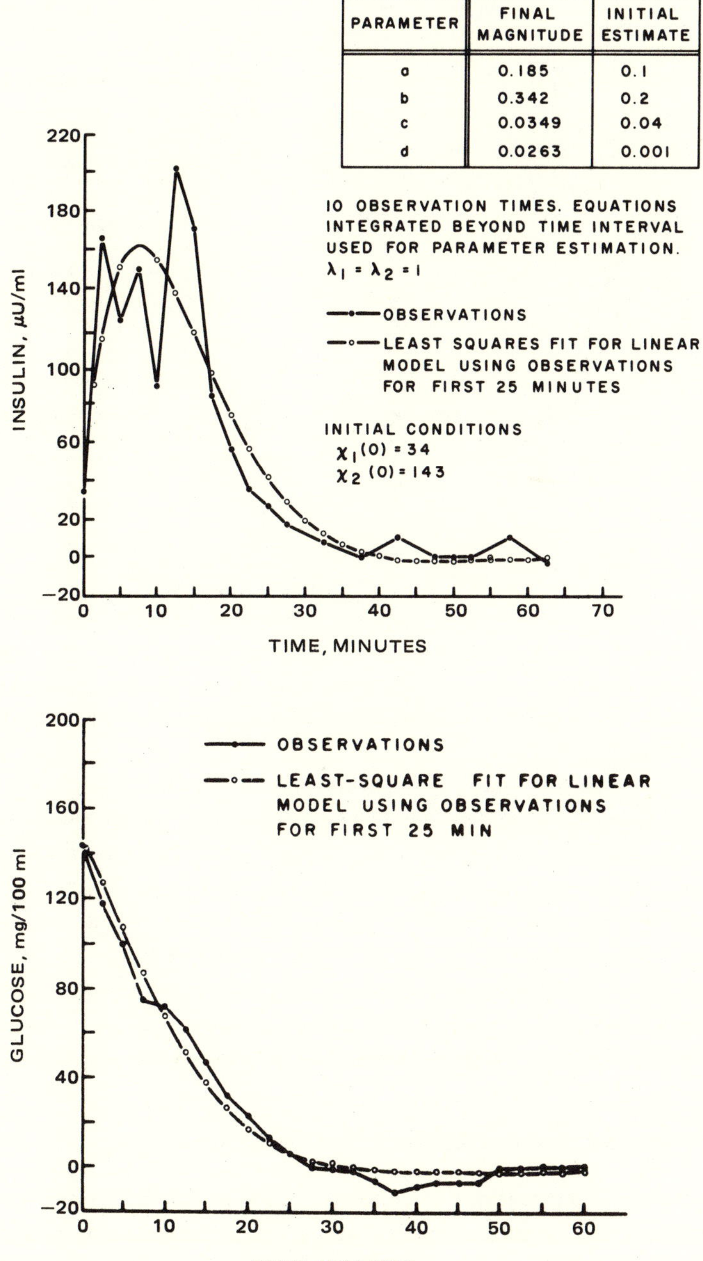

Figure 7.2. Least-squares fit for data set number 2. (a) Insulin response; (b) glucose response (Reference 6).

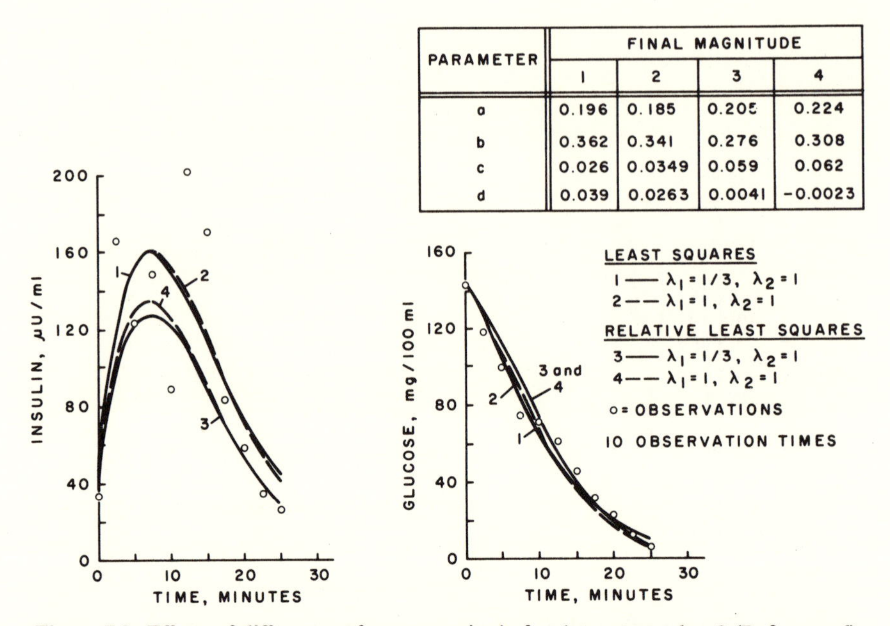

PARAMETER	FINAL MAGNITUDE			
	1	2	3	4
a	0.196	0.185	0.205	0.224
b	0.362	0.341	0.276	0.308
c	0.026	0.0349	0.059	0.062
d	0.039	0.0263	0.0041	-0.0023

Figure 7.3. Effects of different performance criteria for data set number 2 (Reference 6).

$\lambda_1 = 1$ to 0.276 for $\lambda_1 = 1/3$. The magnitude of d is negative for $\lambda_1 = 1$, but this did not appear to affect the shape of the curves, when comparing curves 3 and 4.

All four sets of curves and parameters in Figure 7.3 were obtained using the Gauss–Newton method. The same parameter values were obtained to three significant digits for curves 3 and 4 using the quasilinearization and modified quasilinearization methods. The quasilinearization and modified quasilinearization methods were not run for curves 1 and 2 because they were programmed for the relative least-squares performance criterion. However, it is expected that if these methods were utilized with the same performance criterion and weighting matrix given by equation (7.15), the same results would be obtained as for the Gauss–Newton method.

7.1.5. Discussion and Conclusions

A relatively good fit of the empirical data was obtained using the minimal model. A more complex model is needed to model the biphasic response. The minimal model may be perfectly adequate, however, for computing optimal inputs and for differentiating between diabetic and normal subjects.

In all cases, the final parameter values given in Figure 7.3 yielded complex eigenvalues, indicating that the responses are oscillatory. The damping ratios were between 0.8 and 0.9.

Minimizing the sum of the squares of the relative deviations appeared to have more effect on the model response curves than did varying the constant weighting factors λ_1 and λ_2.

From a biomedical point of view, this section makes available parameter estimates for the canine blood glucose regulation process. From an analytical and computational point of view this section shows that the methods used can provide a good fit to empirical data for a minimal model.

7.2. Fitting of Nonlinear Models of Drug Metabolism to Experimental Data

In recent years it has been shown that the biotransformation of various drugs in the therapeutic dose range is not adequately described by linear models. For enzymatically mediated reactions, e.g., the phenomena of saturation and substrate depletion lead to nonlinear effects.

A basic task is to fit nonlinear theoretical models to observed data for various metabolic processes. This is done in this section for a process involving Michaelis and Menten kinetics. A numerical experiment (Reference 14) is described and various extensions are suggested.

7.2.1. Introduction

In many circumstances drug distributions throughout the body can be described by means of linear ordinary differential equations (References 15, 16). Nevertheless, it is well documented that for certain drugs, e.g., ethanol, undergoing enzymatically mediated reactions, saturation and substrate depletion may arise. These processes are described by means of nonlinear differential equations (Reference 17).

A basic task is the fitting of the theoretical differential equations to observed experimental kinetic data. In this section it will be shown how quasilinearization may be employed for this purpose and the results of some numerical experiments will be described (References 18–21).

7.2.2. A Model Employing Michaelis and Menten Kinetics for Metabolism

Consider a saturable metabolic process. Let u be the concentration of drug in plasma at time t, v be the amount of drug in urine at time t, w

be the concentration of metabolite in plasma at time t, z be the amount of metabolite in urine at time t, and V be the apparent volume of distribution of drug in plasma. Further assume that the rate of transformation of the drug into the metabolite is given by $au/(b+u)$; i.e., Michaelis and Menten kinetics obtain. The differential equations for the process are

$$\dot{u} = -\frac{au}{b+u} - pu, \tag{7.27}$$

$$\dot{v} = pVu, \tag{7.28}$$

$$\dot{w} = \frac{au}{b+u} - qw, \tag{7.29}$$

$$\dot{z} = qVv, \qquad 0 \leq t \leq T, \tag{7.30}$$

where a, b, p, and q are certain rate constants. The initial conditions are

$$u(0) = u_0, \tag{7.31}$$

$$v(0) = 0, \tag{7.32}$$

$$w(0) = 0, \tag{7.33}$$

$$z(0) = 0. \tag{7.34}$$

If the rate constants a, b, p, and q and the initial concentration u_0 are known, then it is a simple matter to integrate equations (7.27)–(7.34) numerically, using modern analog, digital, or hybrid computers. Of greater difficulty, and correspondingly greater significance, is the problem of estimating the rate constants in the differential equations on the basis of experimental observations of such a metabolic process.

7.2.3. An Estimation Problem

Let us imagine that the values of the rate constants p and q are known, as is the initial concentration u_0. Suppose that at various times $t_1, t_2, \ldots, t_N$ the amounts of the drug and the metabolite in the urine are observed; let b_i be the observed amount of drug in urine at time t_i, and c_i be the observed amount of metabolite in urine at time t_i, $i = 1, 2, \ldots, N$. The task is to convert these empirical observations into estimates of the Michaelis and Menten constants a and b so that the theoretical predictions of the model are in closest agreement with the observations. More precisely, we wish to minimize the sum S, where

$$S = \sum_{i=1}^{N} \{[v(t_i) - b_i]^2 + [z(t_i) - c_i]^2\}, \tag{7.35}$$

through proper choice of the constants a and b. The quantities $b_1, b_2, \ldots, b_N$ and $c_1, c_2, \ldots, c_N$ are observed; the quantities $v(t_1)$, $v(t_2)$, $\ldots$, $v(t_N)$ and $z(t_1)$, $z(t_2)$, $\ldots$, $z(t_N)$ are determined from equations 7.27–7.34 for any choice of the parameters a and b.

7.2.4. Quasilinearization

Many techniques are available for minimizing the sum of the squares of the deviations S. One of the most effective is quasilinearization (Reference 12). This technique, a successive approximation scheme, will now be described.

Let us suppose that initial approximations to the minimizing values of the parameters a and b are available. Call these a^0 and b^0. By using these values in equations (7.27)–(7.34) we can produce numerically the approximate time histories $u^0(t)$, $v^0(t)$, $w^0(t)$, and $z^0(t)$ for $0 \leq t \leq t_N$. The values of a^0, b^0, $v(t_1)$, $v(t_2)$, $\ldots$, $v(t_N)$, and $z(t_1)$, $z(t_2)$, $\ldots$, $z(t_N)$ are stored in the memory of the computer, a total of $2N + 2$ values.

The technique for obtaining the next approximations will now be described. First, though, it is convenient to rewrite equations (7.27)–(7.30) in the compact form

$$\dot{u} = f_1, \tag{7.36}$$

$$\dot{v} = f_2, \tag{7.37}$$

$$\dot{w} = f_3, \tag{7.38}$$

$$\dot{z} = f_4, \tag{7.39}$$

so that

$$f_1 = -au/(b + u) - pu, \tag{7.40}$$

$$f_2 = pVu, \tag{7.41}$$

$$f_3 = au/(b + u) - qw, \tag{7.42}$$

$$f_4 = qVw. \tag{7.43}$$

The differential equations for u, v, w, and z are linearized about the current approximations by expanding the functions f_1, f_2, f_3, and f_4 in power series about the current approximations and retaining only the linear terms. We take as the linear differential equation for the new approximation functions u^1, v^1, w^1, and z^1

$$\dot{u}^1 = (f_1)^0 + (a^1 - a^0)\left(\frac{\partial f_1}{\partial a}\right)^0 + (b^1 - b^0)\left(\frac{\partial f_1}{\partial b}\right)^0 + (u^1 - u^0)\left(\frac{\partial f_1}{\partial u}\right)^0, \tag{7.44}$$

$$\dot{v}^1 = pVu^0 + (u^1 - u^0)pV, \tag{7.45}$$

$$\dot{w}^1 = (f_3)^0 + (a^1 - a^0)\left(\frac{\partial f_3}{\partial a}\right)^0 + (b^1 - b^0)\left(\frac{\partial f_3}{\partial b}\right)^0 + (u^1 - u^0)\left(\frac{\partial f_3}{\partial u}\right)^0 + (w^1 - w^0)\left(\frac{\partial f_3}{\partial w}\right)^0, \tag{7.46}$$

$$\dot{z}^1 = qVw^0 + (w^1 - w^0)qV. \tag{7.47}$$

The superscript-zero notation indicates that the quantity is evaluated for the current approximation. The initial conditions are

$$u^1(0) = u_0, \tag{7.48}$$

$$v^1(0) = 0, \tag{7.49}$$

$$w^1(0) = 0, \tag{7.50}$$

$$z^1(0) = 0. \tag{7.51}$$

The general solution equations (7.44)–(7.51) can be written in the form

$$u^1(t) = p_1(t) + a^1q_1(t) + b^1r_1(t), \tag{7.52}$$

$$v^1(t) = p_2(t) + a^1q_2(t) + b^1r_2(t), \tag{7.53}$$

$$w^1(t) = p_3(t) + a^1q_3(t) + b^1r_3(t), \tag{7.54}$$

$$z^1(t) = p_4(t) + a^1q_4(t) + b^1r_4(t). \tag{7.55}$$

The functions p_1, p_2, p_3, and p_4 are solutions of the initial-value problem

$$\dot{p}_1 = (f_1)^0 - a^0\left(\frac{\partial f_1}{\partial a}\right)^0 - b^0\left(\frac{\partial f_1}{\partial b}\right)^0 + (p_1 - u^0)\left(\frac{\partial f_1}{\partial u}\right)^0, \tag{7.56}$$

$$\dot{p}_2 = pVu^0 + (p_1 - u^0)pV, \tag{7.57}$$

$$\dot{p}_3 = (f_3)^0 - a^0\left(\frac{\partial f_3}{\partial a}\right)^0 - b^0\left(\frac{\partial f_3}{\partial b}\right)^0 + (p_1 - u^0)\left(\frac{\partial f_3}{\partial u}\right)^0 + (p_3 - w^0)\left(\frac{\partial f_3}{\partial w}\right)^0, \tag{7.58}$$

$$\dot{p}_4 = qVw^0 + qV(p_3 - w^0), \tag{7.59}$$

and the initial conditions

$$p_1(0) = u_0, \tag{7.60}$$

$$p_2(0) = 0, \tag{7.61}$$

$$p_3(0) = 0, \tag{7.62}$$

$$p_4(0) = 0. \tag{7.63}$$

The functions q_1, q_2, q_3, and q_4 are solutions of the initial-value problem

$$\dot{q}_1 = \left(\frac{\partial f_1}{\partial a}\right)^0 + \left(\frac{\partial f_1}{\partial u}\right)^0 q_1 \tag{7.64}$$

$$\dot{q}_2 = pVq_1 \tag{7.65}$$

$$\dot{q}_3 = \left(\frac{\partial f_3}{\partial a}\right)^0 + q_1\left(\frac{\partial f_3}{\partial u}\right)^0 + q_3\left(\frac{\partial f_3}{\partial w}\right)^0 \tag{7.66}$$

$$\dot{q}_4 = q_3 qV \tag{7.67}$$

and

$$q_i(0) = 0, \qquad i = 1, 2, 3, 4. \tag{7.68}$$

The functions r_1, r_2, r_3, and r_4 are solutions of the initial-value problem

$$\dot{r}_1 = \left(\frac{\partial f_1}{\partial b}\right)^0 + \left(\frac{\partial f_1}{\partial u}\right)^0 r_1, \tag{7.69}$$

$$\dot{r}_2 = pVr_1, \tag{7.70}$$

$$\dot{r}_3 = \left(\frac{\partial f_3}{\partial b}\right)^0 + \left(\frac{\partial f_3}{\partial u}\right)^0 r_1 + \left(\frac{\partial f_3}{\partial w}\right)^0 r_3, \tag{7.71}$$

$$\dot{r}_4 = qVr_3, \tag{7.72}$$

and

$$r_i(0) = 0, \qquad i = 1, 2, 3, 4. \tag{7.73}$$

The two constants a^1 and b^1 are determined by the condition that they should minimize the sum S_1, where

$$S_1 = \sum_{i=1}^{N} \{[p_2(t_i) + a^1 q_2(t_k) + b^1 r_2(t_i) - b_i]^2 + [p_4(t_i) + a^1 q_4(t_i) + b^1 r_4(t_i) - c_i]^2\}. \tag{7.74}$$

The conditions for this are that

$$\frac{\partial S_1}{\partial a^1} = 0 \tag{7.75}$$

and

$$\frac{\partial S_1}{\partial b^1} = 0. \tag{7.76}$$

These equations reduce to two linear algebraic equations for the unknowns a^1 and b^1, in view of the fact that the sum S_1 is quadratic in a^1 and b^1.

In more detail these equations are

$$\alpha_{11}a^1 + \alpha_{12}b^1 = \beta_1, \tag{7.77}$$

$$\alpha_{21}a^1 + \alpha_{22}b^1 = \beta_2, \tag{7.78}$$

where

$$\alpha_{11} = \sum_{i=1}^{N} [q_2(t_i)q_2(t_i) + q_4(t_i)q_4(t_i)], \tag{7.79}$$

$$\alpha_{12} = \sum_{i=1}^{N} [r_2(t_i)q_2(t_i) + q_4(t_i)r_4(t_i)], \tag{7.80}$$

$$\alpha_{21} = \sum_{i=1}^{N} [q_2(t_i)r_2(t_i) + q_4(t_i)r_4(t_i)], \tag{7.81}$$

$$\alpha_{22} = \sum_{i=1}^{N} [r_2(t_i)r_2(t_i) + r_4(t_i)r_4(t_i)], \tag{7.82}$$

$$\beta_1 = \sum_{i=1}^{N} [b_i - p_2(t_i)]q_2(t_i) + \sum_{i=1}^{N} [c_i - p_4(t_i)]q_4(t_i), \tag{7.83}$$

and

$$\beta_2 = \sum_{i=1}^{N} [b_i - p_2(t_i)]r_2(t_i) + \sum_{i=1}^{N} [c_i - p_4(t_i)]r_4(t_i). \tag{7.84}$$

As a typical cycle of the calculation is entered, values of a^0, b^0, and $u^0(t)$, $v^0(t)$, $w^0(t)$, and $z^0(t)$ for $0 \leq t \leq t_N$ are stored in the machine. The 12 differential equations for the p's, q's, and r's (with known initial conditions) are integrated from $t = 0$ to $t = t_N$. As each point $t = t_i$, $i = 1, 2, \ldots, N$, is passed, the sums in equations (7.79)–(7.84) are updated. When $t = t_N$ is reached, the values of the coefficients in equations (7.77) and (7.78) are known. These linear algebraic equations for the constants a^1

and b^1 are then solved. The final step in a cycle is to integrate equations (7.44)–(7.47) using the computed values of a^1 and b^1 and the known initial conditions in equations (7.48)–(7.51) from $t = 0$ to $t = t_N$. In this way the values of a^1, b^1, and $u^1(t)$, $v^1(t)$, $w^1(t)$ and $z^1(t)$, $0 \leq t \leq t_N$ are determined, and we are ready to enter a new cycle.

The method outlined is quadratically convergent, if convergent at all. This means that ultimately the number of correct digits obtained with each additional cycle will be doubled. If the initial approximations are too poor, no convergence will result. Ordinarily this is not too important, as the example shows. Generally about five complete cycles are permitted, or the calculation is stopped when a sufficiently small percentage change takes place from one cycle to the next.

7.2.5. Numerical Results

Suppose that a metabolic process is described by the differential equations and initial conditions given above. Assume that it is known that

$$p = 0.05, \tag{7.85}$$

$$q = 1.0, \tag{7.86}$$

$$V = 10^3, \tag{7.87}$$

$$u_0 = 5.0. \tag{7.88}$$

Furthermore, observations on the amounts in the urine from time 0 until time 4 have yielded the values in Table 7.3.

It is assumed initially that

$$a^0 = 1.1 \tag{7.89}$$

and

$$b^0 = 2.2. \tag{7.90}$$

We wish to find the values of a and b that give the closest fit of the theoretical differential equations to the experimental data.

By carrying out the quasilinearization scheme described earlier, it was found that the following approximations were obtained. (See Table 7.4.) The true values of the constants a and b are 1.00000 and 2.00000, respectively. The data in Table 7.3 were generated by integrating the differential equations in Section 7.2.2 with these numerical values for the constants a and b, and the other parameters as given above.

Table 7.3. Observed Amounts of the Drug and Metabolite in the Urine at Various Times[a]

Time	v	z
0.100	0.2475961 E 2	0.3448866 E 1
0.200	0.4904146 E 2	0.1332865 E 2
0.300	0.7284995 E 2	0.2898885 E 2
0.400	0.9618947 E 2	0.4983992 E 2
0.500	0.1190645 E 3	0.7534818 E 2
0.600	0.1414794 E 3	0.1050297 E 3
0.700	0.1634388 E 3	0.1384459 E 3
0.800	0.1849471 E 3	0.1751991 E 3
0.900	0.2060088 E 3	0.2149288 E 3
1.000	0.2266284 E 3	0.2573076 E 3
1.100	0.2468105 E 3	0.3020391 E 3
1.200	0.2665596 E 3	0.3488538 E 3
1.300	0.2858801 E 3	0.3975071 E 3
1.400	0.3047771 E 3	0.4477769 E 3
1.500	0.3232549 E 3	0.4994614 E 3
1.600	0.3413181 E 3	0.5523770 E 3
1.700	0.3589717 E 3	0.6063567 E 3
1.800	0.3762200 E 3	0.6612485 E 3
1.900	0.3930679 E 3	0.7169143 E 3
2.000	0.4095203 E 3	0.7732275 E 3
2.100	0.4255818 E 3	0.8300732 E 3
2.200	0.4412573 E 3	0.8873462 E 3
2.300	0.4565515 E 3	0.9449504 E 3
2.400	0.4714695 E 3	0.1002798 E 4
2.500	0.4860159 E 3	0.1060809 E 4
2.600	0.5001958 E 3	0.1118908 E 4
2.700	0.5140139 E 3	0.1177029 E 4
2.800	0.5274753 E 3	0.1235109 E 4
2.900	0.5405852 E 3	0.1293091 E 4
3.000	0.5533484 E 3	0.1350922 E 4
3.100	0.5657698 E 3	0.1408552 E 4
3.200	0.5778547 E 3	0.1465938 E 4
3.300	0.5896082 E 3	0.1523036 E 4
3.400	0.6010352 E 3	0.1579809 E 4
3.500	0.6121411 E 3	0.1636219 E 4
3.600	0.6229309 E 3	0.1692235 E 4
3.700	0.6334099 E 3	0.1747823 E 4
3.800	0.6435833 E 3	0.1802955 E 4
3.900	0.6534563 E 3	0.1857603 E 4
4.000	0.6630342 E 3	0.1911741 E 4

[a] Numbers following E denote powers of ten.

Table 7.4. Approximations to the Optimal Values of the Michaelis and Menten Constants for a Hypothetical Process

Cycle number	0	1	2	3	4
a	1.10000	1.00315	0.99831	0.99960	0.99999
b	2.20000	2.03348	2.00290	1.99758	2.00000

The execution time on the IBM System 360, Mod 44, at USC is in the order of one minute for the entire experiment. An Adams–Moulton integration scheme of fourth order with a step size of 0.1 was employed.

A second experiment yields an interesting result. Let the parameter values in equations (7.27)–(7.34) be

$$a = 1.0, \tag{7.91}$$

$$b = 20.0, \tag{7.92}$$

$$p = 0.05, \tag{7.93}$$

$$q = 1.0, \tag{7.94}$$

$$V = 4.5, \tag{7.95}$$

and

$$u_0 = 0.22222. \tag{7.96}$$

Numerical integration of these equations from $t = 0$ to $t = 4$ yields the data for v and z shown in Table 7.5. The quasilinearization scheme for estimating the values of a and b from these data is applied, and the results are shown in Table 7.6.

Apparently the process is not convergent. Notice, though, that toward the end the ratio of a to b is about 1/20, the true ratio. The explanation may lie in the observation that the initial concentration is small and

$$\frac{au}{b+u} \cong \frac{a}{b} u \tag{7.97}$$

during the entire process. Since the zero-order kinetics are not important during the portion of the process observed, only the ratio of a to b is important. When a higher initial concentration is employed, convergence is obtained.

Table 7.5. Observed Amounts of the Drug and Metabolite in the Urıne at Various Times

Time	v	z
0.100	0.4975207 E−2	0.2384238 E−3
0.200	0.9901173 E−2	0.9201004 E−3
0.300	0.1477839 E−1	0.1998288 E−2
0.400	0.1960733 E−1	0.3430717 E−2
0.500	0.2438848 E−1	0.5179223 E−2
0.600	0.2912229 E−1	0.7209294 E−2
0.700	0.3380924 E−1	0.9489745 E−2
0.800	0.3844981 E−1	0.1199241 E−1
0.900	0.4304442 E−1	0.1469183 E−1
1.000	0.4759355 E−1	0.1756502 E−1
1.100	0.5209763 E−1	0.2059123 E−1
1.200	0.5655712 E−1	0.2375172 E−1
1.300	0.6097245 E−1	0.2702956 E−1
1.400	0.6534404 E−1	0.3040948 E−1
1.500	0.6967235 E−1	0.3387772 E−1
1.600	0.7395780 E−1	0.3742184 E−1
1.700	0.7820076 E−1	0.4103064 E−1
1.800	0.8240169 E−1	0.4469403 E−1
1.900	0.8656102 E−1	0.4840289 E−1
2.000	0.9067917 E−1	0.5214905 E−1
2.100	0.9475654 E−1	0.5592511 E−1
2.200	0.9879351 E−1	0.5972444 E−1
2.300	0.1027905	0.6354105 E−1
2.400	0.1067479	0.6736958 E−1
2.500	0.1106660	0.7120520 E−1
2.600	0.1145454	0.7504362 E−1
2.700	0.1183863	0.7888097 E−1
2.800	0.1221892	0.8271372 E−1
2.900	0.1259543	0.8653879 E−1
3.000	0.1296822	0.9035343 E−1
3.100	0.1333731	0.9415513 E−1
3.200	0.1370274	0.9794170 E−1
3.300	0.1406456	0.1017112
3.400	0.1442278	0.1054617
3.500	0.1477746	0.1091919
3.600	0.1512862	0.1129003
3.700	0.1547630	0.1165857
3.800	0.1582053	0.1202471
3.900	0.1616135	0.1238835
4.000	0.1649879	0.1274942

Table 7.6. Approximations to the Optimal Estimates of the Michaelis and Menten Constants for the Second Process

Cycle number	0	1	2	3	4	5
a	1.10000	0.04709	0.04906	0.09418	0.09337	0.21368
b	22.00000	0.78125	0.81250	1.72312	1.70494	4.14118

7.2.6. Discussion

The major purpose of this section has been the presentation of a feasible method for handling the inverse problems which arise in nonlinear models of drug distribution. In addition to the Michaelis and Menten constants it is desirable to be able to estimate the apparent volume of distribution and the other parameters. The modifications in the method to do this are obvious and will not be described here.

Inevitably the observations are corrupted with errors. The occurrence of these errors adversely affects our ability to estimate the unknown constants. Controlled numerical experiments can be conducted to determine quantitatively the extent of these effects (References 18, 19).

The approach to system identification presented yields estimates of the system parameters characteristic of the individual that produced the observations. In conjunction with the theory of optimal control processes (Reference 22) it may be possible to devise schemes for the optimal administration of drugs to the individual.

Exercises

1. Derive the quasilinearization equations for the identification of the unknown parameter a in the van der Pol equation

$$\ddot{x} + a(x^2 - 1)\dot{x} + bx = y(t),$$

where $x(0)$ and $\dot{x}(0)$ are the known initial conditions, b is a known parameter, and $y(t)$ is a known forcing function. Assume both x and $\dot{x}$ are observed.

2. Derive the quasilinearization equations in Exercise 1 if both a and b are unknown parameters.

3. Derive the equations for the identification of the unknown parameters in Exercises 1 and 2 using the Gauss–Newton method.

4. Consider the motion of a mass attached to a nonlinear spring

$$\ddot{x} + \omega_0^2 x + a x^3 = y(t).$$

Derive the quasilinearization equations for estimating the unknown parameter a. Assume that $x(0)$ and $\dot{x}(0)$ are known initial conditions, b is a known parameter, and $y(t)$ is a known forcing function. Assume that only x is observed.

5. Derive the quasilinearization equations in Exercise 4 if both a and ω_0^2 are unknown parameters.

6. Derive the equations for the identification of the unknown parameters in Exercises 4 and 5 using the Gauss–Newton method.

PART IV

Optimal Inputs for System Identification

8

Optimal Inputs

The accuracy of parameter estimates is increased by the use of optimal inputs. In this chapter, a historical background of optimal inputs is given first. This is followed by the design of optimal inputs for linear systems in Section 8.2 and nonlinear systems in Section 8.3. An improved method for the numerical determination of optimal inputs and multiparameter optimal inputs is discussed in Chapter 9.

8.1. Historical Background

System parameters are usually identified by applying a known input signal and observing the system response. The estimation accuracy is enhanced by the use of optimal inputs. A survey (Reference 1) of optimal input design in the literature credits Levin (Reference 2) in 1960 with the first systematic attempt at obtaining optimal inputs. Levin considers the optimal estimation of the impulse response of a discrete-time linear system in the presence of noise. Levadi (Reference 3) in 1966 considers the estimation of gain parameters in continuous-time systems. Nahi and Wallis (Reference 4) in 1969 consider the design of optimal inputs for continuous-time systems from the point of view of the Cramér–Rao lower bound and its inverse, the Fisher information matrix. The Cramér–Rao lower bound is equal to the minimum estimation variance of the unbiased system parameter estimator. Optimal inputs for probing dynamic systems are formulated through maximization of the Fisher information matrix in the form of a conventional integral-quadratic-criterion optimal control problem with amplitude constraints. For a vector-valued parameter, the trace of the information matrix is maximized.

Aoki and Staley (Reference 5) in 1970 consider the optimal input sequences to minimize the inverse of the trace of the Fisher information matrix M. A sufficient condition is given for the input sequences to maximize the smallest eigenvalue λ_1 of the information matrix and let it approach infinity asymptotically. By definition the trace of the matrix M is equal to the sum of its eigenvalues

$$\operatorname{Tr} M = \lambda_1 + \lambda_2 + \cdots + \lambda_k.$$

The estimation error covariance matrix goes to zero asymptotically as λ_1 approaches infinity.

Nahi and Napjus (Reference 6) in 1971 consider the optimality criteria to be used for the design of inputs for vector parameter estimation of continuous-time systems with amplitude constraints. The intuitive sensitivity criterion for scalar parameter estimation

$$J = \int_0^T g_a^2(t, a)\, dt, \qquad g_a = \frac{\partial g}{\partial a},$$

where g_a is the sensitivity of the measurements to the parameter a, is justified by the Cramér–Rao lower bound M^{-1}. The estimation error covariance is

$$\operatorname{cov}(a) \geq M^{-1},$$

where M is the Fisher information matrix. For scalar observations, the Fisher information matrix agrees with J to within a constant. If $\mathbf{a}$ is a k-dimensional parameter vector, a suitable scalar function of the Fisher information matrix must be chosen. For Gaussian estimation error a k-dimensional ellipsoid surface of constant probability density can be defined. The volume of the ellipsoid is proportional to the determinant of M^{-1} for efficient estimation. Thus the maximization of the det M is a possible criterion for the design of optimal inputs since it is equivalent to the minimization of the volume of the ellipsoid. A criterion which is more convenient to implement, however, is the trace of the information matrix or the weighted trace.

Goodwin (Reference 7) in 1971 considers optimal inputs for nonlinear system initial state and parameter vector estimation of discrete-time systems with amplitude and state constraints. A scalar loss function plus the trace of the covariance matrix is minimized. The method requires an initial estimate of the parameters and initial states, which can be obtained from any nonoptimal input–output record of the system.

Perhaps one of the most prolific researchers in the design of optimal inputs for system identification is Mehra (References 1, 8–16). Mehra (References 8, 9), and Mehra and Stepner (Reference 10) show that the design of optimal inputs for linear systems involves the solution of two-point boundary-value problems with homogeneous boundary conditions. Nontrivial solutions exist at certain values of the Lagrange multiplier which correspond to the boundary-value problem eigenvalues. Mehra's method is described in this text in Section 8.2.4. Mehra (Reference 11) and Gupta, Mehra, and Hall (Reference 12) consider the design of optimal inputs in the frequency domain. Mehra (References 1, 13–15) and Mehra and Gupta (Reference 16) extend these results for time and frequency-domain syntheses of optimal inputs to multi-input–multi-output systems with process noise and D-optimality criterion (minimization of the determinant of the information matrix).

Kalaba and Spingarn (References 17–22) consider the design of optimal inputs for linear and nonlinear system parameter estimation with nonzero boundary conditions. In Reference 21 they consider an improved method for the numerical determination of optimal inputs. In Reference 22 they consider the sensitivity of the parameter estimates to observations. References 17–22 form the basis for Chapter 8 and part of Chapter 9 in this text.

Additional results on optimal inputs are given in References 23–25.

8.2. Linear Optimal Inputs

Optimal inputs for linear system identification are discussed in this section. The optimal input is determined such that the sensitivity of the system output to an unknown parameter is maximized. A quadratic performance criterion subject to an input energy constraint is used. In most cases the methods of solution are illustrated by means of simple examples.

The performance criterion is maximized by finding the solution of a two-point boundary-value problem. The two-point boundary-value equations are derived using either the Euler–Lagrange equations or Pontryagin's maximum principle. The numerical solution to these equations is obtained using either the method of complementary functions or the analytical solution.

In Section 8.2.1 the linear dynamic system example

$$\dot{x} = -ax + y, \qquad x(0) = c$$

is considered, where $x(t)$ is a scalar state variable, $y(t)$ is a scalar input, and

a is an unknown system parameter to be estimated. The performance criterion is

$$J = \max_{y} \int_0^T [x_a^2(t, a) - qy^2] \, dt,$$

where

$$x_a(t, a) = \frac{\partial x(t, a)}{\partial a}.$$

The method of quasilinearization is used for parameter estimation. The sensitivity of parameter estimates to observations is considered in Section 8.2.2 using the same linear example as above.

In Section 8.2.3 the second-order linear dynamic system example

$$\ddot{x} = -ax - b\dot{x} + y, \qquad x(0) = c_1, \qquad \dot{x}(0) = c_2$$

is considered, where a is the unknown system parameter to be estimated. The parameter b is assumed to be known.

Mehra's method for the determination of optimal inputs for linear system identification is discussed in Section 8.2.4. Equations are derived using Pontryagin's maximum principle. The critical time length corresponding to an eigenvalue of the two-point boundary-value equations with homogeneous boundary conditions, is obtained by integrating the matrix Riccati equation. The solution is then obtained using the transition matrix.

A comparison of optimal inputs for homogeneous and nonhomogeneous boundary conditions is discussed in Section 8.2.5. The linear dynamic system

$$\dot{x} = -ax + y, \qquad x(0) = c$$

is considered as an example, where a is the unknown parameter. Mehra's method, described in Section 8.2.4, is used to obtain the solution for homogeneous boundary conditions. For nonhomogeneous boundary conditions, the solution is obtained using the method of complementary functions.

8.2.1. Optimal Inputs and Sensitivities for Parameter Estimation

Observed data for parameter estimation are often both difficult and expensive to obtain. Thus when an experiment is conducted, the input to the system should be such that the sensitivity to the parameter being estimated is maximized. Evaluation of the optimal inputs and sensitivities (Reference 17) requires the solution to a two-point boundary-value problem. The Gauss–Newton method or the method of quasilinearization may then be used for parameter estimation. Numerical results are given for a simple example utilizing both optimal and nonoptimal inputs. The results clearly show the advantages of utilizing an optimal input.

Introduction

Consider the dynamical system given by the linear or nonlinear differential equation

$$\dot{x} = f(x, y, a, t) \tag{8.1}$$

and initial condition

$$x(0) = c, \tag{8.2}$$

where $x(t)$ is a scalar state variable, $y(t)$ is a scalar control function or input, a is an unknown system parameter, and t is the time. It is desired to estimate the unknown system parameter a. This can be accomplished by minimizing the sum S through the proper choice of a:

$$S = \sum_{i=1}^{N} [x(t_i, a) - b_i]^2, \tag{8.3}$$

where b_i is the observed value of the dependent variable at time t_i, $i = 1, 2, \ldots, N$, and

$$x(t_i, a) = x(t_i).$$

In performing an experiment, it is desired to obtain the maximum benefit from the observations. In particular, the input to the system should be such that it maximizes the sensitivity of the state variable $x(t, a)$ to the parameter a. Furthermore, the magnitude of the input should be constrained so that it does not become excessively large. The performance index is

$$M = \max_{y} \int_0^{T_f} x_a^2 \, dt \tag{8.4}$$

subject to the input energy constraint

$$E = \int_0^{T_f} y^2 \, dt. \tag{8.5}$$

The above is equivalent to maximizing the performance index

$$J = \int_0^{T_f} [x_a^2(t, a) - qy^2(t)] \, dt, \tag{8.6}$$

where

$$x_a(t, a) = \frac{\partial x(t, a)}{\partial a} \tag{8.7}$$

is the sensitivity and q is a constant Lagrange multiplier. The control function $y(t)$ is determined such that J is maximized. It can easily be shown that the $y(t)$ obtained via the above procedure for any given value of q is optimal for all inputs with energy less than or equal to E. The problem is to maximize the sensitivity

$$\max_{y(t)} \int_0^{T_f} x_a^2(t, a)\, dt \tag{8.8}$$

subject to the constraint

$$\int_0^{T_f} y^2(t)\, dt \leq E, \tag{8.9}$$

where E is a constant. For a linear system, the inequality constraint should be satisfied as an equality. Substitute $y(t)$ by $\alpha y(t)$; then x_a will be multiplied by α. Thus

$$\alpha^2 \int_0^{T_f} x_a^2(t, a)\, dt \tag{8.10}$$

is increased for $\alpha > 1$. Then the sensitivity is maximized for all

$$\int_0^{T_f} y^2(t)\, dt \leq E \tag{8.11}$$

when

$$\int_0^{T_f} y^2(t)\, dt = E. \tag{8.12}$$

The above concepts may be applied to any type of system, but are of particular interest for biological systems where the number of experiments that can easily be performed is limited. The maximum amount of information is desired from each set of observations. However, before these concepts are applied to a biological system some indication of the magnitudes of the input and the sensitivity should be obtained. This is the subject of this section utilizing a simple example. The example is given for illustrative purposes in order to demonstrate numerically the advantages of utilizing the preferred input. Later sections will be concerned with the generalization to the vector case and to nonlinear problems.

Example. The example to be considered is given by the differential equation

$$\dot{x} = -ax + y, \qquad 0 \leq t \leq T_f \tag{8.13}$$

with initial condition

$$x(0) = c. \tag{8.14}$$

Differentiating equation (8.13) with respect to a, the sensitivity is

$$\dot{x}_a = -x - ax_a. \tag{8.15}$$

Introduce the notation

$$x_1 = x, \qquad x_2 = x_a. \tag{8.16}$$

Then we have

$$\dot{x}_1 = -ax_1 + y, \tag{8.17}$$

$$\dot{x}_2 = -x_1 - ax_2. \tag{8.18}$$

If we utilize the Lagrange multipliers, the performance index becomes

$$J' = \int_0^{T_f} F\, dt, \tag{8.19}$$

where

$$F = x_2^2 - qy^2 + \lambda_1(t)(\dot{x}_1 + ax_1 - y) + \lambda_2(t)(\dot{x}_2 + x_1 + ax_2) \tag{8.20}$$

and

$$x_1(0) = c, \qquad x_2(0) = 0. \tag{8.21}$$

The optimal input is the control function that maximizes integral equation (8.19) and is obtained via the Euler–Lagrange equations.

Euler–Lagrange Equations for the Optimal Input

The Euler–Lagrange equations and associated transversality conditions are

$$\frac{\partial F}{\partial x_1} - \frac{d}{dt}\frac{\partial F}{\partial \dot{x}_1} = 0, \qquad \left.\frac{\partial F}{\partial \dot{x}_1}\right|_{t=T_f} = 0, \tag{8.22}$$

$$\frac{\partial F}{\partial x_2} - \frac{d}{dt}\frac{\partial F}{\partial \dot{x}_2} = 0, \qquad \left.\frac{\partial F}{\partial \dot{x}_2}\right|_{t=T_f} = 0, \tag{8.23}$$

$$\frac{\partial F}{\partial y} - \frac{d}{dt}\frac{\partial F}{\partial \dot{y}} = 0, \qquad \left.\frac{\partial F}{\partial \dot{y}}\right|_{t=T_f} = 0, \tag{8.24}$$

augmented by the constraint equations (8.17) and (8.18). Thus, for the given example the Euler–Lagrange equations along with equality constraints (8.17) and (8.18) lead to the two-point boundary-value equations and associated boundary conditiosn

$$\dot{x}_1 = -ax_1 - (1/2q)\lambda_1, \qquad x_1(0) = c, \tag{8.25}$$

$$\dot{x}_2 = -x_1 - ax_2, \qquad x_2(0) = 0, \tag{8.26}$$

$$\dot{\lambda}_1 = a\lambda_1 + \lambda_2, \qquad \lambda_1(T_f) = 0, \tag{8.27}$$

$$\dot{\lambda}_2 = 2x_2 + a\lambda_2, \qquad \lambda_2(T_f) = 0, \tag{8.28}$$

where

$$y = -(1/2q)\lambda_1. \tag{8.29}$$

These equations can be expressed in the matrix form

$$\begin{bmatrix} \dot{x}_1 \\ \dot{x}_2 \\ \dot{\lambda}_1 \\ \dot{\lambda}_2 \end{bmatrix} = \begin{bmatrix} -a_1 & 0 & -1/2q & 0 \\ -1 & -a & 0 & 0 \\ 0 & 0 & a & 1 \\ 0 & 2 & 0 & a \end{bmatrix} \begin{bmatrix} x_1 \\ x_2 \\ \lambda_1 \\ \lambda_2 \end{bmatrix}. \tag{8.30}$$

The solution can be obtained numerically utilizing the method of complementary functions.

Complementary Functions

Using the method of complementary functions the homogeneous differential equations are

$$\dot{H} = FH = \begin{bmatrix} a_1 & b_1 & c_1 & d_1 \\ a_2 & b_2 & c_2 & d_2 \\ a_3 & b_3 & c_3 & d_3 \\ a_4 & b_4 & c_4 & d_4 \end{bmatrix} \begin{bmatrix} h_{11} & h_{21} & h_{31} & h_{41} \\ h_{12} & h_{22} & h_{32} & h_{42} \\ h_{13} & h_{23} & h_{33} & h_{43} \\ h_{14} & h_{24} & h_{34} & h_{44} \end{bmatrix} \tag{8.31}$$

with the initial condition

$$H(0) = I. \tag{8.32}$$

All the h's are functions of time. Comparing the coefficients in (8.30) with the coefficients in (8.31) it is obvious that

$$\begin{aligned} &a_1 = b_2 = -a, \qquad && c_1 = -1/2q, \\ &a_2 = -1, && c_3 = d_4 = a, \\ &b_4 = 2, && d_3 = 1, \end{aligned} \tag{8.33}$$

and all the rest of the coefficients are zero. The solution is given by the linear combination

$$\begin{bmatrix} x_1 \\ x_2 \\ \lambda_1 \\ \lambda_2 \end{bmatrix} = A_1 \begin{bmatrix} h_{11} \\ h_{12} \\ h_{13} \\ h_{14} \end{bmatrix} + A_2 \begin{bmatrix} h_{21} \\ h_{22} \\ h_{23} \\ h_{24} \end{bmatrix} + A_3 \begin{bmatrix} h_{31} \\ h_{32} \\ h_{33} \\ h_{34} \end{bmatrix} + A_4 \begin{bmatrix} h_{41} \\ h_{42} \\ h_{43} \\ h_{44} \end{bmatrix}, \tag{8.34}$$

where A_1, A_2, A_3, and A_4 are constants to be determined by the boundary conditions. Utilizing the boundary conditions at $t = 0$ and $t = T_f$, four linear algebraic equations in terms of A_1, A_2, A_3, and A_4 are readily obtained:

$$x_1(0) = A_1 h_{11}(0) + A_2 h_{21}(0) + A_3 h_{31}(0) + A_4 h_{41}(0), \tag{8.35}$$

$$x_2(0) = A_1 h_{12}(0) + A_2 h_{22}(0) + A_3 h_{32}(0) + A_4 h_{42}(0), \tag{8.36}$$

$$\lambda_1(T_f) = A_1 h_{13}(T_f) + A_2 h_{23}(T_f) + A_3 h_{33}(T_f) + A_4 h_{43}(T_f), \tag{8.37}$$

$$\lambda_2(T_f) = A_1 h_{14}(T_f) + A_2 h_{24}(T_f) + A_3 h_{34}(T_f) + A_4 h_{44}(T_f), \tag{8.38}$$

where

$$x_1(0) = c, \qquad x_2(0) = \lambda_1(T_f) = \lambda_2(T_f) = 0. \tag{8.39}$$

The optimal trajectory is obtained by integrating (8.31) with initial condition (8.32) from $t = 0$ to $t = T_f$, storing the values of $H(t)$ at the end points and at the instants where a solution printout is desired. The linear algebraic equations (8.35)–(8.38) are then solved for A_1, A_2, A_3, and A_4 and the solution is obtained from equations (8.34).

Analytical Solution

In order to utilize the optimal input for parameter estimation, it would be desirable to express $y(t)$ analytically. Expressing (8.30) in the form

$$\dot{\mathbf{x}} = F\mathbf{x} \tag{8.40}$$

the eigenvalues can be calculated from the determinantal equation

$$| mI - F | = \begin{vmatrix} m + a & 0 & 1/2q & 0 \\ 1 & m + a & 0 & 0 \\ 0 & 0 & m - a & -1 \\ 0 & -2 & 0 & m - a \end{vmatrix} = 0. \tag{8.41}$$

Then the eigenvalues are

$$m_1, m_2 = \pm(a^2 + 1/q^{1/2})^{1/2}, \tag{8.42}$$

$$m_3, m_4 = \pm(a^2 - 1/q^{1/2})^{1/2}. \tag{8.43}$$

For $a^2 < 1/q^{1/2}$, two of the eigenvalues are real and two are imaginary. Assuming that the solution is of the form

$$x_1(t) = c_1 e^{m_1 t} + c_2 e^{-m_1 t} + c_3 \cos m_3 t + c_4 \sin m_3 t, \tag{8.44}$$

then by differentiation and algebraic manipulation, equations (8.25)–(8.28) can be solved for x_2, λ_1, and λ_2 in terms of c_1, c_2, c_3, and c_4. Utilizing the boundary conditions at $t = 0$ and $t = T_f$, the four linear algebraic equations can be solved for c_1, c_2, c_3, and c_4. From (8.29) the optimal input is

$$y(t) = c_1(a + m_1)e^{m_1 t} + c_2(a - m_1)e^{-m_1 t} + c_3(a \cos m_3 t - m_3 \sin m_3 t) + c_4(a \sin m_3 t + m_3 \cos m_3 t). \tag{8.45}$$

System Identification via Quasilinearization

The method of quasilinearization is utilized for minimizing the sum S given by equation (8.3). Differential equation (8.1) is first linearized around the kth approximation by expanding the function f in a power series and retaining only the linear terms. The $k + 1$ approximation is

$$\dot{x}_{k+1} = f_k + (a_{k+1} - a_k)\left.\frac{\partial f}{\partial a}\right|_k + (x_{k+1} - x_k)\left.\frac{\partial f}{\partial x}\right|_k, \qquad \dot{x}_{k+1}(0) = c. \tag{8.46}$$

The general solution of equation (8.46) is

$$x_{k+1}(t) = p(t) + a_{k+1}h(t). \tag{8.47}$$

The function $p(t)$ is the solution of the initial-value equation

$$\dot{p}(t) = f_k - a_k\left.\frac{\partial f}{\partial a}\right|_k + (p - x_k)\left.\frac{\partial f}{\partial x}\right|_k, \qquad p(0) = c \tag{8.48}$$

and the function $h(t)$ is the solution of the initial-value equation

$$\dot{h}(t) = \left.\frac{\partial f}{\partial a}\right|_k + \left.\frac{\partial f}{\partial x}\right|_k h(t), \qquad h(0) = 0. \tag{8.49}$$

For the given example we have

$$\left.\frac{\partial f}{\partial a}\right|_k = -x_k, \qquad \left.\frac{\partial f}{\partial x}\right|_k = -a_k. \tag{8.50}$$

Substituting equation (8.47) into (8.3) yields

$$S = \sum_{i=1}^{N} [p(t_i) + a_{k+1}h(t_i) - b_i]^2. \tag{8.51}$$

The condition for minimum S is given by

$$\frac{\partial S}{\partial a_k} = 2 \sum_{i=1}^{N} h(t_i)[p(t_i) + a_{k+1}h(t_i) - b_i] = 0. \tag{8.52}$$

Solving for a_{k+1} gives

$$a_{k+1} = \frac{\sum_{i=1}^{N} [b_i - p(t_i)]h(t_i)}{\sum_{i=1}^{N} h^2(t_i)}. \tag{8.53}$$

In a typical cycle, the two initial-value equations for p and h are integrated and stored as a function of time from $t = 0$ to $t = t_N$. As each point t_i, $i = 1, 2, \ldots, N$ is passed, the sums in (8.53) are updated. Equation (8.53) is solved for the constant a_{k+1} when $t = t_N$ is reached. Equation (8.47) is then evaluated for $x_{k+1}(t)$ and the latter is stored as a function of time. Utilizing the new value of a and the stored values of $x(t)$, the process is then repeated to obtain the next approximation, etc.

Numerical Results and Discussion

The optimal input is evaluated via the method of complementary functions. To obtain the numerical results, a fourth-order Runge–Kutta method with grid intervals of 1/100 sec was utilized for numerical integration.

Numerical results are shown in Figure 8.1 for a as a parameter with values from 0.1 to 2 and

$$q = 1, \qquad x(0) = c = 1, \qquad T_f = 1 \text{ sec}.$$

Figures 8.1a and 8.1b show that the optimal input magnitudes and sensitivities increase as a decreases. Furthermore, the sensitivity curves show that x_a has a maximum for time $0 \leq t \leq T_f$ although this maximum may be increased in some cases if T_f is extended. The time at which the maximum

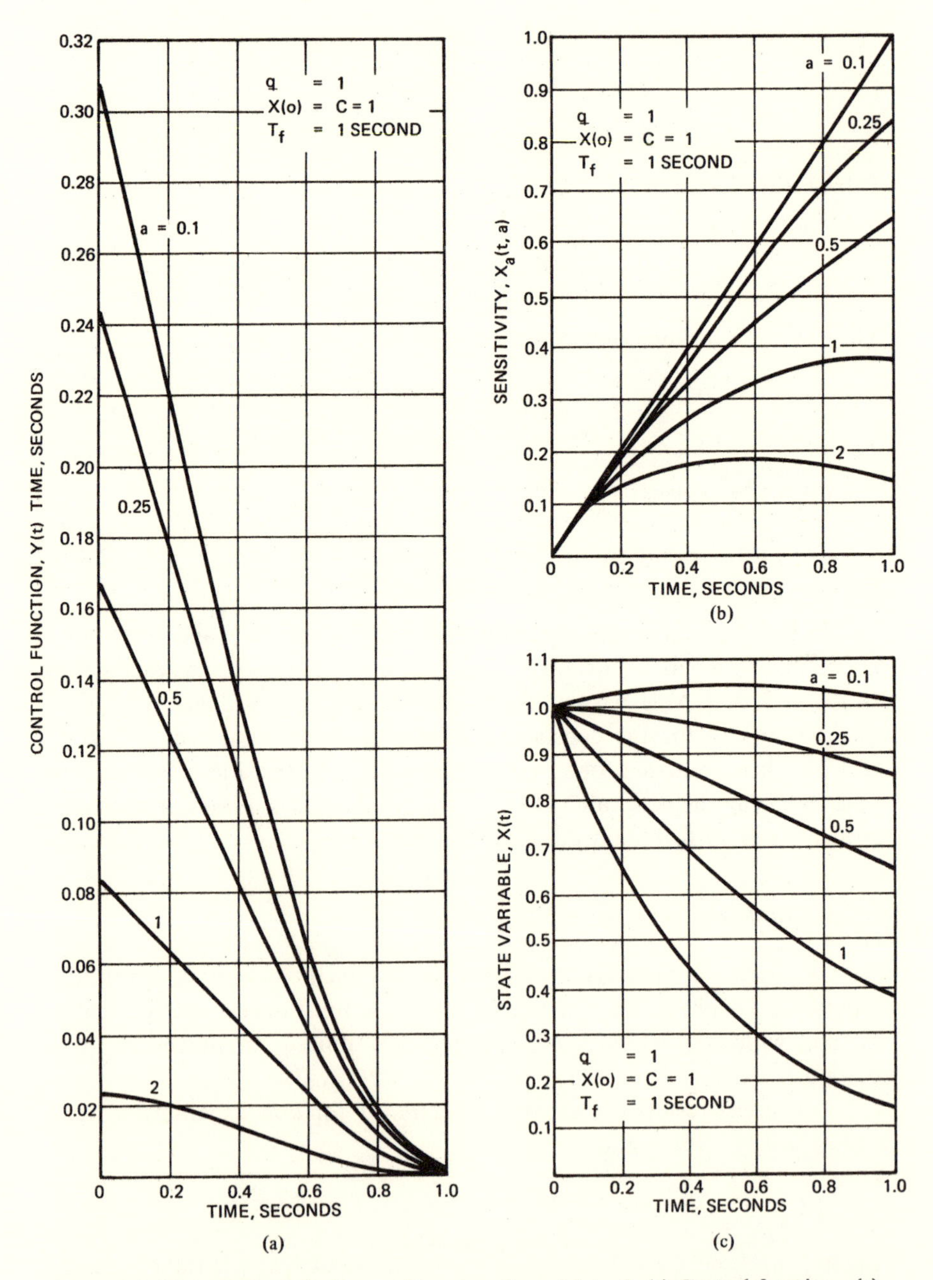

Figure 8.1. Numerical results for a with values from 0.1 to 2. (a) Control function $y(t)$ versus time. (b) Sensitivity $x_a(t, a)$ versus time. (c) State variable $x(t)$ versus time. (Reference 17.)

occurs is a function of a. The state variable x as a function of time is shown in Figure 8.1c.

In order to demonstrate the advantages of utilizing the optimal input, the parameter estimation accuracy is compared with and without the optimal input in the presence of observation noise. The analytical solution for the optimal input is first evaluated for

$$q = 0.08, \qquad x(0) = 1, \qquad T_f = 1 \text{ sec}, \qquad a = 0.1.$$

The coefficients of the optimal input (8.45) are

$$\begin{aligned} c_1 &= 1.17187, & m_1 &= 1.88296, \\ c_2 &= -7.27024, & m_3 &= 1.87764. \\ c_3 &= 7.09837, \\ c_4 &= 9.16885, \end{aligned}$$

For the nonoptimal input, a sine wave with angular frequency $\omega = 0.1$ rad/sec is utilized. Then for the same average power input as for the optimal input we have

$$\begin{aligned} y(t) &= K \sin \omega t, \\ K &= 291.322. \end{aligned} \tag{8.54}$$

Figure 8.2a shows both the optimal and nonoptimal inputs as a function of time. The nonoptimal input is approximately a ramp rather than a sine wave because of the small terminal time, T_f. Figure 8.2b shows the optimal and nonoptimal state variables and sensitivities as a function of time, corresponding to the inputs given in Figure 8.2a. It is seen that the sensitivity for the optimal input is almost double that for the nonoptimal input at the terminal time, $t = T_f$.

The parameter estimation problem is solved via the method of quasilinearization. First it is assumed that the true value of the parameter a is equal to 0.1. Then for the conditions given, the optimal input is that shown in Figure 8.2a. The observations are formed by adding noise to the optimal input response, $x(t)$, shown in Figure 8.2b:

$$b_i = x(t_i) + v(t_i), \tag{8.55}$$

where b_i is the observation at time t_i, $x(t_i)$ is the state variable, and $v(t_i)$ is the white Gaussian noise with zero mean and standard deviation equal to 0.3. Starting with the initial condition, which is assumed to be known

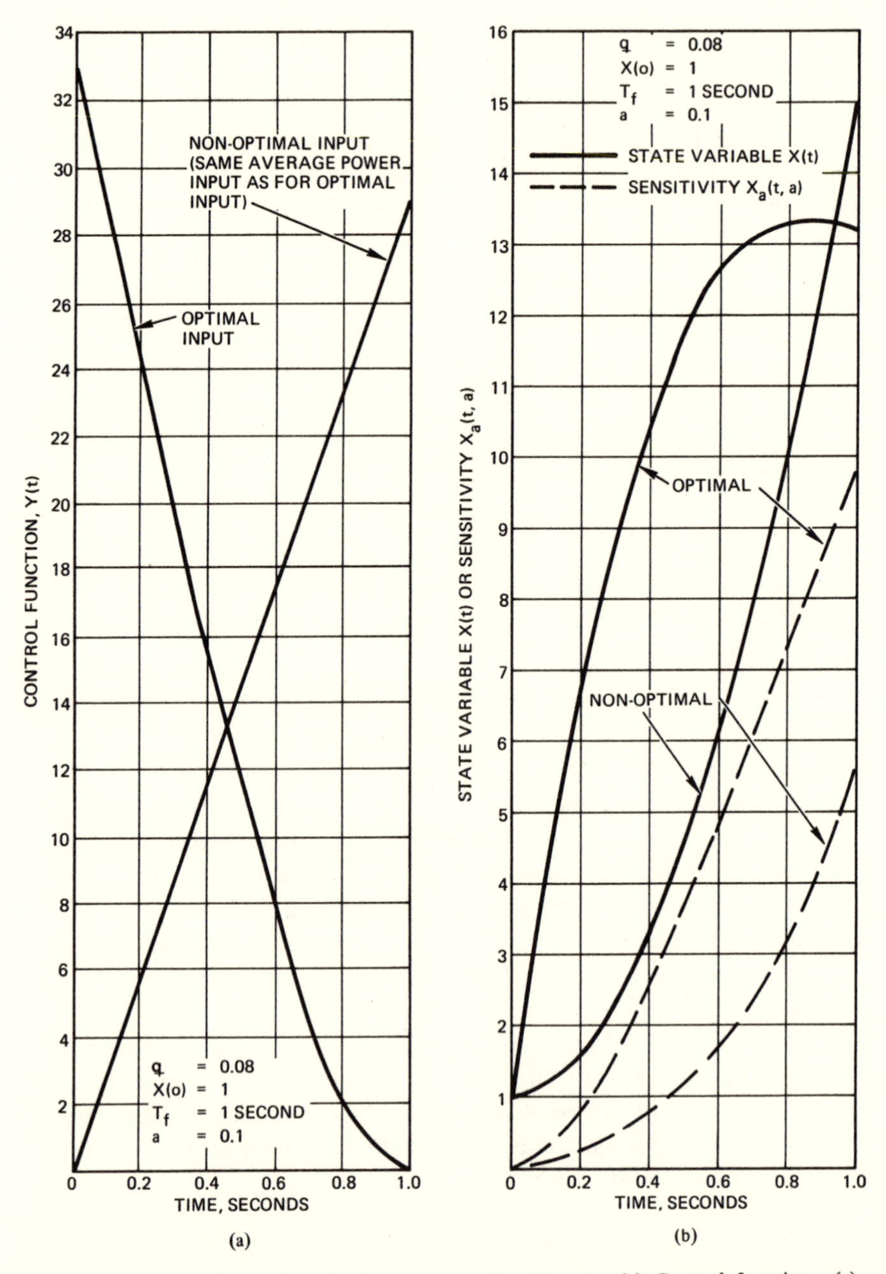

Figure 8.2. Comparison of optimal and nonoptimal inputs. (a) Control function $y(t)$ versus time. (b) State variable $x(t)$ and sensitivity $x_a(t, a)$ versus time. (Reference 17.)

exactly at time $t = 0$, ten additional observations, equation (8.55), are made at 0.1-sec intervals. The first approximation of the parameter a is assumed to be

$$a = 0.15.$$

Usually about four cycles are sufficient for the parameter estimates to converge. A fourth-order Runge–Kutta integration scheme with grid intervals of 1/100 sec was utilized for the quasilinearization program.

The procedure was repeated for the nonoptimal input with the same average power input. The nonoptimal input is shown in Figure 8.2a. The observations are formed according to equation (8.55) by adding noise to the nonoptimal input response, $x(t)$, shown in Figure 8.2b. As previously, ten observations are made, and the first approximation of the parameter a is assumed to be 0.15.

Table 8.1 shows the parameter estimates for a typical run for both the optimal and nonoptimal inputs. Note that the additive noise was identical for both the optimal and nonoptimal inputs. The percent error in the final estimate of a was 0.78% for the optimal input and 5.5% for the nonoptimal input.

Table 8.2 shows the average estimates and standard deviations of ten runs for both the optimal and nonoptimal inputs. While the average estimates are nearly the same, the standard deviation of the nonoptimal input is 72% greater than that of the optimal input. The results clearly show the advantages of utilizing the optimal input.

In an actual experiment where the true value of the parameter to be estimated is unknown, the optimal input would be found for an initial

Table 8.1. Estimates of the Parameter a for a Typical Run (True Value of $a = 0.1$)

Cycle number	Estimate of a	
	Optimal input	Nonoptimal input
0	0.15	0.15
1	0.099717	0.093076
2	0.10079	0.094526
3	0.10078	0.094508
4	0.10078	0.094508
Percent error (cycle 4)	0.78%	5.5%

Table 8.2. Estimate of the Parameter a. Average of 10 Runs (True Value of $a = 0.1$)

	Optimal input	Nonoptimal input
Average estimate	0.10168	0.10182
Standard deviation	0.01492	0.02571

approximation of a. Then, using quasilinearization, a new estimate of a would be found. Using this estimate, the optimal input would be obtained and the process repeated until a satisfactory estimate is obtained.

8.2.2. Sensitivity of Parameter Estimates to Observations

The sensitivity of parameter estimates to observation noise or disturbances affects the accuracy of the parameter estimates in system identification. The parameter estimate/observation sensitivity is derived for a scalar nonlinear differential system. In the design of optimal inputs, as in the previous section, the sensitivity of the state variable to the unknown parameter is maximized. It is shown that the parameter estimate/observation sensitivity tends to be lowered for optimal inputs (Reference 22).

Introduction

System identification is concerned with the estimation of the unknown parameters from the information obtained from the observations. For accurate results the sensitivity of the parameter estimates to observation noise or disturbances should be small (Reference 26). In this section the parameter estimate/observation sensitivity for a scalar nonlinear differential system is derived. It is shown that the parameter estimate/observation sensitivity can be lowered by increasing the sensitivity of the state variable to the unknown parameter.

In the design of optimal inputs for system identification the sensitivity of the state variable to the unknown parameter or the sensitivity of the observation to the unknown parameter is maximized. The justification for this approach is the Cramér–Rao lower bound, which provides a lower bound for the estimation error covariance. The parameter estimate/observation sensitivity tends to be lowered for optimal inputs. A numerical example is given.

Parameter Estimate/Observation Sensitivity

Given the nonlinear dynamical system

$$\dot{x} = f(x, y, a, t), \qquad x(0) = c, \tag{8.56}$$

the unknown system parameter a is estimated by minimizing the sum S:

$$S = \sum_{i=1}^{N} [x(t_i, a) - b_i]^2, \tag{8.57}$$

where $x(t)$ is a scalar state variable, $y(t)$ is a control function or input, a is an unknown system parameter, c is the initial condition, and b_i is the observed value of $x(t_i, a)$ at time t_i, $i = 1, 2, \ldots, N$. It is desired to compute the sensitivity of the parameter estimate, $\hat{a}$, to observation, b_j. For system identification the sum in equation (8.57) is minimized by differentiating equation (8.57) and setting the result equal to zero:

$$S_a = \sum_{i=1}^{N} 2[x(t_i, a) - b_i]x_a(t_i, a) = 0, \tag{8.58}$$

where x_a is obtained by differentiating equation (8.56) with respect to a:

$$\dot{x}_a = \frac{\partial \dot{x}}{\partial a} = f_x x_a + f_a, \qquad x_a(0) = 0. \tag{8.59}$$

Now S_a is a function of the parameter estimate, $\hat{a}$, and the observations, b_i:

$$S_a = S_a(\hat{a}, b_1, b_2, \ldots, b_N), \tag{8.60}$$

$$\hat{a} = \hat{a}(b_1, b_2, \ldots, b_N). \tag{8.61}$$

Differentiating equation (8.58) with respect to b_j yields

$$S_{aa} \frac{\partial \hat{a}}{\partial b_j} + S_{ab_j} = 0, \tag{8.62}$$

where

$$S_{aa} = \sum_{i=1}^{N} 2x_a^2(t_i, a) + \sum_{i=1}^{N} 2[x(t_i, a) - b_i]x_{aa}(t_i, a), \tag{8.63}$$

$$S_{ab_j} = -2x_a(t_j, a). \tag{8.64}$$

Solving equation (8.62) for the parameter estimate/observation sensitivity yields

$$\frac{\partial \hat{a}}{\partial b_j} = -\frac{S_{ab_j}}{S_{aa}} = \frac{x_a(t_j, a)}{\sum_{i=1}^{N} x_a^2(t_i, a) + \sum_{i=1}^{N} [x(t_i, a) - b_i]x_{aa}(t_i, a)}, \tag{8.65}$$

where x, x_a, and x_{aa} are the solutions of the differential equations

$$\dot{x} = f(x, y, a, t), \qquad x(0) = c, \tag{8.66}$$

$$\dot{x}_a = f_x x_a + f_a, \qquad x_a(0) = 0, \tag{8.67}$$

$$\dot{x}_{aa} = f_{xa} x_a + f_x x_{aa} + f_{aa} + (f_{xx} x_a + f_{ax}) x_a, \qquad x_{aa}(0) = 0. \tag{8.68}$$

Assume that $x(t_i, a)$ is close to its observed value, b_j, so that the second term in the denominator of equation (8.65) is small. Then we have

$$\frac{\partial \hat{a}}{\partial b_j} \doteq \frac{x_a(t_j, a)}{\sum_{i=1}^{N} x_a^2(t_i, a)}. \tag{8.69}$$

Equation (8.69) shows that increasing the state variable/unknown parameter sensitivity, $x_a(t)$, decreases the parameter estimate/observation sensitivity, $\partial \hat{a}/\partial b_j$. The parameter estimate/observation sensitivity should be small in order to minimize the effects of observation noise or disturbances on the parameter estimate. (See Exercise 1.)

In the design of optimal inputs for system identification, the state variable/unknown parameter sensitivity is usually maximized. Thus the parameter estimate/observation sensitivity tends to be lowered for optimal inputs.

Example. Consider as an example the linear differential equation

$$\dot{x} = -ax + y, \qquad 0 \leq t \leq T, \qquad x(0) = c, \tag{8.70}$$

with nonoptimal sine wave input

$$y(t) = K \sin \omega t. \tag{8.71}$$

Given the observations

$$b_i = x(t_i) + v(t_i), \tag{8.72}$$

where $v(t)$ is a zero mean Gaussian white noise process, the unknown system parameter, a, can be estimated using the method of quasilinearization for system identification. The parameter estimate/observation sensitivity can be computed using equation (8.65).

For comparison, the parameter estimate/observation sensitivity can be computed using the optimal input derived in Section 8.2.1.

Numerical Results

The parameter estimate/observation sensitivity for the sine wave input was evaluated for the following parameter values in equations (8.70) and (8.71):

$$a = 0.1, \qquad x(0) = 1, \qquad T = 1 \text{ sec}, \tag{8.73}$$

$$\omega = 0.1 \text{ rad/sec}, \qquad K = 291.322. \tag{8.74}$$

Table 8.3 shows the parameter estimate/observation sensitivity, equation (8.65), as a function of time for the sine wave input. Also shown are $x_a(t)$ and $x_{aa}(t)$. The observation noise was assumed to be zero so that the second term in the denominator of equation (8.65) is zero.

Table 8.4 shows the parameter estimate/observation sensitivity for the optimal input with the parameter values given in equation (8.73) and $q = 0.08$, which as derived in Section 8.2.1 corresponds to the same input energy as for the sine wave input. Also given is the ratio of the $\partial\hat{a}/\partial b_j$ for the sine wave input to the $\partial\hat{a}/\partial b_j$ for the optimal input. The ratio varies from 1.2 to 2.5.

Using the parameter estimate/observation sensitivity, it should be possible to predict the change in the parameter estimate to a change in the observation, b_j. Starting with the initial condition, eleven observations were made at 0.1-sec intervals. The first approximation of the parameter a was assumed to be $\hat{a} = 0.15$. Using a fourth-order Runge–Kutta integration method with grid intervals of 1/100 sec for the quasilinearization system

Table 8.3. Sine Wave Input Sensitivities

t_j	$x_a(t_j)$	$x_{aa}(t_j)$	$\partial\hat{a}/\partial b_j$
0	0	0	0
0.1	−0.103836	1.01418 E−2	−1.53992 E−3
0.2	−0.234496	4.30459 E−2	−3.47764 E−3
0.3	−0.420274	0.106653	−6.23278 E−3
0.4	−0.688894	0.2144	−1.02165 E−2
0.5	−1.06751	0.385048	−1.58315 E−2
0.6	−1.58273	0.642523	−2.34723 E−2
0.7	−2.26058	1.01575	−3.35251 E−2
0.8	−3.12657	1.53849	−4.63680 E−2
0.9	−4.20565	2.24919	−6.23709 E−2
1	−5.52221	3.19083	−8.18960 E−2

Table 8.4. Optimal Input Sensitivities

t_j	$x_a(t_j)$	$x_{aa}(t_j)$	$\partial\hat{a}/\partial b_j$	$\left(\frac{\partial\hat{a}}{\partial b_j}\right)_{\text{sine wave}} \Big/ \left(\frac{\partial\hat{a}}{\partial b_j}\right)_{\text{optimal}}$
0	0	0	0	—
0.1	−0.256739	2.05280 E−2	−8.50101 E−4	1.81
0.2	−0.793751	0.120702	−2.62822 E−3	1.32
0.3	−1.56162	0.350569	−5.17073 E−3	1.21
0.4	−2.512	0.749935	−8.31760 E−3	1.23
0.5	−3.5986	1.34878	−1.19155 E−2	1.33
0.6	−4.77831	2.16786	−1.58217 E−2	1.48
0.7	−6.01255	3.21959	−1.99084 E−2	1.68
0.8	−7.26874	4.50914	−2.40678 E−2	1.93
0.9	−8.52183	6.03588	−2.82170 E−2	2.21
1	−9.75585	7.79509	−3.23030 E−2	2.54

identification method, $\hat{a}$ converges to 0.1 when the observation noise is zero. Using the optimal input with an observation disturbance, Δb_5, equal to 0.5 at $t = 0.5$ sec, the successive approximations for $\hat{a}$ converge to 0.0941016. The percent change in $\hat{a}$ with the optimal input, with disturbance $\Delta b_5 = 0.5$, is

$$\frac{0.0941016 - 0.1}{0.1} = -5.90\%.$$

The predicted parameter estimate is

$$0.1 + \frac{\partial \hat{a}}{\partial b_5} \Delta b_5 = 0.1 + (-0.0119155)(0.5) = 0.0940,$$

which is correct to almost four decimal places.

Using the sine wave or nonoptimal input with an observation disturbance, Δb_5, equal to 0.5, the successive approximations for $\hat{a}$ converge to 0.0921356. The percent change in $\hat{a}$ with the sine wave input, with disturbance $\Delta b_5 = 0.5$, is

$$\frac{0.0921356 - 0.1}{0.1} = -7.86\%.$$

The predicted parameter estimate is

$$0.1 + \frac{\partial \hat{a}}{\partial b_5} \Delta b_5 = 0.1 + (-0.0158315)(0.5) = 0.0921,$$

which is correct to four decimal places.

The results show that for this example the optimal input lowers the parameter estimate/observation sensitivity, decreasing the sensitivity of the parameter estimate to observation noise or disturbances. The $\partial \hat{a}/\partial b_j$ can be used to estimate the change in the parameter estimate due to a change in the noise disturbance. Similar results were obtained with the noise disturbance Δb_5 added to a zero mean white Gaussian observation noise, equation (8.72), with $\sigma = 0.1$.

8.2.3. Optimal Inputs for a Second-Order Linear System

The optimal inputs for a second-order linear system are derived in this section. To obtain the optimal inputs, the second-order system requires the solution of eight two-point boundary-value equations (Reference 18). The existence of a critical time length for the terminal time is demonstrated.

Example. Consider the second-order linear differential equation

$$\ddot{x} + b\dot{x} + ax = y(t) \tag{8.75}$$

with initial conditions

$$x(0) = c_1, \qquad \dot{x}(0) = c_2. \tag{8.76}$$

The optimal input is to be determined such that the sensitivity to the parameter a is maximized. Equations (8.75) and (8.76) can be expressed in the form

$$\dot{x}_1 = x_2, \qquad x_1(0) = c_1, \tag{8.77}$$

$$\dot{x}_2 = -bx_2 - ax_1 + y, \qquad x_2(0) = c_2. \tag{8.78}$$

The optimal return is defined by the equation

$$J = \max_{y(t)} \int_0^{T_f} [x_{1a}^2(t, a) - qy^2(t)]\, dt, \tag{8.79}$$

where

$$x_{1a} = \frac{\partial x_1}{\partial a}, \qquad q > 0. \tag{8.80}$$

Differentiating equations (8.77) and (8.78) with respect to a and defining

$$x_3 = x_{1a}, \tag{8.81}$$

$$x_4 = x_{2a} \tag{8.82}$$

four differential equations are obtained when combined with equations (8.77) and (8.78):

$$\dot{x}_1 = x_2, \qquad x_1(0) = c_1, \tag{8.83}$$

$$\dot{x}_2 = -bx_2 - ax_1 + y, \qquad x_2(0) = c_2, \tag{8.84}$$

$$\dot{x}_3 = x_4, \qquad x_3(0) = 0, \tag{8.85}$$

$$\dot{x}_4 = -bx_4 - ax_3 - x_1, \qquad x_4(0) = 0. \tag{8.86}$$

Utilizing the Lagrange multipliers, equation (8.79) subject to equality constraints (8.83)–(8.86) becomes

$$J' = \max_{y(t)} \int_0^{T_f} \phi\, dt, \tag{8.87}$$

where

$$\phi = x_3^2 - qy^2 + \lambda_1(\dot{x}_1 - x_2) + \lambda_2(\dot{x}_2 + bx_2 + ax_1 - y) + \lambda_3(\dot{x}_3 - x_4) + \lambda_4(\dot{x}_4 + bx_4 + ax_3 + x_1). \quad (8.88)$$

The Euler–Lagrange equations and transversality conditions yield the additional equations

$$\dot{\lambda}_1 = a\lambda_2 + \lambda_4, \qquad \lambda_1(T_f) = 0, \quad (8.89)$$

$$\dot{\lambda}_2 = -\lambda_1 + b\lambda_2, \qquad \lambda_2(T_f) = 0, \quad (8.90)$$

$$\dot{\lambda}_3 = 2x_3 + a\lambda_4, \qquad \lambda_3(T_f) = 0, \quad (8.91)$$

$$\dot{\lambda}_4 = -\lambda_3 + b\lambda_4, \qquad \lambda_4(T_f) = 0, \quad (8.92)$$

and

$$y(t) = -(1/2q)\lambda_2. \quad (8.93)$$

Equations (8.83)–(8.86) and (8.89)–(8.93) can be expressed in the vector form

$$\dot{\mathbf{x}} = F\mathbf{x} \quad (8.94)$$

with boundary conditions

$$x_i(0) = c_i, \qquad i = 1, 2, \quad (8.95)$$

$$x_j(0) = 0, \qquad j = 3, 4, \quad (8.96)$$

$$\lambda_k(T_f) = 0, \qquad k = 1, 2, 3, 4, \quad (8.97)$$

where

$$F = \begin{bmatrix} 0 & 1 & 0 & 0 & 0 & 0 & 0 & 0 \\ -a & -b & 0 & 0 & 0 & -1/2q & 0 & 0 \\ 0 & 0 & 0 & 1 & 0 & 0 & 0 & 0 \\ -1 & 0 & -a & -b & 0 & 0 & 0 & 0 \\ 0 & 0 & 0 & 0 & 0 & a & 0 & 1 \\ 0 & 0 & 0 & 0 & -1 & b & 0 & 0 \\ 0 & 0 & 2 & 0 & 0 & 0 & 0 & a \\ 0 & 0 & 0 & 0 & 0 & 0 & -1 & b \end{bmatrix}, \quad (8.98)$$

$$\mathbf{x} = (x_1, x_2, x_3, x_4, \lambda_1, \lambda_2, \lambda_3, \lambda_4)^T. \quad (8.99)$$

The numerical solution of equation (8.94) can be obtained via the method of complementary functions or from the explicit analytical solution.

Complementary Functions

Using the method of complementary functions, the homogeneous differential equation is

$$\dot{H} = FH, \qquad H(0) = I. \tag{8.100}$$

For the given example both F and H are 8×8 matrices. The solution to equation (8.94) is given by

$$\mathbf{x} = H\mathbf{c}_A, \tag{8.101}$$

where the vector of constants

$$\mathbf{c}_A = (A_1, A_2, \ldots, A_8)^T \tag{8.102}$$

is determined from the equation

$$\begin{bmatrix} x_1(0) \\ \vdots \\ x_4(0) \\ \lambda_1(T_f) \\ \vdots \\ \lambda_4(T_f) \end{bmatrix} = \begin{bmatrix} h_{11}(0) & \cdots & h_{81}(0) \\ \vdots & & \vdots \\ h_{14}(0) & \cdots & h_{84}(0) \\ h_{15}(T_f) & \cdots & h_{85}(T_f) \\ \vdots & & \vdots \\ h_{18}(T_f) & \cdots & h_{88}(T_f) \end{bmatrix} \begin{bmatrix} A_1 \\ \\ \vdots \\ \\ A_8 \end{bmatrix}. \tag{8.103}$$

The optimal trajectory is obtained by integrating equation (8.100) from time $t = 0$ to $t = T_f$ and storing the values of $H(t)$ at the boundaries. The vector $\mathbf{c}_A$ is evaluated and utilized to obtain the full set of initial conditions of $\mathbf{x}(t)$ from equation (8.101). Equation (8.94) is then integrated from time $t = 0$ to $t = T_f$ to obtain $\mathbf{x}(t)$.

Analytical Solution

In order to obtain the analytical expression for the optimal input $y(t)$, the eigenvalues of the matrix F are evaluated from the characteristic equation

$$| mI - F | = 0. \tag{8.104}$$

For the given example, equation (8.104) yields an eighth-order equation whose roots are the eigenvalues. The eight eigenvalues are

$$m_1, m_2, m_3, m_4 = \pm\left\{\frac{-(2a - b^2) \pm [(2a - b^2)^2 - 4(a^2 - K)]^{1/2}}{2}\right\}^{1/2}, \tag{8.105}$$

$$m_5, m_6, m_7, m_8 = \pm\left\{\frac{-(2a - b^2) \pm [(2a - b^2)^2 - 4(a^2 + K)]^{1/2}}{2}\right\}^{1/2}, \tag{8.106}$$

where K is a positive constant given by

$$K = 1/q^{1/2}. \tag{8.107}$$

Assuming that $a > b^2/4$, all the eigenvalues are complex for

$$0 < K < \frac{(4a - b^2)b^2}{4}. \tag{8.108}$$

Four of the eigenvalues are imaginary and four are complex for

$$\frac{(4a - b^2)b^2}{4} < K < a^2 \tag{8.109}$$

and two of the eigenvalues are real, two are imaginary, and four are complex for

$$K > a^2. \tag{8.110}$$

Assuming that the solution is desired for the last condition, then substituting equation (8.107) into inequality (8.110) gives

$$q < 1/a^4. \tag{8.111}$$

The optimal trajectory, $x_1(t)$, is of the form

$$\begin{aligned} x_1(t) = {} & C_1e^{\alpha_1 t} + C_2e^{-\alpha_1 t} + C_3 \cos \omega_1 t + C_4 \sin \omega_1 t + C_5e^{\alpha t} \cos \omega t \\ & + C_6e^{\alpha t} \sin \omega t + C_7e^{-\alpha t} \cos \omega t + C_8e^{-\alpha t} \sin \omega t, \end{aligned} \tag{8.112}$$

where $C_1, C_2, \ldots, C_8$ are constants and the eigenvalues are

$$m_1, m_2 = \pm\alpha_1, \tag{8.113}$$

$$m_3, m_4 = \pm j\omega_1, \tag{8.114}$$

$$m_5, m_6 = \alpha \pm j\omega, \tag{8.115}$$

$$m_7, m_8 = -\alpha \pm j\omega. \tag{8.116}$$

Utilizing equation (8.112), then by differentiation and algebraic manipulation, equation (8.94) can be solved for $x_2, \ldots, x_4$ and $\lambda_1, \ldots, \lambda_4$ in terms of $C_1, C_2, \ldots, C_8$. The boundary conditions at $t = 0$ and $t = T_f$ yield eight linear algebraic equations which can be solved for C_1, C_2, $\ldots$, C_8. The optimal input, $y(t)$, is then obtained from equation (8.93):

$$\begin{aligned} y(t) = {} & C_1 a_0 e^{\alpha_1 t} + C_2 b_0 e^{-\alpha_1 t} + C_3(a_1 \cos \omega_1 t - b_1 \sin \omega_1 t) \\ & + C_4(a_1 \sin \omega_1 t + b_1 \cos \omega_1 t) + C_5 e^{\alpha t}(a_2 \cos \omega t - b_2 \sin \omega t) \\ & + C_6 e^{\alpha t}(a_2 \sin \omega t + b_2 \cos \omega t) + C_7 e^{-\alpha t}(a_3 \cos \omega t - b_3 \sin \omega t) \\ & + C_8 e^{-\alpha t}(a_3 \sin \omega t + b_3 \cos \omega t), \end{aligned} \tag{8.117}$$

where $a_0, a_1, \ldots, a_3$ and $b_0, b_1, \ldots, b_3$ are constants which are functions of the eigenvalues.

Numerical Results and Discussion

The optimal input was evaluated numerically via the method of complementary functions and verified via the analytical solution. For the former, a fourth-order Runge–Kutta method with grid intervals of 1/100 sec was utilized for numerical integration.

The parameters for the linear example given by equation (8.75) are assumed to be

$$\begin{aligned} a &= 4, \\ b &= 2, \\ x(0) &= 1, \\ \dot{x}(0) &= 0. \end{aligned} \tag{8.118}$$

Figure 8.3 shows the optimal return as a function of the terminal time, T_f, with q as a parameter. The curves show that for a given value of q, the return is initially very small, but increases rapidly as the terminal time is increased to some critical time length. The existence of a critical length for the case where the boundary conditions are homogeneous is discussed in Section 8.2.4. The determination of the critical length is not trivial; however, it is easily determined numerically from the optimal return versus terminal time curve. The slope of the curve is extremely steep in the vicinity of the critical length. Furthermore, for the two curves shown in Figure 8.3, it was found that if the critical length is just slightly exceeded, i.e., by 0.01 sec, the return assumes a large negative value. The initial magnitude of the input is then also negative instead of positive. This solution and all solutions obtained for terminal times greater than the critical length are not optimal.

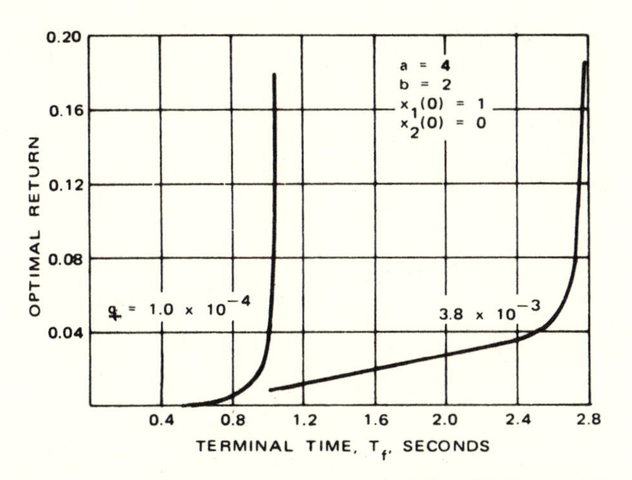

Figure 8.3. Optimal return versus terminal time (Reference 18).

For $q = 1.0 \times 10^{-4}$, the eigenvalues are

$$\begin{aligned} m_1, m_2 &= \pm 2.72, \\ m_3, m_4 &= \pm j3.37, \\ m_5, m_6 &= 2.09 \pm j2.53, \\ m_7, m_8 &= -2.09 \pm j2.53. \end{aligned} \tag{8.119}$$

Figure 8.3 shows that the critical length is 1.04 sec. Increasing q increases the penalty for using a large input, $y(t)$. For $q = 3.8 \times 10^{-3}$, the eigenvalues are

$$\begin{aligned} m_1, m_2 &= \pm 0.234, \\ m_3, m_4 &= \pm j2.01, \\ m_5, m_6 &= 1.36 \pm j1.96, \\ m_7, m_8 &= -1.36 \pm j1.96. \end{aligned} \tag{8.120}$$

Figure 8.3 shows that the critical length is now 2.77 sec.

Figure 8.4 shows the numerical results for the optimal input, $y(t)$, sensitivity, $x_{1a}(t)$, and state variable, $x_1(t)$, as functions of time, for $q = 3.8 \times 10^{-3}$ and terminal time, T_f, as a parameter. Figure 8.4a shows that the magnitude of the optimal input is increased at all instants of time when the terminal time, T_f, is increased. Figures 8.4b and 8.4c show that the sensitivity, $x_{1a}(t)$ and the state variable, $x_1(t)$, also increase when the terminal time is increased. The largest increases occur when the terminal time

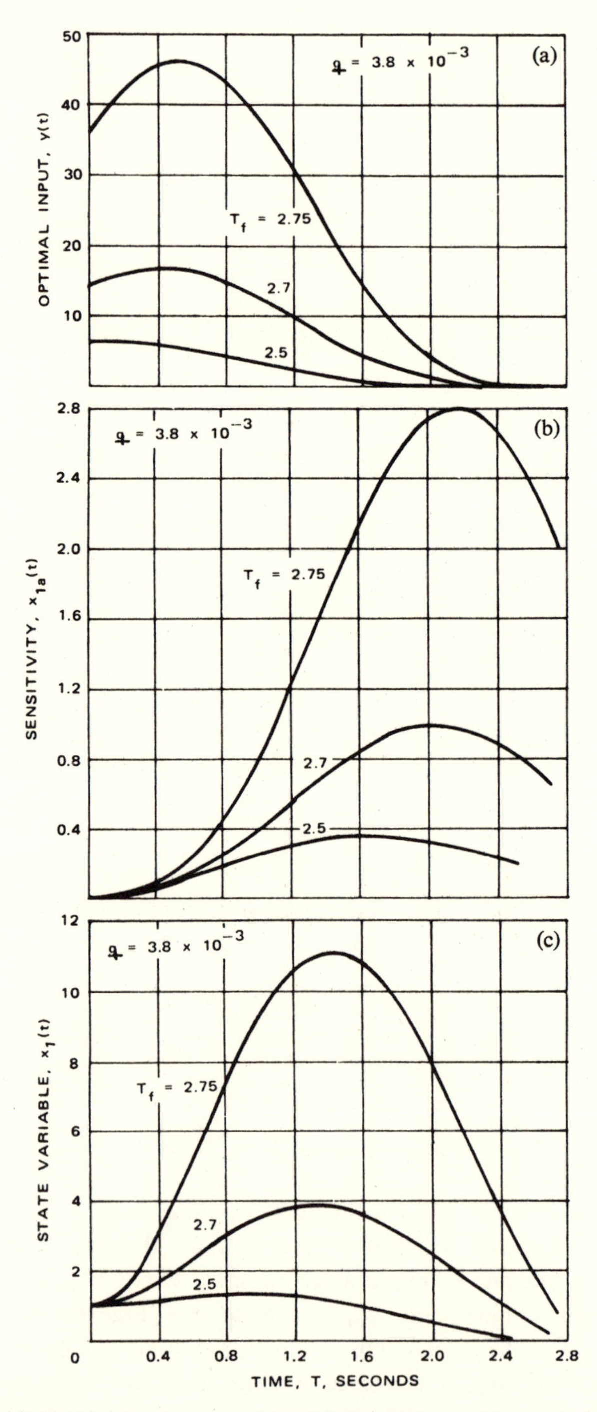

Figure 8.4. (a) Optimal input versus time. (b) Sensitivity $x_{1a}(t)$ versus time. (c) State variable $x_1(t)$ versus time. (Reference 18.)

Table 8.5. Comparison of Optimal and Step Input Returns with the Same Average Power Input ($q = 3.8 \times 10^{-3}$)

Terminal time T_f, (sec)	Return		Average power input $1/T_f \int_0^{T_f} y^2(t)\,dt$
	Optimal input	Step input	
2.5	0.0403	−0.0139	12.6
2.7	0.0652	−0.3674	109.0
2.75	0.117	−3.19	850.0

is near the critical length. In Table 8.5, the optimal return is compared with the return obtained for a step input with the same average power input as for the optimal input. The step input return is positive for terminal times, T_f, less than 2.5 sec, but becomes negative when the terminal time is increased. Table 8.5 also shows that average power of the optimal input increases when the terminal time is increased. For terminal times greater than the critical length, equation (8.94) does not give the optimal return and the return obtained for a step input with the same average power input may actually be greater.

In the above example, it was assumed that the true value of the parameter a was known for deriving the optimal input. In an actual experiment where the true value of the parameter to be estimated is unknown, the optimal input would be found for an initial estimate of a. Then, utilizing parameter estimation techniques as described in Chapters 5 and 6, a new estimate of a would be found. Using this estimate, the optimal input would be obtained and the process repeated until a satisfactory estimate is obtained.

8.2.4. Optimal Inputs Using Mehra's Method

Mehra's method for obtaining optimal inputs for linear system identification is discussed in this section. Mehra considers the case where the initial conditions of the state vector are zero. The Fisher information matrix and the Cramér–Rao lower bound are introduced into the formulation. The two-point boundary-value equations are derived in matrix form using Pontryagin's maximum principle. The matrix Riccati equation is derived for obtaining the critical time length corresponding to the largest eigenvalue of the boundary-value problem. The solution to the two-point boundary-value problem is shown to exist at the eigenvalue.

Derivation of the Two-Point Boundary-Value Equations

Mehra (Reference 9) has shown that the design of optimal inputs for linear system identification involves the solution of two-point boundary-value problems with homogeneous boundary conditions. Nontrivial solutions exist for certain values of the multiplier, q, which correspond to the boundary-value problem eigenvalues. The optimal input is obtained for the largest eigenvalue.

Utilizing Mehra's approach, the problem may be stated as follows. Consider the time-invariant linear system

$$\dot{\mathbf{x}}(t) = F\mathbf{x}(t) + G\mathbf{y}(t), \tag{8.121}$$

$$\mathbf{z}(t) = H\mathbf{x}(t) + \mathbf{v}(t), \tag{8.122}$$

where $\mathbf{x}$ is an $n \times 1$ state vector, $\mathbf{y}$ is an $m \times 1$ control vector, and $\mathbf{z}$ is a $p \times 1$ measurement vector. F, G, and H are, respectively, $n \times n$, $n \times m$, and $p \times n$ matrices. The vector $\mathbf{v}$ is a zero mean Gaussian white noise process:

$$E[\mathbf{v}(t)] = 0, \tag{8.123}$$

$$E[v(t)v^T(\tau)] = R\delta(t - \tau). \tag{8.124}$$

Let a denote an unknown parameter in the above system. Then it is desired to determine the optimal input such that the Fisher information matrix (a scalar in this case)

$$M = \int_0^{T_f} \mathbf{x}_a{}^T H^T R^{-1} H \mathbf{x}_a \, dt \tag{8.125}$$

is maximized subject to the input energy constraint

$$\int_0^{T_f} \mathbf{y}^T\mathbf{y} \, dt = E, \tag{8.126}$$

where

$$\mathbf{x}_a = \frac{\partial \mathbf{x}}{\partial a}. \tag{8.127}$$

The inverse of the Fisher information matrix, M^{-1}, is the Cramér–Rao lower bound for an unbiased estimator of a. Consider a to be an unknown parameter in the matrix F. Then

$$\dot{\mathbf{x}}_a = F\mathbf{x}_a + F_a\mathbf{x}, \tag{8.128}$$

where

$$F_a = \frac{\partial}{\partial a}[F]. \tag{8.129}$$

Let $\mathbf{x}_A$ equal the augmented state vector

$$\mathbf{x}_A = \begin{bmatrix} \mathbf{x} \\ \mathbf{x}_a \end{bmatrix}. \tag{8.130}$$

Then we have

$$\dot{\mathbf{x}}_A = F_A \mathbf{x}_A + G_A \mathbf{y}, \tag{8.131}$$

where

$$F_A = \begin{bmatrix} F & 0 \\ F_a & F \end{bmatrix}, \tag{8.132}$$

$$G_A = \begin{bmatrix} G \\ 0 \end{bmatrix}. \tag{8.133}$$

The performance index can be expressed as the return, J

$$J = \max_{\mathbf{y}} \frac{1}{2} \int_0^{T_f} [\mathbf{x}_A{}^T H_A{}^T R^{-1} H_A \mathbf{x}_A - q\mathbf{y}^T\mathbf{y}]\, dt, \tag{8.134}$$

where H_A is the $p \times 2n$ matrix

$$H_A = [0 \quad H]. \tag{8.135}$$

Utilizing Pontryagin's maximum principle, the Hamiltonian function is

$$\mathscr{H} = \frac{1}{2}[-\mathbf{x}_A{}^T H_A{}^T R^{-1} H_A \mathbf{x}_A + q\mathbf{y}^T\mathbf{y}] + \boldsymbol{\lambda}^T[F_A \mathbf{x}_A + G_A \mathbf{y}]. \tag{8.136}$$

The costate vector $\boldsymbol{\lambda}(t)$ is the solution of the vector differential equation

$$\begin{aligned} \dot{\boldsymbol{\lambda}} &= -\left[\frac{\partial \mathscr{H}}{\partial \mathbf{x}_A}\right]^T \\ &= H_A{}^T R^{-1} H_A \mathbf{x}_A - F_A{}^T \boldsymbol{\lambda}. \end{aligned} \tag{8.137}$$

The vector $\mathbf{y}(t)$ that maximizes $\mathscr{H}$ is

$$\begin{aligned} \frac{\partial \mathscr{H}}{\partial \mathbf{y}} &= q\mathbf{y} + G_A{}^T \boldsymbol{\lambda} = 0, \\ \mathbf{y} &= -(1/q) G_A{}^T \boldsymbol{\lambda}. \end{aligned} \tag{8.138}$$

The two-point boundary-value problem is then given by

$$\begin{bmatrix} \dot{\mathbf{x}}_A \\ \dot{\boldsymbol{\lambda}} \end{bmatrix} = \begin{bmatrix} F_A & -(1/q)G_A G_A{}^T \\ H_A{}^T R^{-1} H_A & -F_A{}^T \end{bmatrix} \begin{bmatrix} \mathbf{x}_A \\ \boldsymbol{\lambda} \end{bmatrix} \tag{8.139}$$

with the boundary conditions

$$\mathbf{x}_A = \begin{bmatrix} \mathbf{x}(0) \\ 0 \end{bmatrix}, \qquad \boldsymbol{\lambda}(T_f) = 0. \tag{8.140}$$

Equation (8.139) can also be expressed in the form

$$\dot{\mathbf{x}}_B = A\mathbf{x}_B, \tag{8.141}$$

where

$$\mathbf{x}_B = \begin{bmatrix} \mathbf{x}_A \\ \boldsymbol{\lambda} \end{bmatrix}, \tag{8.142}$$

$$A = \begin{bmatrix} F_A & -(1/q)G_A G_A{}^T \\ H_A{}^T R^{-1} H_A & -F_A{}^T \end{bmatrix}. \tag{8.143}$$

The derivations are given above for a single unknown parameter, a, but are extended to the case where several parameters are unknown in Section 9.2.

As in the previous sections, since a is unknown, an initial estimate of a must be utilized in order to obtain the optimal input. Then utilizing the parameter estimation techniques described in Chapters 5 and 6, a new estimate of a is found and the process repeated until a satisfactory estimate of a is obtained.

Solution for Homogeneous Boundary Conditions

Mehra's method of solution is based on the assumption that the boundary conditions in equations (8.140) are homogeneous, i.e., the initial conditions vector, $\mathbf{x}(0) = 0$. The solution is trivial except for certain values of q which are the eigenvalues of the two-point boundary-value problem. To obtain the optimal input, the Riccati matrix and the transition matrix are defined as follows.

The Riccati matrix, $P(t)$, is defined by the relation

$$\mathbf{x}_A(t) = P(t)\boldsymbol{\lambda}(t). \tag{8.144}$$

Differentiating equation (8.144), substituting from equation (8.139), and rearranging yields

$$\dot{P} = F_A P + P F_A^T - P H_A^T R^{-1} H_A P - (1/q) G_A G_A^T, \tag{8.145}$$

$$P(0) = 0. \tag{8.146}$$

Let $\Phi(t; q)$ denote the transition matrix of equation (8.139) for a particular value of q. Then we have

$$\begin{bmatrix} \mathbf{x}_A(T_f) \\ \boldsymbol{\lambda}(T_f) \end{bmatrix} = \begin{bmatrix} \Phi_{xx}(T_f; q) & \Phi_{x\lambda}(T_f; q) \\ \Phi_{\lambda x}(T_f; q) & \Phi_{\lambda\lambda}(T_f; q) \end{bmatrix} \begin{bmatrix} \mathbf{x}_A(0) \\ \boldsymbol{\lambda}(0) \end{bmatrix}. \tag{8.147}$$

The second equation in equation (8.147) along with the boundary conditions gives

$$\boldsymbol{\lambda}(T_f) = \Phi_{\lambda\lambda}(T_f; q)\boldsymbol{\lambda}(0) = 0. \tag{8.148}$$

For a nontrivial solution, the determinant of the matrix must equal zero:

$$| \Phi_{\lambda\lambda}(T_f; q) | = 0. \tag{8.149}$$

The optimal input is then obtained as follows:

(a) The matrix Riccati equation (8.145) with initial condition (8.146) is integrated forward in time for a particular value of q. When the elements of $P(t)$ become very large, the critical length, T_{crit}, has been reached. The terminal time, T_f, is set equal to T_{crit}.

(b) The transition matrix is obtained by integrating

$$\dot{\Phi}(t; q) = A\Phi(t; q) \tag{8.150}$$

$$\Phi(0; q) = I \tag{8.151}$$

from $t = 0$ to $t = T_f = T_{\text{crit}}$. The matrix A is defined by equation (8.143).

(c) The initial costate vector, $\boldsymbol{\lambda}(0)$, is obtained from equations (8.148) and (8.149) as an eigenvector of $\Phi_{\lambda\lambda}(T_f; q)$ corresponding to the zero eigenvalue. A unique value of $\boldsymbol{\lambda}(0)$ is found by using the normalization condition of the input energy constraint, equation (8.126).

(d) Equation (8.139) is integrated forward in time using $\boldsymbol{\lambda}(0)$ obtained above. The optimal input is obtained utilizing equation (8.138).

If a particular terminal time, T_f, is desired, then the matrix Riccati equation must be integrated several times with different values of q, in order to determine the value of q corresponding to the desired T_f.

Some numerical errors are introduced in the solution because the critical length, T_{crit}, and the eigenvector of $\Phi_{\lambda\lambda}(T_f; q)$ associated with the zero eigenvalue cannot be determined exactly. The accuracy of the solution can be improved (References 8, 9), by integrating P^{-1} when appropriate and finding the eigenvector associated with the smallest eigenvalue of $\Phi_{\lambda\lambda}^T \Phi_{\lambda\lambda}$.

8.2.5. Comparison of Optimal Inputs for Homogeneous and Nonhomogeneous Boundary Conditions

Mehra's method for the design of optimal inputs for linear system identification described in Section 8.2.4 involves the solution of homogeneous linear differential equations with homogeneous boundary conditions. Sections 8.2.1 to 8.2.3 have considered the solution for problems with nonhomogeneous boundary conditions. In this section the methods of solution for homogeneous and nonhomogeneous boundary conditions are compared (Reference 19).

Solution for Nonhomogeneous Boundary Conditions

When the boundary conditions are nonhomogeneous, the two-point boundary-value problem given by equations (8.139) and (8.140) can be solved using the method of complementary functions. The homogeneous differential equations are

$$\dot{H} = AH, \qquad H(0) = I, \tag{8.152}$$

where H is a $4n \times 4n$ matrix (not the measurement matrix defined previously) and A is the matrix defined by equation (8.143). The solution to equation (8.141) is given by

$$\mathbf{x}_B = H\mathbf{c}_B, \tag{8.153}$$

where the vector of constants

$$\mathbf{c}_B = \begin{bmatrix} \mathbf{c}_1 \\ \mathbf{c}_2 \end{bmatrix} \tag{8.154}$$

is determined from the equation

$$\begin{bmatrix} \mathbf{x}_A(0) \\ \boldsymbol{\lambda}(T_f) \end{bmatrix} = \begin{bmatrix} H_{11}(0) & H_{12}(0) \\ H_{21}(T_f) & H_{22}(T_f) \end{bmatrix} \begin{bmatrix} \mathbf{c}_1 \\ \mathbf{c}_2 \end{bmatrix}. \tag{8.155}$$

The optimal trajectory is obtained by integrating equation (8.152) from time

$t = 0$ to time $t = T_f$ and storing the values of $H(t)$ at the boundaries. The vector $\mathbf{c}_B$ is evaluated and utilized to obtain the full set of initial conditions of $\mathbf{x}_B$ from equation (8.153). Equation (8.141) is then integrated from time $t = 0$ to $t = T_f$ and the optimal input is obtained utilizing equation (8.138).

For a given value of q, the optimal return, J [given by equation (8.134)], will increase as the critical length is approached. The input energy also increases and becomes infinite at the critical length. Thus the desired input energy can be obtained for a given value of q by evaluating equations (8.153) and (8.141) for several different values of the terminal time, T_f, where $T_f < T_{\text{crit}}$. The desired input energy is then obtained by plotting a curve of input energy versus T_f, and finding the terminal time corresponding to the desired input energy. Alternately, a different procedure may be utililized by finding the value of q corresponding to the desired input energy for a given value of T_f.

Example. Consider the scalar linear differential equation

$$\dot{x} = -ax + y, \tag{8.156}$$

$$z = x + v, \tag{8.157}$$

where v is a zero mean Gaussian white noise process with variance r. The initial condition is given by

$$x(0) = c. \tag{8.158}$$

The optimal input is to be determined such that the return is maximized. The two-point boundary-value problem is given by equation (8.139), where

$$F_A = \begin{bmatrix} -a & 0 \\ -1 & -a \end{bmatrix}, \tag{8.159}$$

$$G_A = \begin{bmatrix} 1 \\ 0 \end{bmatrix}, \tag{8.160}$$

$$H_A = [0 \quad 1]. \tag{8.161}$$

Equation (8.139) then becomes

$$\begin{bmatrix} \dot{x} \\ \dot{x}_a \\ \dot{\lambda}_1 \\ \dot{\lambda}_2 \end{bmatrix} = \begin{bmatrix} -a & 0 & -1/q & 0 \\ -1 & -a & 0 & 0 \\ 0 & 0 & a & 1 \\ 0 & 1 & 0 & a \end{bmatrix} \begin{bmatrix} x \\ x_a \\ \lambda_1 \\ \lambda_2 \end{bmatrix} \tag{8.162}$$

with boundary conditions

$$x(0) = c, \qquad x_a(0) = 0, \tag{8.163}$$

$$\lambda_1(T_f) = 0, \qquad \lambda_2(T_f) = 0. \tag{8.164}$$

Numerical Results

To obtain numerical results, a fourth-order Runge–Kutta method with grid intervals of 0.01 sec was utilized for the numerical integrations.

The following parameters are assumed to be given:

$$a = 0.1,$$
$$q = 0.075,$$
$$r = 1.$$

Then for the homogeneous boundary conditions, matrix Riccati equation (8.145) is integrated until overflow occurs. This occurs at the critical length, $T_{\text{crit}} = 1.02$ sec. The transition matrix is obtained by integrating equation (8.150) from $t = 0$ to $t = T_{\text{crit}}$. The normalized eigenvector of $\Phi_{\lambda\lambda}(T_f; q)$ corresponding to the zero eigenvalue is then determined and is given by

$$\boldsymbol{\lambda}(0) = \begin{bmatrix} 1 \\ -1.373 \end{bmatrix}. \tag{8.165}$$

To obtain the desired input energy, the above equation is multiplied by an appropriate scalar constant. Equation (8.139) is then integrated forward in time to obtain the optimal input and state variables.

For the nonhomogeneous boundary conditions, it is assumed that

$$x(0) = c = 0.1. \tag{8.166}$$

Then the optimal input can be found using either the method of complementary functions or the analytical solution. The optimal return as a function of the terminal time, T_f, is shown in Figure 8.5. The optimal return increases as the terminal time approaches the critical length, T_{crit}. The energy input also increases as shown in Figure 8.6. Figure 8.7 shows the optimal input as a function of time for $T_f = 1$ sec. Also shown is the optimal input for the homogeneous case with $T_f = 1.02$ sec and the same input energy as for the nonhomogeneous case. It is seen that the solutions are almost identical.

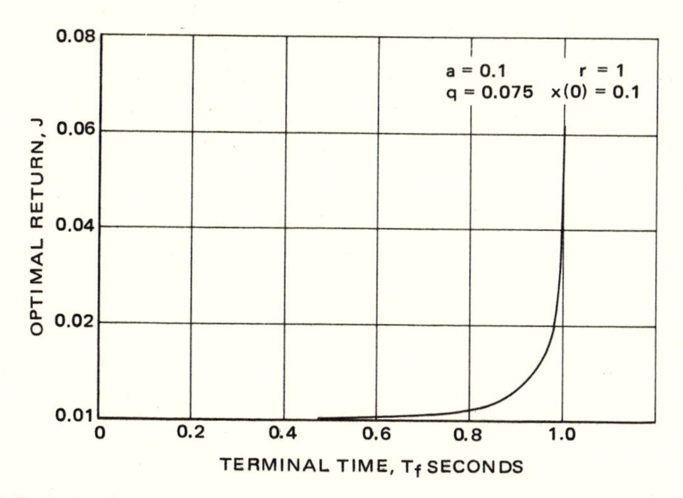

Figure 8.5. Optimal return versus terminal time for nonhomogeneous boundary conditions (Reference 19).

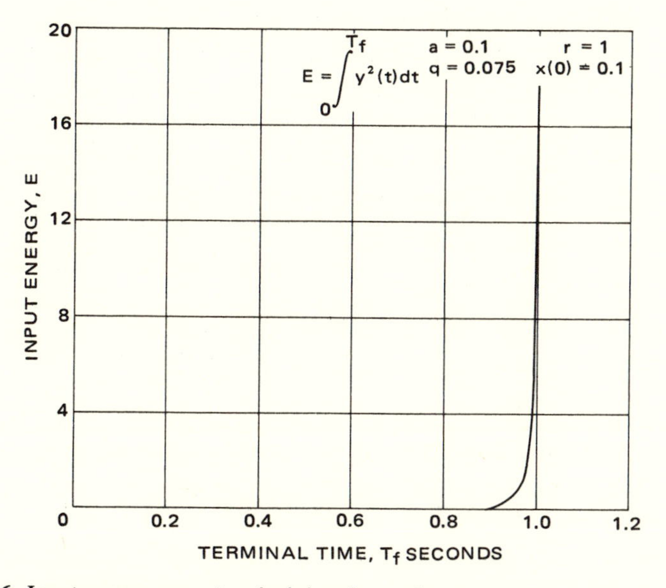

Figure 8.6. Input energy versus terminal time for nonhomogeneous boundary conditions (Reference 19).

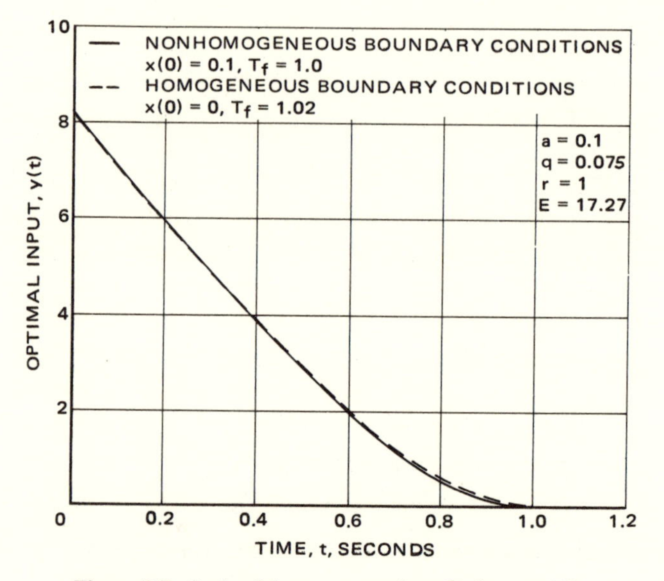

Figure 8.7. Optimal input versus time (Reference 19).

8.3. Nonlinear Optimal Inputs

Optimal inputs for nonlinear system identification are discussed in this section. In Section 8.3.1 the nonlinear dynamic system example (Reference 20)

$$\dot{x} = -ax - bx^2 + y, \qquad x(0) = c$$

is considered where a is the unknown parameter to be estimated. The performance index is

$$J = \max_{y} \frac{1}{2} \int_0^T [x_a^2(t) - qy^2(t)]\, dt.$$

The two-point boundary-value equations are derived using Pontryagin's maximum principle. The numerical results are obtained using either quasilinearization or the Newton–Raphson method.

General equations for optimal inputs for nonlinear process parameter estimation are discussed in Section 8.3.2. The general equations are derived (Reference 18) using the Euler–Lagrange method. The solution is obtained via quasilinearization.

8.3.1. Optimal Input System Identification for Nonlinear Dynamic Systems

The estimation accuracy for a nonlinear dynamic system is increased by the use of optimal inputs. The performance criterion is selected such that the sensitivity of the measured state variables to the unknown parameters is maximized. The application of Pontryagin's maximum principle yields a nonlinear two-point boundary-value problem. In this section the boundary-value problem for a simple nonlinear example is solved using two different methods: the method of quasilinearization and the Newton–Raphson method. The estimation accuracy is discussed in terms of the Cramér–Rao lower bound.

Introduction

The design of optimal inputs for linear system identification has been investigated by many authors in the literature. Few examples however are given for nonlinear systems. Goodwin (Reference 7) gives several examples with magnitude constraints on the inputs. Mehra (Reference 1) derives equations for nonlinear systems in general. In this section, the optimal input for a simple nonlinear dynamic system is determined and curves are given showing the effects of increasing nonlinearity (Reference 20).

The performance index is selected such that the sensitivity of the measured state variables to the unknown parameters is maximized subject to an input energy constraint. The performance index is maximized using Pontryagin's maximum principle. The solution requires the evaluation of a nonlinear two-point boundary-value problem. This is done using two different methods: (1) quasilinearization and (2) the Newton–Raphson method. The relationship of the performance index to the Fisher information matrix and the Cramér–Rao lower bound is also discussed.

Consider the nonlinear dynamic system given by

$$\dot{x} = -ax - bx^2 + y \tag{8.167}$$

with initial condition

$$x(0) = c, \tag{8.168}$$

where $x(t)$ is a scalar state variable, $y(t)$ is a scalar input, a is an unknown system parameter, and b is a known parameter. It is desired to maximize the sensitivity of the state variable, $x(t)$, to the unknown parameter a. Measurements of $x(t)$ utilized to estimate the parameter a will then yield maximum estimation accuracy.

Using a quadratic performance criterion it is desired to determine the optimal input such that the integral

$$M_1 = \int_0^{T_f} x_a^2(t)\,dt \tag{8.169}$$

is maximized, subject to the input energy constraint

$$\int_0^{T_f} y^2(t)\,dt, \tag{8.170}$$

where

$$x_a(t) = \frac{\partial x(t)}{\partial a}. \tag{8.171}$$

The performance index can be expressed as the return

$$J = \max_y \frac{1}{2}\int_0^{T_f} [x_a^2(t) - qy^2(t)]\,dt. \tag{8.172}$$

Differentiating equations (8.167) and (8.168) yields

$$\dot{x}_a = -x - ax_a - 2bxx_a, \qquad x_a(0) = 0. \tag{8.173}$$

Utilizing Pontryagin's maximum principle, the Hamiltonian function is

$$\begin{aligned}\mathscr{H} = \tfrac{1}{2}[-x_a^2 + qy^2] + \lambda_1(-ax - bx^2 + y)\\ + \lambda_2(-x - ax_a - 2bxx_a).\end{aligned} \tag{8.174}$$

The costate variables, λ_1 and λ_2, are the solutions of the differential equations

$$\begin{aligned}\dot{\lambda}_1 &= -\frac{\partial \mathscr{H}}{\partial x} = (a + 2bx)\lambda_1 + (1 + 2bx_a)\lambda_2,\\ \dot{\lambda}_2 &= -\frac{\partial \mathscr{H}}{\partial x_a} = x_a + (a + 2bx)\lambda_2.\end{aligned} \tag{8.175}$$

The input $y(t)$ that maximizes $\mathscr{H}$ is

$$\frac{\partial \mathscr{H}}{\partial y} = qy + \lambda_1 = 0, \qquad y = -\frac{1}{q}\lambda_1. \tag{8.176}$$

The nonlinear two-point boundary-value problem and associated boundary conditions are then given by

$$\begin{aligned}
\dot{x} &= -ax - bx^2 - (1/q)\lambda_1, & x(0) &= c, \\
\dot{x}_a &= -x - ax_a - 2bxx_a, & x_a(0) &= 0, \\
\dot{\lambda}_1 &= (a + 2bx)\lambda_1 + (1 + 2bx_a)\lambda_2, & \lambda_1(T_f) &= 0, \\
\dot{\lambda}_2 &= x_a + (a + 2bx)\lambda_2, & \lambda_2(T_f) &= 0.
\end{aligned} \tag{8.177}$$

The solution to these equations is obtained using the method of quasilinearization or the Newton–Raphson method.

Quasilinearization

Equations (8.177) can be expressed in the vector form

$$\dot{\mathbf{x}} = \mathbf{f}(\mathbf{x}, t), \qquad \mathbf{x}(0) = \mathbf{c}, \tag{8.178}$$

where

$$\begin{aligned}
\mathbf{x} &= (x, x_a, \lambda_1, \lambda_2)^T \\
\mathbf{f} &= (f_1, f_2, f_3, f_4)^T.
\end{aligned} \tag{8.179}$$

The components of $\mathbf{f}$ are given by the right sides of differential equations (8.177). Using the method of quasilinearization (Reference 20), differential equation (8.178) is first linearized about the kth approximation by expanding the function $\mathbf{f}$ in a Taylor series and retaining only the linear terms. The $(k + 1)$ approximation is

$$\dot{\mathbf{x}}_{k+1} = \mathbf{f}(\mathbf{x}_k, t) + J_k(\mathbf{x}_{k+1} - \mathbf{x}_k), \tag{8.180}$$

where the Jacobian matrix, J_k, is defined by

$$J_k = J_k[\mathbf{f}(\mathbf{x}_k, t)] \tag{8.181}$$

with the ij element

$$J_k|_{ij} = \left.\frac{\partial f_i(\mathbf{x})}{\partial x_j}\right|_{\mathbf{x}=\mathbf{x}_k(t)}. \tag{8.182}$$

Equation (8.180) can be expressed in the form

$$\dot{\mathbf{x}}_{k+1} = A_k(t)\mathbf{x}_{k+1} + \mathbf{u}_k(t), \tag{8.183}$$

where

$$\begin{aligned}
A_k(t) &= J_k \\
\mathbf{u}_k(t) &= \mathbf{f}(\mathbf{x}_k, t) - J_k\mathbf{x}_k.
\end{aligned} \tag{8.184}$$

The numerical solution to linear equation (8.183) is obtained using the method of complementary functions. The solution is given by the linear combination

$$\mathbf{x}_{k+1}(t) = \mathbf{p}_{k+1}(t) + H_{k+1}(t)\mathbf{c}_A, \tag{8.185}$$

where the 4×1 vector, $\mathbf{p}_{k+1}$, and the 4×4 matrix, H_{k+1}, are the solutions of the initial-value equations

$$\dot{\mathbf{p}}_{k+1}(t) = A_k(t)\mathbf{p}_{k+1}(t) + \mathbf{u}_k(t), \qquad \mathbf{p}_{k+1}(0) = 0, \tag{8.186}$$

$$\dot{H}_{k+1}(t) = A_k(t)H_{k+1}(t), \qquad H_{k+1}(0) = I. \tag{8.187}$$

The vector of constants

$$\mathbf{c}_A = (\gamma_1, \gamma_2, \gamma_3, \gamma_4)^T \tag{8.188}$$

is obtained by substituting the boundary conditions into equation (8.185) and solving for $\mathbf{c}_A$.

For the given example, $A_k(t)$ and $\mathbf{u}_k(t)$ in equation (8.183) are given by

$$A_k(t) = \begin{bmatrix} -(a + 2bx) & 0 & -1/q & 0 \\ -(1 + 2bx_a) & -(a + 2bx) & 0 & 0 \\ 2b\lambda_1 & 2b\lambda_2 & a + 2bx & 1 + 2bx_a \\ 2b\lambda_2 & 1 & 0 & a + 2bx \end{bmatrix}, \tag{8.189}$$

$$\mathbf{u}_k(t) = \begin{bmatrix} bx^2 \\ 2bxx_a \\ -2b(x\lambda_1 + x_a\lambda_2) \\ -2bx\lambda_2 \end{bmatrix}. \tag{8.190}$$

To obtain a numerical solution, an initial guess is made for the vector $\mathbf{x}(t)$, i.e., assume $\mathbf{x}(t) = \mathbf{x}_0(t)$. Equations (8.186) and (8.187) are integrated from time $t = 0$ to $t = T_f$. The values of $p(t)$ and $H(t)$ at the boundaries are utilized to evaluate the vector $\mathbf{c}_A$ from equation (8.185). The full set of initial conditions of the first approximation, $\mathbf{x}_1(t)$, is equal to $\mathbf{c}_A$. Equation (8.183) is then integrated from time $t = 0$ to $t = T_f$ to obtain $\mathbf{x}_1(t)$. The values of $\mathbf{x}_1(t)$ are stored and the above sequence is repeated to obtain the second approximation, etc. This method requires the integration of 24 differential equations for each approximation.

In an actual experiment where the true value of the parameter to be estimated is unknown, the optimal input is found for an initial estimate of the parameter a. Then, using the method of quasilinearization as above or the Newton–Raphson method, a new estimate of a is found. Using this estimate, another optimal input is obtained and the process is repeated until a satisfactory estimate is obtained.

Newton–Raphson Method

Using the Newton–Raphson method the unknown initial conditions in equations (8.177) are assumed to be given by

$$\begin{aligned} \lambda_1(0) &= c_1, \\ \lambda_2(0) &= c_2. \end{aligned} \tag{8.191}$$

Expanding the boundary conditions

$$\begin{aligned} \lambda_1(T_f, c_1, c_2) &= 0, \\ \lambda_2(T_f, c_1, c_2) &= 0 \end{aligned} \tag{8.192}$$

in a Taylor series around the kth approximation and retaining only the linear terms gives

$$\begin{aligned} \lambda_1(T_f, c_1^k, c_2^k) + (c_1^{k+1} - c_1^k)\lambda_{1c_1} + (c_2^{k+1} - c_2^k)\lambda_{1c_2} &= 0, \\ \lambda_2(T_f, c_1^k, c_2^k) + (c_1^{k+1} - c_1^k)\lambda_{2c_1} + (c_2^{k+1} - c_1^k)\lambda_{2c_2} &= 0, \end{aligned} \tag{8.193}$$

where

$$\lambda_{ic_j} = \frac{\partial \lambda_i}{\partial c_j}. \tag{8.194}$$

Solving for the $(k + 1)$ approximation yields

$$\mathbf{c}^{k+1} = \mathbf{c}^k - \Lambda_c^{-1}\boldsymbol{\lambda}(T_f, \mathbf{c}^k), \tag{8.195}$$

where

$$\begin{aligned} \mathbf{c} &= \begin{bmatrix} c_1 \\ c_2 \end{bmatrix}, \\ \Lambda_c &= \begin{bmatrix} \lambda_{1c_1} & \lambda_{1c_2} \\ \lambda_{2c_1} & \lambda_{2c_2} \end{bmatrix}, \\ \boldsymbol{\lambda} &= \begin{bmatrix} \lambda_1 \\ \lambda_2 \end{bmatrix}. \end{aligned} \tag{8.196}$$

The equations for the components of matrix Λ_c are obtained as follows. Differentiating equations (8.177) with respect to c_1 yields

$$\begin{aligned}
\dot{x}_{c_1} &= -ax_{c_1} - 2bxx_{c_1} - (1/q)\lambda_{1c_1}, \\
\dot{x}_{ac_1} &= -x_{c_1} - ax_{ac_1} - 2b(xx_{ac_1} + x_a x_{c_1}), \\
\dot{\lambda}_{1c_1} &= (a + 2bx)\lambda_{1c_1} + 2bx_{1c_1}\lambda_1 + (1 + 2bx_a)\lambda_{2c_1} + 2bx_{ac_1}\lambda_2, \\
\dot{\lambda}_{2c_1} &= x_{ac_1} + (a + 2bx)\lambda_{2c_1} + 2bx_{c_1}\lambda_2.
\end{aligned} \tag{8.197}$$

The initial conditions are

$$\begin{aligned}
x_{c_1}(0) &= 0, \qquad \lambda_{1c_1}(0) = 1, \\
x_{ac_1}(0) &= 0, \qquad \lambda_{2c_1}(0) = 0.
\end{aligned} \tag{8.198}$$

Differentiating equations (8.177) with respect to c_2 yields

$$\begin{aligned}
\dot{x}_{c_2} &= -ax_{c_2} - 2bxx_{c_2} - (1/q)\lambda_{1c_2}, \\
\dot{x}_{ac_2} &= -x_{c_2} - ax_{ac_2} - 2b(xx_{ac_2} + x_a x_{c_2}), \\
\dot{\lambda}_{1c_2} &= (a + 2bx)\lambda_{1c_2} + 2bx_{c_2}\lambda_1 + (1 + 2bx_a)\lambda_{2c_2} + 2bx_{ac_2}\lambda_2, \\
\dot{\lambda}_{2c_2} &= x_{ac_2} + (a + 2bx)\lambda_{2c_2} + 2bx_{c_2}\lambda_2.
\end{aligned} \tag{8.199}$$

The initial conditions are

$$\begin{aligned}
x_{c_2}(0) &= 0, \qquad \lambda_{1c_2}(0) = 0, \\
x_{ac_2}(0) &= 0, \qquad \lambda_{2c_2}(0) = 1.
\end{aligned} \tag{8.200}$$

To obtain a numerical solution, an initial guess is made for initial conditions (8.191). Initial-value equations (8.177), (8.197), and (8.199) are then integrated from time $t = 0$ to $t = T_f$. A new value for $\mathbf{c}$ is calculated from equation (8.195) and the above sequence is repeated to obtain the second approximation, etc. This method requires the integration of 12 differential equations for each approximation.

Fisher Information Matrix

The performance criterion given by equation (8.169) maximizes the sensitivity of the state variable, $x(t)$, to the unknown parameter a. Assume now that the measurements include additive white noise

$$z(t) = x(t) + v(t), \tag{8.201}$$

where $z(t)$ is the scalar measurement and $v(t)$ is a zero mean Gaussian white noise process such that:

$$E[v(t)] = 0,$$
$$Ev(t)v^T(\tau) = R\delta(t-\tau) = \sigma_n^2\delta(t-\tau). \tag{8.202}$$

The Fisher information matrix (Reference 12) is given by

$$M = \int_0^{T_f} x_a{}^T R^{-1} x_a\, dt = \frac{1}{\sigma_n^2}\int_0^{T_f} x_a{}^2(t)\, dt, \tag{8.203}$$

where M is a scalar in this case since both x_a and R are scalars. The plant equation may be either linear or nonlinear. If $\sigma_n = 1$, equations (8.169) and (8.203) are identical.

The inverse of the Fisher information matrix, M^{-1}, is the Cramér–Rao lower bound. Assume that $\hat{a}$ is an estimate of a. Then the covariance of the estimate $\hat{a}$ is

$$\text{cov}(a) = E[(\hat{a}-a)(\hat{a}-a)^T] \geq M^{-1}. \tag{8.204}$$

Substituting equation (8.203) into equation (8.204) yields

$$\text{cov}(a) \geq \frac{\sigma_n^2}{\int_0^{T_f} x_a^2\, dt}. \tag{8.205}$$

The optimal input is dependent upon the measurement noise variance. In the discussions which follow, it is assumed that $\sigma_n = 1$.

Numerical Results

Numerical results were obtained using both the method of quasilinearization and the Newton–Raphson method. A fourth-order Runge–Kutta method was utilized for integration with grid intervals of 1/100 sec.

The parameters of equation (8.167) were assumed to be as follows:

$$a = 0.1, \qquad x(0) = 0.1,$$
$$q = 0.075, \qquad T_f = 1.$$

The nonlinearity parameter, b, was varied from 0 to 0.1. For $b = 0$, the system is linear. It was shown in Section 8.2.5 that for this case the optimal return, J, increases as the terminal time approaches a critical length, T_{crit}. The terminal time, T_f, should be close to the critical length for maximum optimal return. If a particular terminal time is desired, the value of q must be found by trial and error. The critical length for linear systems with homogeneous boundary conditions may be determined using the matrix Riccati equation. It was shown in Section 8.2.5 that $T_{\text{crit}} = 1.02$ sec for $q = 0.075$.

Utilizing quasilinearization, the initial approximation, $\mathbf{x}_0(t)$, was obtained by assuming that the components of $\mathbf{x}(t)$ are equal to their known boundary conditions throughout the time interval 0 to T_f, i.e.,

$$\left.\begin{array}{ll} x(t) = 0.1, & \lambda_1(t) = 0 \\ x_a(t) = 0, & \lambda_2(t) = 0 \end{array}\right\} \quad 0 \leq t \leq T_f.$$

The optimal input was found for the linear case first using the above set of parameters. The effects of increasing nonlinearity were then determined by using progressively larger values of b.

Figure 8.8 shows the optimal return, J, defined by equation (8.172),

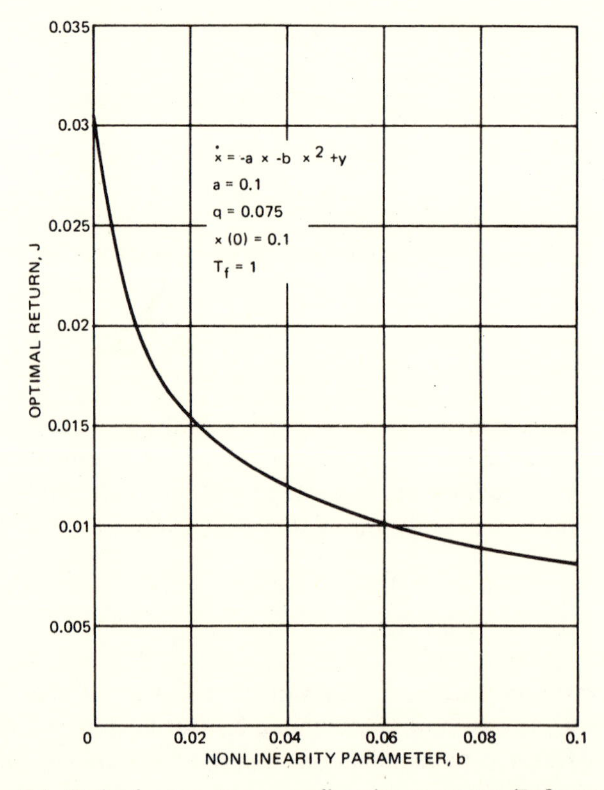

Figure 8.8. Optimal return versus nonlinearity parameter (Reference 20).

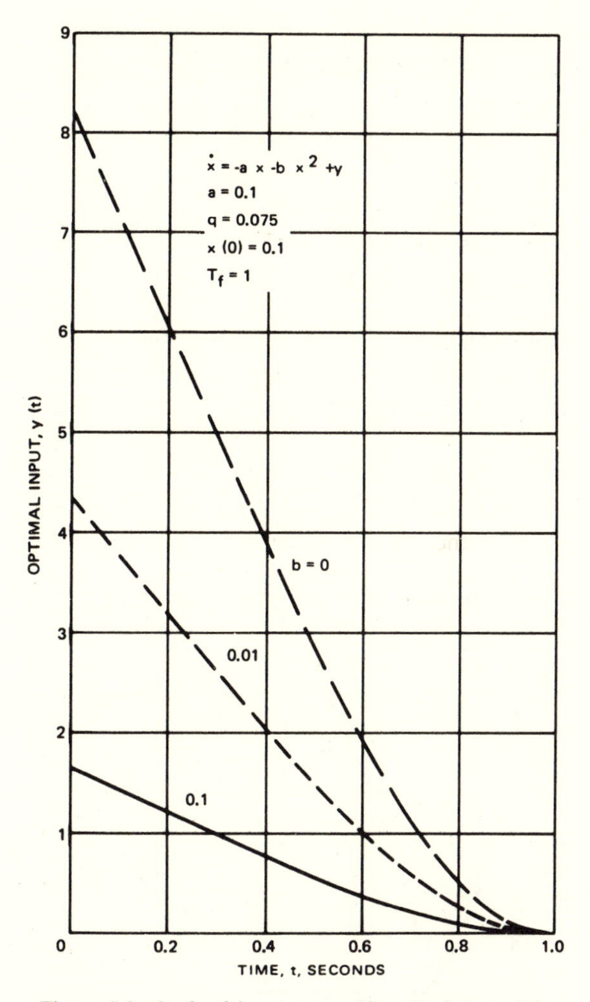

Figure 8.9. Optimal input versus time (Reference 20).

as a function of the nonlinearity parameter, b. The optimal return decreases as b increases. Figure 8.9 shows the optimal input, $y(t)$, as a function of time with b as a parameter equal to 0, 0.01, and 0.1. Figures 8.10 and 8.11 show the corresponding curves for the state variable, $x(t)$, and the sensitivity, $x_a(t)$.

For the linear case, the optimal input is obtained in one iteration of the quasilinearization program. For the nonlinear case, the optimal input is obtained in approximately five iterations (to within approximately five or six digit accuracy).

Utilizing the Newton–Raphson method, the unknown initial conditions, $\lambda_1(0)$ and $\lambda_2(0)$, were assumed to be equal to their known boundary values at time T_f:

$$\lambda_1(0) = 0, \qquad \lambda_2(0) = 0.$$

The initial conditions for the state variable, $x(t)$, and sensitivity, $x_a(t)$, are known:

$$x(0) = 0.1, \qquad x_a(0) = 0.$$

The Newton–Raphson method gives the same solution as the method of quasilinearization to within five or six digits. For the linear case, the optimal input is obtained in one iteration of the Newton–Raphson program. For

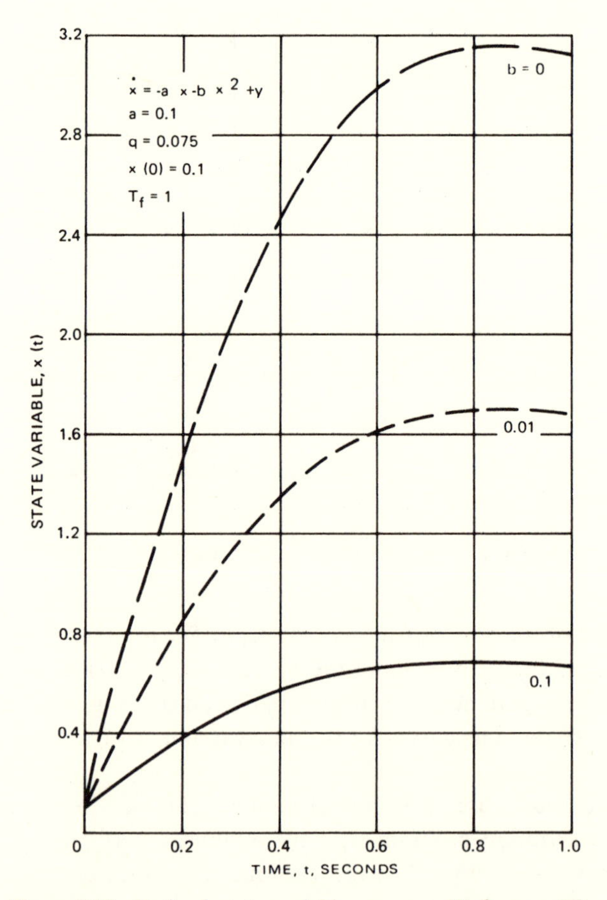

Figure 8.10. Optimal state variable response (Reference 20).

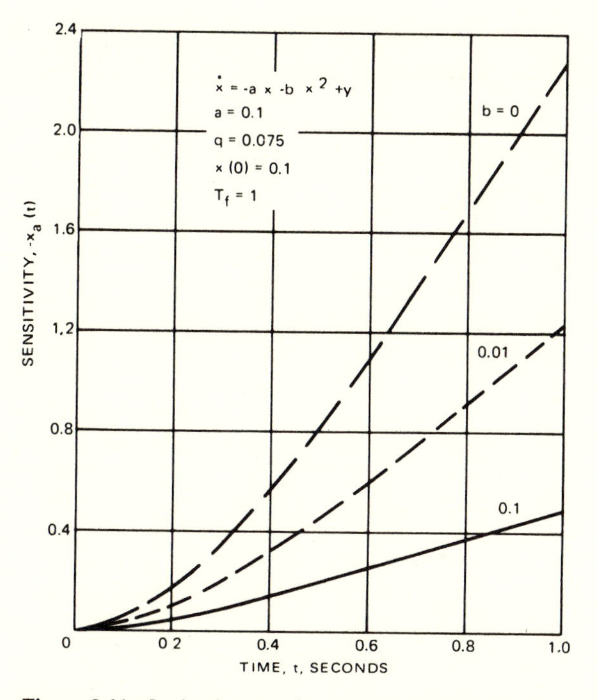

Figure 8.11. Optimal sensitivity response (Reference 20).

the nonlinear case the optimal input is obtained in approximately six iterations. The total computer times for the method of quasilinearization and the Newton–Raphson method were approximately the same.

For $b = 0$, the Fisher information matrix is

$$M = \frac{1}{\sigma_n^2} \int_0^{T_f} x_a^2 \, dt = \frac{1.36}{\sigma_n^2}.$$

The estimation accuracy is then given by the Cramér–Rao lower bound in equation (8.205):

$$\text{cov}(a) \geq \sigma_n^2/1.36.$$

For a nonoptimal sine wave input with the same average power input as the optimal input and angular frequency $\omega = 0.1$ rad/sec

$$y(t) = K \sin \omega t, \qquad K = 72.0601,$$

the estimation accuracy given by the Cramér–Rao lower bound is

$$\text{cov}(a) \geq \sigma_n^2/0.234.$$

The improvement in the estimation variance of the optimal input over the sine wave input is $1/(1.36/0.234) = 1/5.81$.

8.3.2. General Equations for Optimal Inputs for Nonlinear Process Parameter Estimation

In this section equations are derived for obtaining the optimal input for a set of n first-order nonlinear differential equations with unknown parameter, a (Reference 18). The optimal input requires the solution of $4n$ two-point boundary-value equations. The nonlinear process is assumed to be described by the differential equations

$$\begin{aligned} \dot{x}_1 &= f_1(x_1, x_2, \ldots, x_n, a, t), & x_1(0) &= c_1, \\ \dot{x}_2 &= f_2(x_1, x_2, \ldots, x_n, a, t), & x_2(0) &= c_2, \\ &\vdots \\ \dot{x}_n &= f_n(x_1, x_2, \ldots, x_n, y, a, t), & x_n(0) &= c_n, \end{aligned} \tag{8.206}$$

where a is an unknown system parameter to be estimated, y is the input, t is time, and

$$\begin{aligned} x_1 &= x_1(t, a), \\ x_2 &= x_2(t, a), \\ &\vdots \\ x_n &= x_n(t, a). \end{aligned}$$

Observations are made of $x_1(t, a)$ at discrete times, $t_1, t_2, \ldots t_n$. It is desired to maximize the sensitivity of the observations to the parameter a,

$$\max_{y(t)} \int_0^{T_f} x_{1a}^2(t, a)\, dt, \tag{8.207}$$

subject to the energy constraint

$$\int_0^{T_f} y^2(t)\, dt \leq E, \tag{8.208}$$

where

$$x_{1a}(t, a) = \frac{\partial x_1(t, a)}{\partial a},$$

$$E = \text{const.}$$

where the Jacobian matrix J_k is defined by

$$J_k = J[\mathbf{f}(\mathbf{x}_k, t)]$$

with the ijth element

$$\Big[J_k\Big]_{ij} = \frac{\partial f_i(\mathbf{x})}{\partial x_j}\bigg|_{\mathbf{x}=\mathbf{x}_k(t)}.$$

Equation (8.217) can be expressed in the form

$$\dot{\mathbf{x}}_{k+1} = F_k(t)\mathbf{x}_{k+1} + \mathbf{y}_k(t), \tag{8.218}$$

where

$$F_k(t) = J_k,$$
$$\mathbf{y}_k(t) = \mathbf{f}(\mathbf{x}_k, t) - J_k\mathbf{x}_k.$$

The numerical solution to equation (8.218) can be obtained utilizing the method of complementary functions. The particular solution is the solution of the equation

$$\dot{\mathbf{p}}_{k+1} = F_k(t)\mathbf{p}_{k+1} + \mathbf{y}_k(t), \qquad \mathbf{p}_{k+1}(0) = 0, \tag{8.219}$$

where $\mathbf{p}_{k+1}$ is the vector

$$\mathbf{p}_{k+1} = (p_1^{k+1}, p_2^{k+1}, \ldots, p_{4n}^{k+1})^T$$

and the homogeneous solution is the solution of the equation

$$\dot{H}_{k+1} = F_k(t)H_{k+1}, \qquad H_{k+1}(0) = I, \tag{8.220}$$

where H_{k+1} is the matrix

$$H_{k+1} = \begin{bmatrix} h_{11}^{k+1} & h_{21}^{k+1} & \cdots & h_{(4n)1}^{k+1} \\ h_{12}^{k+1} & & & \vdots \\ \vdots & & & \\ h_{1(4n)}^{k+1} & & \cdots & h_{(4n)(4n)}^{k+1} \end{bmatrix}.$$

The solution is then given by the linear combination

$$\mathbf{x}_{k+1}(t) = \mathbf{p}_{k+1}(t) + H_{k+1}(t)\mathbf{c}_A, \tag{8.221}$$

where the vector of constants

$$\mathbf{c}_A = (A_1, A_2, \ldots, A_{4n})^T$$

is determined from the boundary conditions and the matrix equation

$$\begin{bmatrix} x_1(0) \\ x_2(0) \\ \vdots \\ x_{2n}(0) \\ \lambda_1(T_f) \\ \lambda_2(T_f) \\ \vdots \\ \lambda_{2n}(T_f) \end{bmatrix} = \begin{bmatrix} p_1^{k+1}(0) \\ p_2^{k+1}(0) \\ \vdots \\ p_{2n}^{k+1}(0) \\ p_{2n+1}^{k+1}(T_f) \\ p_{2n+2}^{k+1}(T_f) \\ \vdots \\ p_{4n}^{k+1}(T_f) \end{bmatrix} + \begin{bmatrix} h_{11}^{k+1}(0) & h_{21}^{k+1}(0) & \cdots & h_{(4n)1}^{k+1}(0) \\ h_{12}^{k+1}(0) & & & \vdots \\ \vdots & & & \\ h_{1(2n)}^{k+1}(0) & \cdots & & h_{(4n)(2n)}^{k+1}(0) \\ h_{1(2n+1)}^{k+1}(T_f) & h_{2(2n+1)}^{k+1}(T_f) & \cdots & h_{(4n)(2n+1)}^{k+1}(T_f) \\ h_{1(2n+2)}^{k+1}(T_f) & & & \vdots \\ \vdots & & & \\ h_{1(4n)}^{k+1}(T_f) & \cdots & & h_{(4n)(4n)}^{k+1}(T_f) \end{bmatrix} \begin{bmatrix} A_1 \\ A_2 \\ \vdots \\ \vdots \\ A_{4n} \end{bmatrix}. \tag{8.222}$$

The first approximation is obtained by an initial guess for the vector $\mathbf{x}_0(t)$. Equations (8.219) and (8.220) are integrated from time $t = 0$ to $t = T_f$ and values of $\mathbf{p}(t)$ and $H(t)$ are stored. The vector $\mathbf{c}_A$ is then evaluated from matrix equation (8.222) and the solution is obtained from equation (8.221). Alternately, only the boundary values of $\mathbf{p}(t)$ and $H(t)$ need be stored to evaluate the vector $\mathbf{c}_A$ and the full set of initial conditions of $\mathbf{x}_1(t)$ from equation (8.221). Equation (8.218) is then integrated from time $t = 0$ to $t = T_f$ to obtain $\mathbf{x}_1(t)$. The values of $\mathbf{x}_1(t)$ are stored and the above sequence is repeated to obtain the second approximation, etc. If the process is linear, the method of complementary functions becomes a one-sweep method with the elements of the matrix F constant and the particular solution equal to zero. Equations (8.215) and (8.218) are then identical with

$$\begin{aligned} \dot{\mathbf{x}} &= \mathbf{f}(\mathbf{x}, t) \\ &= F\mathbf{x}. \end{aligned} \tag{8.223}$$

Exercises

1. Show that the sum of the squares of the parameter estimate/observation sensitivity given by equation (8.69) is

$$\sum_{j=1}^{N} \left(\frac{\partial \hat{a}}{\partial b_j} \right)^2 = 1 \bigg/ \sum_{i=1}^{N} x_a^2(t_i, a).$$

2. Consider the scalar system

$$\dot{x} = -x + ay, \qquad x(0) = 0,$$
$$z = x + v,$$
$$E[v] = 0, \qquad E[v(t)v(\tau)] = r(t - \tau).$$

Using Mehra's method, show that the optimal input for estimating the parameter a is given by

$$y(t) = -(c_1/q)[\sin \omega t + \omega \cos \omega t],$$

where c_1 is an arbitrary constant and the largest eigenvalue

$$q = \frac{1}{r(1 + \omega^2)}$$

corresponds to the smallest root of the equation

$$\tan \omega T = -\omega.$$

See Mehra, References 8 and 9.

3. Consider the linear system, equation (8.13), used as an example for obtaining optimal inputs. Derive the two-point boundary-value equations using Pontryagin's maximum principle.

4. Derive the system identification equations for Exercise 3 using the Gauss–Newton method.

5. Verify the numerical results in the text for the example given by equation (8.13) using the methods of Exercises 3 and 4.

6. Derive the Euler–Lagrange equations for obtaining the optimal input of the van der Pol equation in Exercise 1, Chapter 7. Assume a is the unknown parameter to be estimated and that only x is observed.

7. Derive the equations in Exercise 6 using Pontryagin's maximum principle.

8. Obtain numerical results in Exercise 6 if the nominal value of a is -3.65, $b = 1.17$, $x(0) = 2$, and $\dot{x}(0) = 0$.

9. Derive the Euler–Lagrange equations for obtaining the optimal input of the mass attached to a nonlinear spring in Exercise 4, Chapter 7. Assume a is the unknown parameter to be estimated.

10. Derive the equations in Exercise 9 using Pontryagin's maximum principle.

11. Obtain numerical results in Exercise 9 if the nominal value of a is 0.1, $\omega_0 = 1$, $x(0) = 1$, and $\dot{x}(0) = 0$.

9

Additional Topics for Optimal Inputs

The determination of the Lagrange multiplier to be used in the calculation of optimal inputs for a quadratic performance criterion is nontrivial. In Section 9.1 an improved method for the numerical determination of optimal inputs is considered in which the Lagrange multiplier is evaluated simultaneously with the optimal input. Multiparameter optimal inputs are considered in Section 9.2. The trace of the information matrix is used as the performance criterion. In Section 9.3 observability, controllability, and identifiability are defined. Optimal inputs for systems with process noise are briefly discussed in Section 9.4, and eigenvalue problems are discussed in Section 9.5.

9.1. An Improved Method for the Numerical Determination of Optimal Inputs

The design of optimal inputs for linear and nonlinear system identification involves the maximization of a quadratic performance index subject to an input energy constraint. In the classical approach, a Lagrange multiplier is introduced whose value is an unknown constant. In the previous chapter, the Lagrange multiplier has been determined by plotting a curve of the Lagrange multiplier as a function of the critical interval length or a curve of the input energy versus the interval length. A new approach is presented in this section in which the Lagrange multiplier is introduced as a state variable and evaluated simultaneously with the optimal input (Reference 1). Numerical results are given for both a linear and a nonlinear dynamic system.

9.1.1. Introduction

The design of optimal inputs for linear system identification and nonlinear system identification has been discussed in Chapter 8. However, the numerical determination of the optimal inputs is far from trivial. The performance index for the optimal input is selected such that the sensitivity of the measured state variables to the unknown parameters is maximized subject to an input energy constraint. The performance index is maximized using Pontryagin's maximum principle. The solution requires the evaluation of a two-point boundary-value problem.

Using a quadratic performance criterion, the optimal input is determined such that the integral

$$M = \int_0^T x_a^2(t)\, dt \tag{9.1}$$

is maximized subject to the input energy constraint

$$E = \int_0^T y^2(t)\, dt, \tag{9.2}$$

where

$$x_a(t) = \frac{\partial x(t)}{\partial a} \tag{9.3}$$

and $x(t)$ is the state variable, $y(t)$ is the input, and a is an unknown system parameter. The performance index is maximized via the classical method by the maximization of the integral

$$J = \max_y \frac{1}{2} \int_0^T \left\{ x_a^2(t) - q\left[y^2(t) - \frac{E}{T} \right] \right\} dt, \tag{9.4}$$

where q is the Lagrange multiplier and is equal to a constant. The magnitude of the Lagrange multiplier must be selected such that the input energy constraint is satisfied.

Using Mehra's method, as described in Section 8.2.4, it was shown that for a linear system with homogeneous boundary conditions, the solution exists only for certain values of q which are the eigenvalues of the two-point boundary-value problem. The eigenvalues q are functions of the interval length T. For a fixed q, the critical length T can be determined by the integration of the Riccati matrix equation. When the elements of the Riccati matrix equation become very large, the critical length has been reached. By integrating for several values of q, a curve relating q to T can be obtained.

In Section 8.2.5 it was shown that for a linear system with nonhomogeneous boundary conditions, the performance index increases as the critical length is approached for a given value of q. The desired input energy is obtained by plotting a curve of input energy versus interval length T and finding the T corresponding to the desired input energy. The optimal input for a system with homogeneous boundary conditions is nearly identical to that of the system with nonhomogeneous boundary conditions when the input energy is the same and the latter has a small initial condition with terminal time near the critical length.

For both homogeneous and nonhomogeneous boundary conditions, the Lagrange multiplier q must be found by trial and error such that the input energy constraint is satisfied. In this section, a new approach for the numerical determination of optimal inputs is presented. The Lagrange multiplier q is introduced as a state variable. The solution simultaneously yields the optimal input and the value of q for which the input energy constraint is satisfied. The new approach is based on the method of solution of the isoperimetric problem (Reference 2) presented in Section 4.2.2. The method is applicable to both linear and nonlinear systems.

The performance criterion given by equation (9.4) maximizes the sensitivity of the state variable, $x(t)$, to the unknown parameter a. Assume that the measurements include additive white noise

$$z(t) = x(t) + v(t), \tag{9.5}$$

where $z(t)$ is the scalar measurement and $v(t)$ is a zero mean Gaussian white noise process:

$$\begin{aligned} E[v(t)] &= 0, \\ E[v(t)v^T(\tau)] &= R\delta(t-\tau) = \sigma_n^2\delta(t-\tau). \end{aligned} \tag{9.6}$$

The Fisher information matrix is given by

$$M = \int_0^T x_a^T R^{-1} x_a \, dt = \frac{1}{\sigma_n^2} \int_0^T x_a^2 \, dt. \tag{9.7}$$

The inverse of the Fisher information matrix, M^{-1}, is the Cramér–Rao lower bound. If $\sigma_n = 1$, as will be assumed in the discussions which follow, then equations (9.1) and (9.7) are identical.

The new approach for obtaining the optimal input will be illustrated using a first-order nonlinear dynamic system as an example. The method can easily be extended to the n-dimensional case, however.

9.1.2. A Nonlinear Example

Consider the nonlinear dynamic system given by

$$\dot{x} = -ax - bx^2 + y \tag{9.8}$$

with initial condition

$$x(0) = c, \tag{9.9}$$

where $x(t)$ is a scalar state variable, $y(t)$ is a scalar input, a is an unknown system parameter, and b is a known parameter. It is desired to maximize sensitivity equation (9.1) subject to input energy constraint (9.2). Performance index (9.3) is unchanged if expressed in the form

$$J = \max_{y} \frac{1}{2} \int_0^T [x_a^2(t) - qy^2(t)]\, dt + \frac{q(T)E}{2}. \tag{9.10}$$

Differentiating equations (9.8) and (9.9) with respect to a yields

$$\dot{x}_a = -x - ax_a - 2bxx_a, \qquad x_a(0) = 0. \tag{9.11}$$

Now the value of the Lagrange multiplier q for which input energy constraint (9.2) is satisfied can be obtained along with the optimal input by adjoining the differential constraint

$$\dot{q}(t) = 0 \tag{9.12}$$

with unknown initial condition and introducing the Lagrange multiplier $p_3(t)$.

Utilizing Pontryagin's maximum principle, the Hamiltonian function for performance index (9.10) subject to differential constraints (9.8), (9.11), and (9.12) is

$$\begin{aligned} H = \tfrac{1}{2}[-x_a^2 + qy^2] + p_1(-ax - bx^2 + y) \\ +p_2(-x - ax_a - 2bxx_a) + p_3 \times 0. \end{aligned} \tag{9.13}$$

The costate variables, p_1, p_2, and p_3, are solutions of the differential equations

$$\begin{aligned} \dot{p}_1 &= -\frac{\partial H}{\partial x} = (a + 2bx)p_1 + (1 + 2bx_a)p_2, \\ \dot{p}_2 &= -\frac{\partial H}{\partial x_a} = x_a + (a + 2bx)p_2, \\ \dot{p}_3 &= -\frac{\partial H}{\partial q} = \frac{-y^2}{2}. \end{aligned} \tag{9.14}$$

The input $y(t)$ that maximizes H is

$$\frac{\partial H}{\partial y} = qy + p_1, \qquad y = -\frac{1}{q} p_1. \tag{9.15}$$

The boundary conditions are

$$\begin{aligned} p_1(T) &= 0, & p_2(T) &= 0, \\ p_3(T) &= \frac{\partial}{\partial q}\left[-\frac{qE}{2}\right]_{t=T} = -\frac{E}{2}, & p_3(0) &= 0. \end{aligned} \tag{9.16}$$

The nonlinear two-point boundary-value equations and associated boundary conditions are then given by

$$\begin{aligned} \dot{x} &= -ax - bx^2 - \frac{1}{q} p_1, & x(0) &= c, \\ \dot{x}_a &= -x - ax_a - 2bxx_a, & x_a(0) &= 0, \\ \dot{q} &= 0, & p_3(0) &= 0, \\ \dot{p}_1 &= (a + 2bx)p_1 + (1 + 2bx_a)p_2, & p_1(T) &= 0, \\ \dot{p}_2 &= x_a + (a + 2bx)p_2, & p_2(T) &= 0, \\ \dot{p}_3 &= -\frac{1}{2q^2} p_1^2, & p_3(T) &= -\frac{E}{2}. \end{aligned} \tag{9.17}$$

These equations differ from the classical equations by the addition of the differential equations $\dot{q}$ and $\dot{p}_3$. The unknown initial conditions are $p_1(0)$, $p_2(0)$, and $q(0)$.

9.1.3. Solution via Newton–Raphson Method

Using the Newton–Raphson method, we assume the unknown initial conditions are given by

$$\begin{aligned} p_1(0) &= c_1, \\ p_2(0) &= c_2, \\ q(0) &= c_3. \end{aligned} \tag{9.18}$$

Expanding the boundary conditions

$$\begin{aligned} p_1(T, c_1, c_2, c_3) &= 0, \\ p_2(T, c_1, c_2, c_3) &= 0, \\ p_3(T, c_1, c_2, c_3) &= -E/2 \end{aligned} \tag{9.19}$$

in a Taylor's series around the kth approximation, retaining only the linear terms, and rearranging yields

$$\begin{bmatrix} c_1^{k+1} \\ c_2^{k+1} \\ c_3^{k+1} \end{bmatrix} = \begin{bmatrix} c_1^{k} \\ c_2^{k} \\ c_3^{k} \end{bmatrix} - \begin{bmatrix} p_{1c_1} & p_{1c_2} & p_{1c_3} \\ p_{2c_1} & p_{2c_2} & p_{2c_3} \\ p_{3c_1} & p_{3c_2} & p_{3c_3} \end{bmatrix}^{-1} \begin{bmatrix} p_1(T, c_1^k, c_2^k, c_3^k) \\ p_2(T, c_1^k, c_2^k, c_3^k) \\ p_3(T, c_1^k, c_2^k, c_3^k) + E/2 \end{bmatrix} \tag{9.20}$$

where

$$p_{ic_j} = \frac{\partial p_i}{\partial c_j}. \tag{9.21}$$

The equations for the p_{ic_j}'s are obtained by differentiating equations (9.17) with respect to c_1, c_2, and c_3:

$$\begin{aligned}
\dot{x}_{c_1} &= -ax_{c_1} - 2bxx_{c_1} - \frac{1}{q} p_{1c_1} + \frac{1}{q^2} q_{c_1} p_1, & x_{c_1}(0) &= 0, \\
\dot{x}_{ac_1} &= -x_{c_1} - ax_{ac_1} - 2bx_{c_1}x_a - 2bxx_{ac_1}, & x_{ac_1}(0) &= 0, \\
\dot{q}_{c_1} &= 0, & q_{c_1}(0) &= 0, \\
\dot{p}_{1c_1} &= (a + 2bx)p_{1c_1} + 2bx_{c_1}p_1 & & \\
&\quad +(1 + 2bx_a)p_{2c_1} + 2bx_{ac_1}p_2, & p_{1c_1}(0) &= 1, \\
\dot{p}_{2c_1} &= x_{ac_1} + (a + 2bx)p_{2c_1} + 2bx_{c_1}p_2, & p_{2c_1}(0) &= 0, \\
\dot{p}_{3c_1} &= -\frac{1}{q^2} p_{1c_1}p_1 + \frac{1}{q^3} q_{c_1}p_1^2, & p_{3c_1}(0) &= 0,
\end{aligned} \tag{9.22}$$

$$\begin{aligned}
\dot{x}_{c_2} &= -ax_{c_2} - 2bxx_{c_2} - \frac{1}{q} p_{1c_2} + \frac{1}{q^2} q_{c_2} p_1, & x_{c_2}(0) &= 0, \\
\dot{x}_{ac_2} &= -x_{c_2} - ax_{ac_2} - 2bx_{c_2}x_a - 2bxx_{ac_2}, & x_{ac_2}(0) &= 0, \\
\dot{q}_{c_2} &= 0, & q_{c_2}(0) &= 0, \\
\dot{p}_{1c_2} &= (a + 2bx)p_{1c_2} + 2bx_{c_2}p_1 & & \\
&\quad +(1 + 2bx_a)p_{2c_2} + 2bx_{ac_2}p_2, & p_{1c_2}(0) &= 0, \\
\dot{p}_{2c_2} &= x_{ac_2} + (a + 2bx)p_{2c_2} + 2bx_{c_2}p_2, & p_{2c_2}(0) &= 1, \\
\dot{p}_{3c_2} &= -\frac{1}{q^2} p_{1c_2}p_1 + \frac{1}{q^3} q_{c_2}p_1^2, & p_{3c_2}(0) &= 0,
\end{aligned} \tag{9.23}$$

$$
\begin{aligned}
\dot{x}_{c_3} &= -ax_{c_3} - 2bxx_{c_3} - \frac{1}{q} p_{1c_3} + \frac{1}{q^2} q_{c_3} p_1, & x_{c_3}(0) &= 0, \\
\dot{x}_{ac_3} &= -x_{c_3} - ax_{ac_3} - 2bx_{c_3}x_a - 2bxx_{ac_3}, & x_{ac_3}(0) &= 0, \\
\dot{q}_{c_3} &= 0, & q_{c_3}(0) &= 1, \\
\dot{p}_{1c_3} &= (a + 2bx)p_{1c_3} + 2bx_{c_3}p_1 \\
&\quad + (1 + 2bx_a)p_{2c_3} + 2bx_{ac_3}p_2, & p_{1c_3}(0) &= 0, \\
\dot{p}_{2c_3} &= x_{ac_3} + (a + 2bx)p_{2c_3} + 2bx_{c_3}p_2, & p_{2c_3}(0) &= 0, \\
\dot{p}_{3c_3} &= -\frac{1}{q^2} p_{1c_3}p_1 + \frac{1}{q^3} q_{c_3} p_1^2, & p_{3c_3}(0) &= 0.
\end{aligned}
\tag{9.24}
$$

To obtain a numerical solution, an initial guess is made for initial conditions (9.18). Initial-value equations (9.17) and (9.22)–(9.24) are then integrated from time $t = 0$ to $t = T$. A new set of values for c_1, c_2, and c_3 are calculated from equation (9.20) and the above sequence is repeated to obtain the second approximation, etc.

9.1.4. Numerical Results and Discussion

Numerical results were obtained using a fourth-order Runge–Kutta method with grid intervals of 1/100 sec.

The parameters of equation (9.8) were assumed to be as follows:

$$
\begin{aligned}
a &= 0.1, & x(0) &= 0.1, \\
q &= 0.075, & T &= 1.
\end{aligned}
\tag{9.25}
$$

For $b = 0$, dynamic system equation (9.8) is linear. From previous numerical results, it is known that (see Section 8.2.5)

$$
E = 17.2742 \qquad \text{for} \quad q = 0.075. \tag{9.26}
$$

Using the Newton–Raphson method, we initialize the algorithm as follows:

$$
\begin{aligned}
p_1(0) &= c_1 = 0, \\
p_2(0) &= c_2 = 0, \\
q(0) &= c_3 = 0.085,
\end{aligned}
\tag{9.27}
$$

with input energy constraint $E = 17.2742$ introduced via boundary condition

$$
p_3(T) = -E/2. \tag{9.28}
$$

Table 9.1. Convergence of q and $E(q)$ for Linear Dynamic System, $b = 0$

Iteration number	q	$E(q)$
0	0.085	5.59959×10^{-3}
1	0.085	5331.31
2	0.0784733	1124.9
3	0.0754061	267.123
4	0.0742463	73.5903
5	0.0742607	28.0558
6	0.0747793	18.372
7	0.0749918	17.2937
8	0.075	17.2742

Final values: $p_1(0) = -0.614874$, $p_2(0) = 0.858588$
$\frac{1}{2}\int_0^T (x_a^2 - qy^2)\,dt = 3.07437 \times 10^{-2}$

At the end of the first iteration, q is reset to 0.085 so that initial estimates of $p_1(0)$ and $p_2(0)$ can be obtained for a given q. Without this initial reset, the algorithm diverges.

Table 9.1 shows the convergence of q and $E(q)$. After eight iterations the solution is exact to within six digits.

For $b = 0.1$, dynamic system equation (9.8) is nonlinear. From previous numerical results, it is known that (see Section 8.3.1)

$$E = 0.700069 \qquad \text{for} \quad q = 0.075. \tag{9.29}$$

Initializing the algorithm as above (9.27) with input energy constraint $E = 0.700069$, the convergence of q and $E(q)$ is shown in Table 9.2. After six iterations, the solution is exact to almost six digits. The final optimal inputs for $b = 0$ and $b = 0.1$ are given in Table 9.3.

The numerical results show that for both linear and nonlinear dynamic systems, the optimal input can be obtained directly for a given input energy E without having to go through the intermediate step of determining the Lagrange multiplier q for a given interval length T. Only an initial estimate of q is required, since the solution converges to the proper q and the optimal input simultaneously. Additional results show that convergence for the nonlinear dynamic system is obtained for an initial estimate of q at least twice the final value of q.

Table 9.2. Convergence of q and $E(q)$ for Nonlinear Dynamic System, $b = 0.1$

Iteration number	q	$E(q)$
0	0.085	5.49326×10^{-3}
1	0.085	13.5278
2	0.0827309	3.79548
3	0.0785015	1.26587
4	0.075788	0.749698
5	0.0750276	0.70042
6	0.0750001	0.700069

Final values: $p_1(0) = -0.125058$, $p_2(0) = 0.18529$
$\frac{1}{2}\int_0^T (x_a^2 - qy^2)\, dt = 8.08180 \times 10^{-3}$

The region of convergence can be increased by using the performance index

$$J_1 = \max_y \frac{1}{2}\int_0^T [(x_a^2 - qy^2)]\, dt + \frac{q(T)E}{2} - k\frac{q^2(0)}{2} \tag{9.30}$$

subject to differential constraints (9.8), (9.11), and (9.12). For the nonlinear dynamic system, convergence is obtained for an initial estimate of q at least 10 times the final value of q when $k = 1$.

Table 9.3. Optimal Inputs, Converged Solution for $y(t)$

Time	Linear system $b = 0$	Nonlinear system $b = 0.1$
0	8.19833	1.66744
0.1	7.1251	1.44049
0.2	6.03681	1.21423
0.3	4.94906	0.990971
0.4	3.88549	0.77483
0.5	2.87637	0.571385
0.6	1.95716	0.387323
0.7	1.16718	0.230114
0.8	0.548439	0.107709
0.9	0.144563	2.82775×10^{-2}
1.0	-4.82978×10^{-7}	9.57485×10^{-8}

9.2. Multiparameter Optimal Inputs

In this section the design of optimal inputs for systems with multiple unknown parameters are considered. The trace of the information matrix is used as the performance criterion. Examples are given for two parameter estimations with a single-input, single-output system; single-input, two-output system; and weighted optimal inputs.

9.2.1. Optimal Inputs for Vector Parameter Estimation

In the design of optimal inputs for estimating more than one parameter, a suitable scalar function of the Fisher information matrix M must be selected as the performance criterion. A criterion which is often used is the trace of the matrix M or the weighted trace (Reference 3)

$$\mathrm{Tr}(M) \text{ or } \mathrm{Tr}(WM) \tag{9.31}$$

wherein the sum of the diagonal elements of the matrix M or WM is maximized. This approach will be discussed in detail here. Other measures of performance are as follows (Reference 4):

A-optimality: $\mathrm{Tr}(M^{-1})$, minimize the average variance of the parameters.

E-optimality: $\lambda_{\max}$, minimize the maximum eigenvalue of M^{-1}.

D-optimality: $\det M^{-1}$, minimize the determinant of M^{-1}.

Using Mehra's approach (Reference 3), consider the time-invariant linear system

$$\dot{\mathbf{x}}(t) = F\mathbf{x}(t) + G\mathbf{y}(t), \tag{9.32}$$

$$\mathbf{z}(t) = H\mathbf{x}(t) + \mathbf{v}(t), \tag{9.33}$$

where $\mathbf{x}$ is an $n\times 1$ state vector, $\mathbf{y}$ is an $m\times 1$ control vector, and $\mathbf{z}$ is a $r\times 1$ measurement vector. F, G, and H are, respectively, $n\times n$, $n\times m$, and $r\times n$ matrices. The vector $\mathbf{v}$ is a zero mean Gaussian white noise process

$$E[\mathbf{v}(t)] = 0, \tag{9.34}$$

$$E[\mathbf{v}(t)\mathbf{v}^T(\tau)] = R\delta(t-\tau). \tag{9.35}$$

Let $\mathbf{p}$ denote a $k\times 1$ vector of unknown parameters. The optimal input is to be determined such that the trace of the $k\times k$ Fisher information matrix

$$\mathrm{Tr}(M) = \int_0^T \mathrm{Tr}\,[X_p{}^T H^T R^{-1} H X_p]\,dt \tag{9.36}$$

is maximized subject to the input energy constraint

$$\int_0^T y^T y\,dt = E, \tag{9.37}$$

where X_p is a $n \times k$ parameter influence coefficient matrix with ij component $\partial x_i/\partial p_j$. In order to maximize $\mathrm{Tr}(M)$, define the augmented $[(k+1)n \times 1]$ state vector $\mathbf{x}_A$

$$\mathbf{x}_A = \begin{bmatrix} \mathbf{x} \\ \mathbf{x}_{p_1} \\ \vdots \\ \mathbf{x}_{p_k} \end{bmatrix} \tag{9.38}$$

where

$$\mathbf{x}_{p_i} = \frac{\partial \mathbf{x}}{\partial p_i}. \tag{9.39}$$

The state equation is then

$$\dot{\mathbf{x}} = F_A \mathbf{x}_A + G_A \mathbf{y}, \tag{9.40}$$

where the $(k+1)n \times (k+1)n$ matrix F_A and $(k+1)n \times m$ matrix G_A are given by

$$F_A = \left[\begin{array}{c|c|c|c} F & 0 & \cdots & 0 \\ \hline F_{p_1} & F & \ddots & \\ \hline \vdots & & \ddots & \\ \hline F_{p_k} & & & F \end{array}\right], \tag{9.41}$$

$$G_A = \begin{bmatrix} G \\ G_{p_1} \\ \vdots \\ G_{p_k} \end{bmatrix}, \tag{9.42}$$

$$F_{p_i} = \frac{\partial F}{\partial p_i}, \qquad G_{p_i} = \frac{\partial G}{\partial p_i}. \tag{9.43}$$

Then the expression for the Tr(M), equation (9.36), can be written in the form

$$\mathrm{Tr}(M) = \int_0^T \mathbf{x}_A{}^T H_A{}^T R_A{}^{-1} H_A \mathbf{x}_A \, dt, \tag{9.44}$$

where the $rk \times (k+1)n$ matrix H_A and the $rk \times rk$ matrix $R_A{}^{-1}$ are given by

$$H_A = \left[\begin{array}{c|c|c|c} 0 & H & 0 & 0 \\ \hline \vdots & & \ddots & \vdots \\ \hline 0 & 0 & 0 & H \end{array}\right], \tag{9.45}$$

$$R_A{}^{-1} = \left[\begin{array}{c|c|c|c} R^{-1} & 0 & \cdots & 0 \\ \hline 0 & R^{-1} & & 0 \\ \hline \vdots & & \ddots & \\ \hline 0 & 0 & \cdots & R^{-1} \end{array}\right]. \tag{9.46}$$

The performance index can then be expressed as the return, J:

$$J = \max_{\mathbf{y}} \frac{1}{2} \int_0^T [\mathbf{x}_A{}^T H_A{}^T R_A{}^{-1} H_A \mathbf{x}_A - q\mathbf{y}^T\mathbf{y}] \, dt. \tag{9.47}$$

Utilizing Pontryagin's maximum principle, the Hamiltonian function is

$$H = \tfrac{1}{2}[-\mathbf{x}_A{}^T H_A{}^T R_A{}^{-1} H_A \mathbf{x}_A + q\mathbf{y}^T\mathbf{y}] + \lambda^T[F_A\mathbf{x}_A + G_A\mathbf{y}]. \tag{9.48}$$

The costate vector $\lambda(t)$ is the solution of the vector differential equation

$$\begin{aligned} \dot{\boldsymbol{\lambda}} &= -\left[\frac{\partial H}{\partial \mathbf{x}_A}\right]^T \\ &= H_A{}^T R_A{}^{-1} H_A \mathbf{x}_A - F_A{}^T \boldsymbol{\lambda}. \end{aligned} \tag{9.49}$$

The input $\mathbf{y}(t)$ that maximizes H is

$$\frac{\partial H}{\partial \mathbf{y}} = q\mathbf{y} + G_A{}^T\boldsymbol{\lambda} = 0, \qquad \mathbf{y} = -\frac{1}{q} G_A{}^T \boldsymbol{\lambda}. \tag{9.50}$$

The two-point boundary-value problem is then

$$\begin{bmatrix} \dot{\mathbf{x}}_A \\ \dot{\boldsymbol{\lambda}} \end{bmatrix} = \begin{bmatrix} F_A & -(1/q)G_A G_A{}^T \\ H_A{}^T R_A{}^{-1} H_A & -F_A{}^T \end{bmatrix} \begin{bmatrix} \mathbf{x}_A \\ \boldsymbol{\lambda} \end{bmatrix} \tag{9.51}$$

with boundary conditions

$$\mathbf{x}_A(0) = \begin{bmatrix} \mathbf{x}(0) \\ 0 \\ \vdots \\ 0 \end{bmatrix}, \qquad \boldsymbol{\lambda}(T) = 0. \tag{9.52}$$

The two-point boundary-value problem can be solved using the method of complementary functions.

9.2.2. Example of Optimal Inputs for Two-Parameter Estimation

Consider the single-input, single-output, second-order system

$$\dot{x}_1 = -ax_1 + bx_2, \qquad x_1(0) = x_{10} \tag{9.53}$$

$$\dot{x}_2 = -cx_1 - dx_2 + y, \qquad x_2(0) = x_{20} \tag{9.54}$$

with measurements

$$z = x_2 + v, \tag{9.55}$$

where v is a zero mean Gaussian white noise process

$$E[v(t)] = 0, \qquad E[v(t)v^T(\tau)] = R\delta(t - \tau) = \sigma^2\delta(t - \tau). \tag{9.56}$$

It is desired to determine the optimal input for estimating parameters b and c.

Equations (9.53) and (9.54) can be expressed in vector form by the time-invariant linear system

$$\dot{\mathbf{x}}(t) = F\mathbf{x}(t) + Gy(t), \tag{9.57}$$

where

$$\mathbf{x}(t) = \begin{bmatrix} x_1(t) \\ x_2(t) \end{bmatrix}, \qquad F = \begin{bmatrix} -a & b \\ -c & -d \end{bmatrix}, \qquad G = \begin{bmatrix} 0 \\ 1 \end{bmatrix}. \tag{9.58}$$

The measurements are

$$z(t) = H\mathbf{x}(t) + v(t), \tag{9.59}$$

where

$$H = [0 \quad 1]. \tag{9.60}$$

The optimal input is to be determined such that the trace of the Fisher information matrix

$$\mathrm{Tr}(M) = \int_0^T \mathrm{Tr}\,[X_p{}^T H^T R^{-1} H X_p]\,dt \tag{9.61}$$

is maximized subject to the input energy constraint

$$\int_0^T y^2\,dt = E, \tag{9.62}$$

where

$$X_p = \begin{bmatrix} \dfrac{\partial x_1}{\partial b} & \dfrac{\partial x_1}{\partial c} \\ \dfrac{\partial x_2}{\partial b} & \dfrac{\partial x_2}{\partial c} \end{bmatrix} = \begin{bmatrix} x_{1b} & x_{1c} \\ x_{2b} & x_{2c} \end{bmatrix}. \tag{9.63}$$

Substituting matrices X_p and H into equation (9.61), and replacing R by σ^2, it can be shown that

$$\mathrm{Tr}(M) = \frac{1}{\sigma^2}\int_0^T \mathrm{Tr}\begin{bmatrix} x_{2b}^2 & x_{2b}x_{2c} \\ x_{2b}x_{2c} & x_{2c}^2 \end{bmatrix} dt. \tag{9.64}$$

Taking the partial derivatives of equation (9.57) with respect to b and c yields

$$\dot{\mathbf{x}}_b = F\mathbf{x}_b + F_b\mathbf{x}, \tag{9.65}$$

$$\dot{\mathbf{x}}_c = F\mathbf{x}_c + F_c\mathbf{x}, \tag{9.66}$$

where

$$F_b = \begin{bmatrix} 0 & 1 \\ 0 & 0 \end{bmatrix}, \quad F_c = \begin{bmatrix} 0 & 0 \\ -1 & 0 \end{bmatrix}. \tag{9.67}$$

Define the augmented 6×1 state vector

$$\mathbf{x}_A = \begin{bmatrix} \mathbf{x} \\ \mathbf{x}_b \\ \mathbf{x}_c \end{bmatrix}. \tag{9.68}$$

The state equation is then

$$\dot{\mathbf{x}}_A = F_A\mathbf{x}_A + G_A y, \tag{9.69}$$

where the 6×6 matrix F_A and the 6×1 vector G_A are given by

$$F_A = \begin{bmatrix} F & 0 & 0 \\ F_b & F & 0 \\ F_c & 0 & F \end{bmatrix} = \begin{bmatrix} -a & b & 0 & 0 & 0 & 0 \\ -c & -d & 0 & 0 & 0 & 0 \\ 0 & 1 & -a & b & 0 & 0 \\ 0 & 0 & -c & -d & 0 & 0 \\ 0 & 0 & 0 & 0 & -a & b \\ -1 & 0 & 0 & 0 & -c & -d \end{bmatrix}, \tag{9.70}$$

$$G_A{}^T = [G \quad 0 \quad 0] = [0 \quad 1 \quad 0 \quad 0 \quad 0 \quad 0]. \tag{9.71}$$

The trace of the information matrix becomes

$$\mathrm{Tr}(M) = \int_0^T \mathbf{x}_A{}^T H_A{}^T R_A{}^{-1} H_A \mathbf{x}_A \, dt, \tag{9.72}$$

where the 2×6 matrix H_A and 2×2 matrix R_A are given by

$$H_A = \begin{bmatrix} 0 & H & 0 \\ 0 & 0 & H \end{bmatrix} = \begin{bmatrix} 0 & 0 & 0 & 1 & 0 & 0 \\ 0 & 0 & 0 & 0 & 0 & 1 \end{bmatrix}, \tag{9.73}$$

$$R_A = \begin{bmatrix} R^{-1} & 0 \\ 0 & R^{-1} \end{bmatrix} = \frac{1}{\sigma^2} \begin{bmatrix} 1 & 0 \\ 0 & 1 \end{bmatrix}. \tag{9.74}$$

Substituting the vector $\mathbf{x}_A$ and matrices H_A and R_A into equation (9.72), it can be shown that as expected

$$\int_0^T \mathbf{x}_A{}^2 H_A{}^T R_A{}^{-1} H_A \mathbf{x}_A \, dt = \frac{1}{\sigma^2} \int_0^T (x_{2b}^2 + x_{2c}^2)\, dt. \tag{9.75}$$

The optimal input is obtained by solving the two-point boundary-value problem given by equations (9.51) and (9.52). By making the appropriate substitutions, equation (9.51) becomes

$$\begin{bmatrix} \dot{x}_1 \\ \dot{x}_2 \\ \dot{x}_{1b} \\ \dot{x}_{2b} \\ \dot{x}_{1c} \\ \dot{x}_{2c} \\ \dot{\lambda}_1 \\ \dot{\lambda}_2 \\ \dot{\lambda}_3 \\ \dot{\lambda}_4 \\ \dot{\lambda}_5 \\ \dot{\lambda}_6 \end{bmatrix} = \left[\begin{array}{cccccc|cccccc} -a & b & 0 & 0 & 0 & 0 & 0 & 0 & 0 & 0 & 0 & 0 \\ -c & -d & 0 & 0 & 0 & 0 & 0 & -1/q & 0 & 0 & 0 & 0 \\ 0 & 1 & -a & b & 0 & 0 & 0 & 0 & 0 & 0 & 0 & 0 \\ 0 & 0 & -c & -d & 0 & 0 & 0 & 0 & 0 & 0 & 0 & 0 \\ 0 & 0 & 0 & 0 & -a & b & 0 & 0 & 0 & 0 & 0 & 0 \\ -1 & 0 & 0 & 0 & -c & -d & 0 & 0 & 0 & 0 & 0 & 0 \\ \hline 0 & 0 & 0 & 0 & 0 & 0 & a & c & 0 & 0 & 0 & 1 \\ 0 & 0 & 0 & 0 & 0 & 0 & -b & d & -1 & 0 & 0 & 0 \\ 0 & 0 & 0 & 0 & 0 & 0 & 0 & 0 & a & c & 0 & 0 \\ 0 & 0 & 0 & 1/\sigma^2 & 0 & 0 & 0 & 0 & -b & d & 0 & 0 \\ 0 & 0 & 0 & 0 & 0 & 0 & 0 & 0 & 0 & 0 & a & c \\ 0 & 0 & 0 & 0 & 0 & 1/\sigma^2 & 0 & 0 & 0 & 0 & -b & d \end{array}\right] \begin{bmatrix} x_1 \\ x_2 \\ x_{1b} \\ x_{2b} \\ x_{1c} \\ x_{2c} \\ \lambda_1 \\ \lambda_2 \\ \lambda_3 \\ \lambda_4 \\ \lambda_5 \\ \lambda_6 \end{bmatrix}. \tag{9.76}$$

Some difficulties in identifying the parameters b and c in the above example are discussed in Section 9.3. Parameter estimation is possible if both state variables, x_1 and x_2, can be measured. The equations for the two-output system are given in the next section followed by an example of weighted optimal inputs.

9.2.3. Example of Optimal Inputs for a Single-Input, Two-Output System

Consider the example of the previous section with two outputs, x_1 and x_2, instead of only one. Equations (9.57) and (9.58) are the same, but the measurements now become

$$\mathbf{z}(t) = H\mathbf{x}(t) + \mathbf{v}(t), \tag{9.77}$$

where

$$H = \begin{bmatrix} 1 & 0 \\ 0 & 1 \end{bmatrix}, \tag{9.78}$$

$$E[\mathbf{v}(t)] = 0, \qquad E[\mathbf{v}(t)\mathbf{v}^T(\tau)] = R\delta(t - \tau), \tag{9.79}$$

$$R = \begin{bmatrix} \sigma_1^2 & 0 \\ 0 & \sigma^2 \end{bmatrix}. \tag{9.80}$$

The trace of the information matrix is

$$\begin{aligned} \mathrm{Tr}(M) &= \int_0^T \mathrm{Tr}\,[X_p^T H^T R^{-1} H X_p]\,dt \\ &= \int_0^T \mathrm{Tr}\begin{bmatrix} \dfrac{x_{1b}^2}{\sigma_1^2} + \dfrac{x_{2b}^2}{\sigma_2^2} & \dfrac{x_{1b}x_{1c}}{\sigma_1^2} + \dfrac{x_{2b}x_{2c}}{\sigma_2^2} \\ \dfrac{x_{1b}x_{1c}}{\sigma_1^2} + \dfrac{x_{2b}x_{2c}}{\sigma_2^2} & \dfrac{x_{1c}^2}{\sigma_1^2} + \dfrac{x_{2c}^2}{\sigma_2^2} \end{bmatrix} dt, \end{aligned} \tag{9.81}$$

where X_p is defined in equation (9.63). The augmented state vector and state equation are defined by equations (9.68)–(9.71). Equation (9.72) then becomes

$$\begin{aligned} \mathrm{Tr}(M) &= \int_0^T \mathbf{x}_A^T H_A^T R_A^{-1} H_A \mathbf{x}_A\,dt \\ &= \int_0^T \left[\frac{x_{1b}^2}{\sigma_1^2} + \frac{x_{2b}^2}{\sigma_2^2} + \frac{x_{1c}^2}{\sigma_1^2} + \frac{x_{2c}^2}{\sigma_2^2}\right] dt, \end{aligned} \tag{9.82}$$

where the 4×6 matrix H_A and the 4×4 matrix R_A^{-1} are given by

$$H_A = \begin{bmatrix} 0 & H & 0 \\ 0 & 0 & H \end{bmatrix} = \begin{bmatrix} 0 & 0 & 1 & 0 & 0 & 0 \\ 0 & 0 & 0 & 1 & 0 & 0 \\ 0 & 0 & 0 & 0 & 1 & 0 \\ 0 & 0 & 0 & 0 & 0 & 1 \end{bmatrix}, \tag{9.83}$$

$$R_A^{-1} = \begin{bmatrix} R^{-1} & 0 \\ 0 & R^{-1} \end{bmatrix} = \begin{bmatrix} 1/\sigma_1^2 & 0 & 0 & 0 \\ 0 & 1/\sigma_2^2 & 0 & 0 \\ 0 & 0 & 1/\sigma_1^2 & 0 \\ 0 & 0 & 0 & 1/\sigma_2^2 \end{bmatrix}. \tag{9.84}$$

Only the matrix $H_A^T R_A^{-1} H_A$ in the two-point boundary-value equation

$$\begin{bmatrix} \dot{\mathbf{x}}_A \\ \dot{\boldsymbol{\lambda}} \end{bmatrix} = \begin{bmatrix} F_A & -(1/q)G_A G_A^T \\ H_A^T R_A^{-1} H_A & -F_A^T \end{bmatrix} \begin{bmatrix} \mathbf{x}_A \\ \boldsymbol{\lambda} \end{bmatrix} \tag{9.85}$$

is changed. Multiplying out the terms in the matrix $H_A^T R_A^{-1} H_A$ gives us

$$H_A^T R^{-1} H_A = \begin{bmatrix} 0 & 0 & 0 & 0 & 0 & 0 \\ 0 & 0 & 0 & 0 & 0 & 0 \\ 0 & 0 & 1/\sigma_1^2 & 0 & 0 & 0 \\ 0 & 0 & 0 & 1/\sigma_2^2 & 0 & 0 \\ 0 & 0 & 0 & 0 & 1/\sigma_1^2 & 0 \\ 0 & 0 & 0 & 0 & 0 & 1/\sigma_2^2 \end{bmatrix}. \tag{9.86}$$

Equation (9.85) then becomes

$$\begin{bmatrix} \dot{x}_1 \\ \dot{x}_2 \\ \dot{x}_{1b} \\ \dot{x}_{2b} \\ \dot{x}_{1c} \\ \dot{x}_{2c} \\ \dot{\lambda}_1 \\ \dot{\lambda}_2 \\ \dot{\lambda}_3 \\ \dot{\lambda}_4 \\ \dot{\lambda}_5 \\ \dot{\lambda}_6 \end{bmatrix} = \left[\begin{array}{cccccc|cccccc} -a & b & 0 & 0 & 0 & 0 & 0 & 0 & 0 & 0 & 0 & 0 \\ -c & -d & 0 & 0 & 0 & 0 & 0 & -1/q & 0 & 0 & 0 & 0 \\ 0 & 1 & -a & b & 0 & 0 & 0 & 0 & 0 & 0 & 0 & 0 \\ 0 & 0 & -c & -d & 0 & 0 & 0 & 0 & 0 & 0 & 0 & 0 \\ 0 & 0 & 0 & 0 & -a & b & 0 & 0 & 0 & 0 & 0 & 0 \\ -1 & 0 & 0 & 0 & -c & -d & 0 & 0 & 0 & 0 & 0 & 0 \\ \hline 0 & 0 & 0 & 0 & 0 & 0 & a & c & 0 & 0 & 0 & 1 \\ 0 & 0 & 0 & 0 & 0 & 0 & -b & d & -1 & 0 & 0 & 0 \\ 0 & 0 & 1/\sigma_1^2 & 0 & 0 & 0 & 0 & 0 & a & c & 0 & 0 \\ 0 & 0 & 0 & 1/\sigma_2^2 & 0 & 0 & 0 & 0 & -b & d & 0 & 0 \\ 0 & 0 & 0 & 0 & 1/\sigma_1^2 & 0 & 0 & 0 & 0 & 0 & a & c \\ 0 & 0 & 0 & 0 & 0 & 1/\sigma_2^2 & 0 & 0 & 0 & 0 & -b & d \end{array}\right] \begin{bmatrix} x_1 \\ x_2 \\ x_{1b} \\ x_{2b} \\ x_{1c} \\ x_{2c} \\ \lambda_1 \\ \lambda_2 \\ \lambda_3 \\ \lambda_4 \\ \lambda_5 \\ \lambda_6 \end{bmatrix}. \tag{9.87}$$

9.2.4. Example of Weighted Optimal Inputs

Using the weighted trace of the information matrix as the performance criterion, equation (9.36) becomes (Reference 5)

$$\mathrm{Tr}(WM) = \mathrm{Tr}\,(W^{1/2}MW^{1/2})$$
$$= \int_0^T \mathrm{Tr}[(X_p W^{1/2})^T H^T R^{-1} X_p W^{1/2}]\,dt, \tag{9.88}$$

where

$$W = \mathrm{diag}(w_1, w_2, \ldots, w_k). \tag{9.89}$$

Since multiplication by $W^{1/2}$ represents a column operation, this is equivalent to replacing

$$\frac{\partial \mathbf{x}}{\partial p_i} \text{ by } w_i^{1/2} \frac{\partial \mathbf{x}}{\partial p_i} \qquad i = 1, \ldots, k. \tag{9.90}$$

Taking the derivative of equation (9.32) with respect to p_i and multiplying both sides of the equation by $w_i^{1/2}$ yields

$$w_i^{1/2} \frac{\partial \dot{\mathbf{x}}}{\partial p_i} = w_i^{1/2} \frac{\partial F}{\partial p_i} \mathbf{x} + w_i^{1/2} \frac{\partial \mathbf{x}}{\partial p_i} F + w_i^{1/2} \frac{\partial G}{\partial p_i} \mathbf{y}. \tag{9.91}$$

Consider the example of Section 9.2.2. The weighted augmented state vector corresponding to equation (9.68) is

$$\mathbf{x}_A = \begin{bmatrix} \mathbf{x} \\ w_1^{1/2}\mathbf{x}_b \\ w_2^{1/2}\mathbf{x}_c \end{bmatrix}. \tag{9.92}$$

The state equation is

$$\dot{\mathbf{x}}_A = F_A \mathbf{x}_A + G_A y$$
$$= \begin{bmatrix} F & 0 & 0 \\ w_1^{1/2} F_b & F & 0 \\ w_2^{1/2} F_c & 0 & F \end{bmatrix} \begin{bmatrix} \mathbf{x} \\ w_1^{1/2}\mathbf{x}_b \\ w_2^{1/2}\mathbf{x}_c \end{bmatrix} + \begin{bmatrix} G \\ 0 \\ 0 \end{bmatrix} y. \tag{9.93}$$

Equation (9.88) can then be expressed in the form

$$\mathrm{Tr}(WM) = \int_0^T \mathbf{x}_A{}^T H_A{}^T R^{-1} H_A \mathbf{x}_A\,dt, \tag{9.94}$$

where

$$H_A = \begin{bmatrix} 0 & H & 0 \\ 0 & 0 & H \end{bmatrix}, \tag{9.95}$$

$$R_A = \begin{bmatrix} R^{-1} & 0 \\ 0 & R^{-1} \end{bmatrix}. \tag{9.96}$$

The two-point boundary-value equation is

$$\begin{bmatrix} \dot{x}_1 \\ \dot{x}_2 \\ w_1^{1/2}\dot{x}_{1b} \\ w_1^{1/2}\dot{x}_{2b} \\ w_2^{1/2}\dot{x}_{1c} \\ w_2^{1/2}\dot{x}_{2c} \\ \dot{\lambda}_1 \\ \dot{\lambda}_2 \\ \dot{\lambda}_3 \\ \dot{\lambda}_4 \\ \dot{\lambda}_5 \\ \dot{\lambda}_6 \end{bmatrix} = \left[\begin{array}{cccccc|cccccc} -a & b & 0 & 0 & 0 & 0 & 0 & 0 & 0 & 0 & 0 & 0 \\ -c & -d & 0 & 0 & 0 & 0 & 0 & -\frac{1}{q} & 0 & 0 & 0 & 0 \\ 0 & w_1^{1/2} & -a & b & 0 & 0 & 0 & 0 & 0 & 0 & 0 & 0 \\ 0 & 0 & -c & -d & 0 & 0 & 0 & 0 & 0 & 0 & 0 & 0 \\ 0 & 0 & 0 & 0 & -a & b & 0 & 0 & 0 & 0 & 0 & 0 \\ -w_2^{1/2} & 0 & 0 & 0 & -c & -d & 0 & 0 & 0 & 0 & 0 & 0 \\ \hline 0 & 0 & 0 & 0 & 0 & 0 & a & c & 0 & 0 & 0 & w_2^{1/2} \\ 0 & 0 & 0 & 0 & 0 & 0 & -b & d & -w_1^{1/2} & 0 & 0 & 0 \\ 0 & 0 & 0 & 0 & 0 & 0 & 0 & 0 & a & c & 0 & 0 \\ 0 & 0 & 0 & \frac{1}{\sigma^2} & 0 & 0 & 0 & 0 & -b & d & 0 & 0 \\ 0 & 0 & 0 & 0 & 0 & 0 & 0 & 0 & 0 & 0 & a & c \\ 0 & 0 & 0 & 0 & 0 & \frac{1}{\sigma^2} & 0 & 0 & 0 & 0 & -b & d \end{array}\right] \begin{bmatrix} x_1 \\ x_2 \\ w_1^{1/2}x_{1b} \\ w_1^{1/2}x_{2b} \\ w_2^{1/2}x_{1c} \\ w_2^{1/2}x_{2c} \\ \lambda_1 \\ \lambda_2 \\ \lambda_3 \\ \lambda_4 \\ \lambda_5 \\ \lambda_6 \end{bmatrix}. \tag{9.97}$$

9.3. Observability, Controllability, and Identifiability

In the previous section, several examples of the equations for multiparameter optimal inputs were given. The performance criterion used for the optimal inputs was the trace of the information matrix M. However, in some cases for long observation intervals the information matrix becomes singular and the parameters are unidentifiable (References 4, 6–8).

It is not the intent to provide a complete discussion of identifiability here. Only some introductory concepts are given. Identifiability is closely related to the concepts of observability and controllability.

System identification is often concerned with both structure determination and parameter identification (Reference 9). In a large class of problems the model structure is given by physical considerations; however, in other cases two or more models may be in contention. For parameter identification the problem of the identifiability of parameters must be considered. A model structure should not be selected in which the parameters are unidentifiable.

Consider the system

$$\dot{\mathbf{x}} = A\mathbf{x} + B\mathbf{y}, \qquad \mathbf{x}(0) = \mathbf{x}_0, \tag{9.98}$$

$$\mathbf{z} = C\mathbf{x}, \tag{9.99}$$

where $\mathbf{x}$ is an $n \times 1$ state vector, $\mathbf{y}$ is an $m \times 1$ control vector, and $\mathbf{z}$ is an $r \times 1$ measurement vector. The system is observable if all of its states are derivable from the measurements of the output (References 10–13). It can be shown that the system is observable if the $n \times nr$ matrix

$$L = [C^T \,|\, A^T C^T \,|\, A^{T^2} C^T \,|\, \cdots \,|\, A^{T^{n-1}} C^T] \tag{9.100}$$

is of rank n. The rank of a square matrix is the order of the largest nonvanishing determinant of the matrix. When L is not square, the rank of L can be determined by finding the rank of the $n \times n$ matrix LL^T. The complete state model of an unobservable system cannot be identified, therefore the parameters belonging to the unobservable states are unidentifiable.

The system is controllable if a control vector can move the system state from any initial state to any other desired state in finite time. It can be shown that the system is controllable if the matrix

$$[B \,|\, AB \,|\, A^2 B \,|\, \cdots \,|\, A^{n-1} B] \tag{9.101}$$

is of rank n.

The system is identifiable if the unknown parameters in the system matrices A, B, and C can be determined from measurements of the state variables. The choice of input signal should be persistently exciting in that it activates all modes of interest of the process throughout the observation interval (Reference 14). If the process is controllable, it should be identifiable.

System equations (9.98) and (9.99) are not uniquely defined by the system's input and output (Reference 15). The system can be described by any equations of the form

$$\dot{\mathbf{w}} = F\mathbf{w} + G\mathbf{y}, \qquad \mathbf{w}(0) = \mathbf{w}_0, \tag{9.102}$$

$$\mathbf{z} = H\mathbf{w}, \tag{9.103}$$

where

$$\mathbf{w} = T\mathbf{x}, \qquad F = TAT^{-1}, \qquad G = TB, \qquad H = CT^{-1} \tag{9.104}$$

and T is a nonsingular $n \times n$ matrix. The parameters F, G, and H contain $n(n + m + r)$ parameters. All the parameters in equations (9.102) and (9.103) cannot be uniquely determined by the input and output. However, the unknown parameters can be uniquely identified if a sufficient number of the remaining parameters are specified.

As an example, consider the system given by equations (9.57)–(9.60). The transfer function is

$$\frac{Z(s)}{Y(s)} = \frac{s + d_0}{s^2 + c_1 s + c_0},$$

where s is the Laplace transform and

$$d_0 = a,$$
$$c_1 = a + d,$$
$$c_0 = ad + bc.$$

The response in this case is specified by the three coefficients, d_0, c_1, and c_0, where c_1 is the trace and c_0 is the determinant of the matrix F. Note that while the parameters a and d can be identified, only the product of the parameters b and c is identifiable in this case.

Using the trace of the information matrix as a criterion for optimal inputs may result in an input which is not persistently exciting (References 6, 7). The information matrix is then nonsingular and the parameters are unidentifiable. However, it can be shown that optimization of the $\mathrm{Tr}(WM)$ is simply a first step to the optimization of criteria such as $\det M^{-1}$ or $\mathrm{Tr}(M^{-1})$ (Reference 6). Minimization of $\det M^{-1}$ or $\mathrm{Tr}(M^{-1})$ can be obtained by solving a sequence of subprograms each involving a maximization of $\mathrm{Tr}(W_i M)$.

More on the concepts of identifiability is given in References 16–19. Some practical difficulties in applying these concepts are discussed in References 20–22.

9.4. Optimal Inputs for Systems with Process Noise

Consider the linear dynamic system

$$\dot{\mathbf{x}} = F\mathbf{x} + G\mathbf{y} + \Gamma\boldsymbol{\eta},$$
$$\mathbf{z} = H\mathbf{x} + \mathbf{v},$$

where $\eta(t)$ is a Gaussian white noise process

$$E[\boldsymbol{\eta}(t)] = 0, \qquad E[\boldsymbol{\eta}(t)\boldsymbol{\eta}^T(\tau)] = Q\delta(t - \tau).$$

The information matrix is given in terms of the expected value of the sensitivity of the smoothed estimate, $\hat{\mathbf{x}}$, of the state vector $\mathbf{x}$ (Reference 3)

$$M = \int_0^T E[\hat{X}_p{}^T H^T R^{-1} \hat{H} X_p] \, dt,$$

where $\hat{X}_p$ is a parameter influence coefficient matrix with ij component $\partial \hat{x}_i / \partial p_j$, and $E[\cdot]$ is the expected value. The smoothed estimate $\hat{\mathbf{x}}$ is obtained from the Kalman filter equations

$$\dot{\hat{\mathbf{x}}} = F\hat{\mathbf{x}} + G\mathbf{y} + K[\mathbf{z} - H\hat{\mathbf{x}}],$$
$$K = PH^T R^{-1}$$
$$\dot{P} = FP + PF^T + \Gamma Q \Gamma - PH^T R^{-1} HP.$$

In this case, both P and K are functions of the unknown parameter vector $\mathbf{p}$.

9.5. Eigenvalue Problems

In the design of multiparameter optimal inputs, a scalar function of the information matrix M is selected as the performance criterion. Usually the sum of the diagonal elements of the matrix M or the weighted matrix MW is maximized. This can be viewed as the conversion of a multiple objective problem to a scalar optimization problem with the solution obtained by standard optimization methods (References 23–25). An important consideration is the determination of the weights for the multiple objective optimization. Mehra (Reference 5) uses the weighted trace as the performance criterion with the weights chosen such that the diagonal elements of the information matrix are nearly equal. This causes the weighted trace to more closely approximate the determinant of M as the performance criterion. For other objectives and in decision theory, the selection of weights is not trivial.

The determination of the weights of belonging of each member to the set is a basic problem in decision theory and in the theory of fuzzy sets (Reference 26). Saaty (References 27, 28) has shown that this problem can be reduced to an eigenvalue problem. Obtaining the eigenvalues of a matrix is of interest not only for fuzzy sets, but for obtaining analytical solutions for two-point boundary-value problems as in Section 3.1.4, obtaining optimal inputs using Mehra's method as in Section 8.2.4, determining the convergence of the Gauss–Seidel method for the solution of linear equations, etc. Therefore this section will be devoted to obtaining the eigenvalues, not by routine methods, but by methods which have been developed only recently.

9.5.1. Convergence of the Gauss–Seidel Method

The solution of linear equations by iterative methods requires for convergence that the absolute magnitudes of all the eigenvalues of the iteration matrix are less than unity. The test for convergence, however, is often difficult to apply because of the computation required. In this section a method for determining the convergence of the Gauss–Seidel iteration is given (Reference 29). The method involves the numerical integration of initial-value differential equations in the complex plane around the unit circle. The Gauss–Seidel method converges if the number of roots inside the unit circle is equal to the order of the iteration matrix.

Introduction

Consider the $n \times n$ system of equations

$$\begin{aligned}
&a_{11}x_1 + a_{12}x_2 + \cdots + a_{1n}x_n = b_1, \\
&a_{21}x_1 + a_{22}x_2 + \cdots + a_{2n}x_n = b_2, \\
&\cdots \\
&a_{n1}x_1 + a_{n2}x_n + \cdots + a_{nn}x_n = b_n.
\end{aligned} \tag{9.105}$$

These equations may be expressed in the matrix form

$$A\mathbf{x} = \mathbf{b}. \tag{9.106}$$

The solution of these linear equations is to be obtained by iterative methods. The iteration equation for Jacobi's method (Reference 30) is easily derived from equation (9.106):

$$\mathbf{x} = B\mathbf{x} + \mathbf{c}. \tag{9.107}$$

By adding subscripts, the recurrence relation is obtained:

$$\mathbf{x}_{k+1} = B\mathbf{x}_k + \mathbf{c}, \tag{9.108}$$

where B is an $n \times n$ matrix

$$B = \begin{bmatrix} b_{11} & b_{12} & \cdots & b_{1n} \\ b_{21} & b_{22} & \cdots & b_{2n} \\ \cdots & & & \\ b_{n1} & b_{n2} & \cdots & b_{nn} \end{bmatrix}. \tag{9.109}$$

Using initial guesses for the components of $\mathbf{x}_k$, equation (9.108) is solved for $\mathbf{x}_{k+1}$. The next iteration starts with the new set of values and so on.

The Gauss–Seidel method (Reference 30–33) is similar to Jacobi's method except that the most recently computed value of each variable is used as soon as it becomes available instead of waiting until each iteration cycle is complete. A necessary and sufficient condition for the convergence of the Gauss–Seidel method is that the absolute magnitudes of all the eigenvalues of the iteration matrix are less than unity, i.e.,

$$|\lambda_i| < 1. \tag{9.110}$$

The matrix for the determination of the eigenvalues of the Gauss–Seidel iteration matrix is obtained from the Jacobi iteration matrix defined by equation (9.109):

$$B(\lambda) = \begin{bmatrix} b_{11} - \lambda & b_{12} & \cdots & b_{1n} \\ b_{21}\lambda & b_{22} - \lambda & \cdots & b_{2n} \\ \cdots & & & \\ b_{n1}\lambda & b_{n2}\lambda & \cdots & b_{nn} - \lambda \end{bmatrix}. \tag{9.111}$$

The eigenvalues are the roots of the equation

$$\det[B(\lambda)] = 0. \tag{9.112}$$

In Reference 34, it has been shown that the problem of finding the roots of an equation of the form given by (9.112) can be reduced to the integration of a system of ordinary differential equations with known initial conditions given by

$$\frac{d}{d\lambda}\,\delta(\lambda) = \mathrm{Tr}[A(\lambda)B_\lambda(\lambda)], \tag{9.113}$$

$$\frac{d}{d\lambda}\,A(\lambda) = \frac{A\,\mathrm{Tr}[AB_\lambda] - AB_\lambda A}{\delta}, \tag{9.114}$$

where

$$\delta = \det[B(\lambda)], \tag{9.115}$$

$$A = \text{adj}[B(\lambda)], \tag{9.116}$$

$$B_\lambda = \frac{dB}{d\lambda}. \tag{9.117}$$

The initial conditions are obtained by evaluating the determinant of B and the adjoint of B at $\lambda = 0$:

$$\delta(0) = \delta_0, \tag{9.118}$$

$$A(0) = A_0. \tag{9.119}$$

The number of roots N within a closed contour C in the complex plane is given by

$$N = \frac{1}{2\pi i} \oint_C \frac{\delta'(\lambda)}{\delta(\lambda)}\, d\lambda, \tag{9.120}$$

where

$$\delta' = \frac{d\delta}{d\lambda}. \tag{9.121}$$

Thus if we integrate around the unit circle in the complex plane, the number of roots N should equal the order of the matrix $B(\lambda)$. This leads to the following theorem.

Theorem. A necessary and sufficient condition for the convergence of the Gauss–Seidel method is that the number of roots N inside the unit circle is equal to n, the order of matrix $B(\lambda)$.

The derivation of the initial-value equations (Reference 34) is given in the next section followed by some numerical results which show that these equations are stable.

Derivation of Initial-Value Equations

The inverse of the matrix $B(\lambda)$ is given by

$$B^{-1} = \frac{\text{adj}\, B}{\det B} \tag{9.122}$$

Premultiplying both sides of equation (9.122) by the matrix B and then multiplying both sides by det B yields

$$I \det B = B \operatorname{adj} B. \tag{9.123}$$

Postmultiplying both sides of equation (9.122) by the matrix B gives

$$I \det B = (\operatorname{adj} B)B. \tag{9.124}$$

Differentiate both sides of equation (9.123) with respect to λ:

$$B_\lambda \operatorname{adj} B + B(\operatorname{adj} B)_\lambda = I(\det B)_\lambda. \tag{9.125}$$

Premultiplying both sides of equation (9.125) by adj B gives

$$(\operatorname{adj} B)B_\lambda \operatorname{adj} B + (\operatorname{adj} B)B(\operatorname{adj} B)_\lambda = (\operatorname{adj} B)(\det B)_\lambda. \tag{9.126}$$

Making use of equation (9.124) in the second term of equation (9.126), we obtain

$$(\operatorname{adj} B)B_\lambda \operatorname{adj} B + (\det B)(\operatorname{adj} B)_\lambda = (\operatorname{adj} B)(\det B)_\lambda. \tag{9.127}$$

Since det B is a scalar, the $(\operatorname{adj} B)_\lambda$ becomes

$$(\operatorname{adj} B)_\lambda = \frac{(\operatorname{adj} B)(\det B)_\lambda - (\operatorname{adj} B)B_\lambda(\operatorname{adj} B)}{\det B}. \tag{9.128}$$

Let b_{ij} equal the element of the ith row and the jth column of the matrix B. Then differentiating the det B with respect λ gives

$$(\det B)_\lambda = \sum_{i,j=1}^{n} \frac{\partial(\det B)}{\partial b_{ij}} \frac{db_{ij}}{d\lambda}. \tag{9.129}$$

However, we have

$$\frac{\partial}{\partial b_{ij}}(\det B) = B_{ij}, \tag{9.130}$$

where B_{ij} is the cofactor of the element in the ith row and the jth column. Substituting equation (9.130) into equation (9.129) yields

$$(\det B)_\lambda = \sum_{i,j=1}^{n} B_{ij} \frac{db_{ij}}{d\lambda}. \tag{9.131}$$

It can be shown that

$$\operatorname{Tr}[(\operatorname{adj} B)B_\lambda] = \sum_{i,j=1}^{n} B_{ij} \frac{db_{ij}}{d\lambda}. \tag{9.132}$$

Then we have

$$(\det B)_\lambda = \mathrm{Tr}[(\mathrm{adj}\ B)B_\lambda]. \tag{9.133}$$

Substituting equation (9.133) into equation (9.128) gives

$$(\mathrm{adj}\ B)_\lambda = \frac{(\mathrm{adj}\ B)\mathrm{Tr}[(\mathrm{adj}\ B)B_\lambda] - (\mathrm{adj}\ B)B_\lambda(\mathrm{adj}\ B)}{\det B}. \tag{9.134}$$

Using the notation of equations (9.115) and (9.116), equations (9.133) and (9.134) become

$$\frac{d}{d\lambda}\,\delta(\lambda) = \mathrm{Tr}[A(\lambda)B_\lambda(\lambda)], \tag{9.135}$$

$$\frac{d}{d\lambda}\,A(\lambda) = \frac{A\,\mathrm{Tr}[AB_\lambda] - AB_\lambda A}{\delta} \tag{9.136}$$

with initial conditions at $\lambda = 0$

$$\delta(0) = \delta_0, \tag{9.137}$$

$$A(0) = A_0. \tag{9.138}$$

To determine the number of roots inside the unit circle, equation (9.120) is evaluated by integrating the differential equation

$$\frac{dN}{d\lambda} = \frac{1}{2\pi i}\,\frac{\mathrm{Tr}[A(\lambda)B_\lambda(\lambda)]}{\delta} \tag{9.139}$$

around the unit circle with the initial condition

$$N = 0. \tag{9.140}$$

Numerical Results

Numerical results were obtained using a fourth-order Runge–Kutta method for integration. Equations (9.135) and (9.136) were integrated with 100 grid intervals along the imaginary axis from $\lambda = 0$ to $\lambda = i$. Equation (9.139) was adjoined at $\lambda = i$ and all the equations were integrated using another 100 intervals around the unit circle. Equation (9.136) involves a square $n \times n$ matrix A. Thus equations (9.135), (9.136), and (9.139) yield $n^2 + 2$ differential equations which must be integrated around the unit circle.

A program for the integration around the unit circle in the complex plane was written in FORTRAN using single precision complex data type. The only inputs required are the order of the matrix A, the initial conditions, $A(0)$ and $\delta(0)$, and the derivative matrix, $B_\lambda(\lambda)$.

Example 1. Assume

$$B(\lambda) = \begin{bmatrix} 1-\lambda & 2 \\ 3\lambda & 4-\lambda \end{bmatrix},$$

from which we obtain

$$A(0) = \text{adj } B(0) = \begin{bmatrix} 4 & -2 \\ 0 & 1 \end{bmatrix},$$

$$\delta(0) = \det B(0) = 4,$$

$$B_\lambda(\lambda) = \begin{bmatrix} -1 & 0 \\ 3 & -1 \end{bmatrix}.$$

The eigenvalues for this example are easily obtained analytically and can be shown to be given by

$$\lambda_1 = 0.3765, \qquad \lambda_2 = 10.6234.$$

Since only one of the roots is inside the unit circle, we expect that $N = 1$ and that the Gauss–Seidel method will not converge. The numerical result obtained was

$$N = 1.0000004 - 0.0000001i,$$

which clearly shows that there is only one root inside the contour.

Example 2. Assume

$$B(\lambda) = \begin{bmatrix} 1-\lambda & -0.25 \\ \lambda & 0.5-\lambda \end{bmatrix},$$

from which we obtain

$$A(0) = \begin{bmatrix} 0.5 & 0.25 \\ 0 & 1 \end{bmatrix},$$

$$\delta(0) = 0.5,$$

$$B_\lambda(\lambda) = \begin{bmatrix} -1 & 0 \\ 1 & -1 \end{bmatrix}.$$

The eigenvalues are

$$\lambda_{1,2} = 0.625 \pm 0.3307i, \qquad |\lambda_1| = |\lambda_2| = 0.707.$$

In this case both roots are inside the unit circle. Thus we expect that $N = 2$ and that the Gauss–Seidel method will converge. The numerical result obtained was

$$N = 2.0000064 + 0.0000017i,$$

which shows that there are two roots inside the contour.

Example 3. As a final example a matrix $B(\lambda)$ was selected such that the eigenvalues are near the unit circle. Assume

$$B(\lambda) = \begin{bmatrix} 0.95 - \lambda & -0.1 \\ \lambda & 1 - \lambda \end{bmatrix},$$

from which we obtain

$$A(0) = \begin{bmatrix} 1 & 0.1 \\ 0 & 0.95 \end{bmatrix},$$

$$\delta(0) = 0.95,$$

$$B_\lambda(\lambda) = \begin{bmatrix} -1 & 0 \\ 1 & -1 \end{bmatrix}.$$

The eigenvalues are

$$\lambda_{1,2} = 0.925 \pm 0.307i, \qquad |\lambda_1| = |\lambda_2| = 0.975.$$

Both roots are inside, but close to the unit circle. The numerical result obtained was

$$N = 1.9716261 + 0.0283000i.$$

While the accuracy is not as good as in Example 2, the result still clearly shows that there are two roots inside the contour. We can determine quantitatively whether we are in the neighborhood of a root by checking at each grid step interval whether the absolute value of the det $B(\lambda)$ is less than some small number ε. For this example, the minimum value of the absolute value of the det $B(\lambda)$ was 0.0161.

9.5.2. Determining the Eigenvalues of Saaty's Matrices for Fuzzy Sets

Saaty has shown that a basic problem in fuzzy set theory can be reduced to an eigenvalue problem. In this section an imbedding method is applied to obtain the largest eigenvalue and eigenvector of a matrix (Reference 26). Numerical results are given for an example and compared with Saaty's results.

An advantage of the imbedding approach is that the number of characteristic roots within any closed contour is obtained, which is of importance in judging the consistency of the estimated relative weight matrices.

Introduction

A basic problem in the theory of fuzzy sets (Reference 35) is the determination of the degree of belonging of each member to the set. In References 27 and 28, Saaty has suggested the following solution to the problem. Let $w_i \geq 0$, $i = 1, 2, \ldots, n$ be the degrees of belonging of the n members. Form the matrix of relative weights whose i, jth element is w_i/w_j. Then Saaty observes that the vector $(w_1, w_2, \ldots, w_n)^T$ is an eigenvector corresponding to the largest eigenvalue [the Perron–Frobenius root (References 36, 37)]. All the other eigenvalues are zero.

The idea is to estimate the matrix of relative weights and then obtain an estimate of the vector $(w_1, w_2, \ldots, w_n)^T$ as an eigenvector corresponding to the largest eigenvalue of the relative weight matrix. To compare a set of n objects in pairs according to their relative weights, Saaty denotes the objects by $A_1, \ldots, A_n$ and their weights by $w_1, \ldots, w_n$. The pairwise comparisons are represented by the matrix

$$A = \begin{array}{c|cccc} & A_1 & A_2 & \cdots & A_n \\ \hline A_1 & \dfrac{w_1}{w_1} & \dfrac{w_1}{w_2} & \cdots & \dfrac{w_1}{w_n} \\ A_2 & \dfrac{w_2}{w_1} & \dfrac{w_2}{w_2} & & \dfrac{w_2}{w_n} \\ \vdots & \vdots & & & \\ A_n & \dfrac{w_n}{w_1} & \dfrac{w_n}{w_2} & & \dfrac{w_n}{w_n} \end{array} \tag{9.141}$$

This matrix, called a reciprocal matrix, has positive entries everywhere and satisfies the reciprocal property $a_{ji} = 1/a_{ij}$. Multiplying this matrix by the vector $w = (w_1, \ldots, w_n)^T$,

$$Aw = nw \tag{9.142}$$

or

$$(A - nI)w = 0. \tag{9.143}$$

This is a system of homogeneous linear equations which has a nontrivial solution if and only if the determinant of $(A - nI)$ vanishes, that is, n is an eigenvalue of A. Matrix A is also consistent; that is, $a_{jk} = a_{ik}/a_{ij}$.

In the general case, the precise values of w_i/w_j are not known and must be estimated. Since the eigenvalues are perturbed by a small perturbation of the coefficients, equation (9.142) becomes

$$A'w' = \lambda_{\max}w', \tag{9.144}$$

where $\lambda_{\max}$ is the largest eigenvalue of A'. To simplify the notation, equation (9.144) is expressed in the form

$$Aw = \lambda_{\max}w, \tag{9.145}$$

where A is Saaty's matrix of pairwise comparisons. The eigenvector associated with the largest eigenvalue is the desired vector of weights.

Numerical methods for obtaining the largest eigenvalue and associated eigenvector are discussed in this section. An imbedding method applied to the eigenvector problem is emphasized. Numerical results are given for an example and compared with Saaty's results.

An Imbedding Method for Eigenvalue Problems

Consider a square matrix, $B(\lambda)$, where the elements of the matrix can be either linear or nonlinear functions of the parameter λ. The eigenvalues are the roots of the determinantal equation

$$\det B(\lambda) = 0. \tag{9.146}$$

It has been shown in Reference 34 that the problem of finding the roots of an equation of the form given by (9.146) can be reduced to the integration of a system of ordinary differential equations with known initial conditions given by

$$\frac{d}{d\lambda}\Delta(\lambda) = \mathrm{Tr}[M(\lambda)B_\lambda(\lambda)], \tag{9.147}$$

$$\frac{d}{d\lambda}M(\lambda) = \frac{M\,\mathrm{Tr}[MB_\lambda] - MB_\lambda M}{\Delta(\lambda)}, \tag{9.148}$$

where

$$\Delta(\lambda) = \det B(\lambda), \tag{9.149}$$

$$M(\lambda) = \mathrm{adj}\, B(\lambda), \tag{9.150}$$

$$B_\lambda = \frac{dB}{d\lambda}. \tag{9.151}$$

The derivations of these equations were given in Section 9.5.1. The initial conditions are obtained by evaluating the determinant of B and the adjoint of B at $\lambda = 0$:

$$\Delta(0) = \Delta_0, \tag{9.152}$$

$$M(0) = M_0. \tag{9.153}$$

For a linear eigenvalue problem, such as that which is of interest in this section, the matrix $B(\lambda)$ has the special form (Reference 34)

$$B(\lambda) = A - \lambda I, \tag{9.154}$$

where A is Saaty's matrix and I is the identity matrix. The form of the initial-value equations can be simplified by introducing the parameter

$$\mu = 1/\lambda. \tag{9.155}$$

Equation (9.154) can be expressed in the form

$$B(\mu) = \mu A - I. \tag{9.156}$$

Differentiating equation (9.156) with respect to μ gives

$$B_\mu(\mu) = A. \tag{9.157}$$

Equations (9.147) and (9.148) then become

$$\frac{d}{d\mu} \Delta(\mu) = \mathrm{Tr}[M(\mu)A], \tag{9.158}$$

$$\frac{d}{d\mu} M(\mu) = \frac{M\,\mathrm{Tr}[MA] - MAM}{\Delta(\mu)}. \tag{9.159}$$

Postmultiplying both sides of equation (9.159) by A and introducing a new matrix C such that

$$C = MA, \tag{9.160}$$

the initial-value equations become

$$\frac{d}{d\mu} \Delta(\mu) = \mathrm{Tr}\, C, \tag{9.161}$$

$$\frac{d}{d\mu} C(\mu) = \frac{C\,\mathrm{Tr}[C] - C^2}{\Delta(\mu)}, \tag{9.162}$$

where

$$\Delta(\mu) = \det B(\mu), \tag{9.163}$$

$$C(\mu) = MA = \operatorname{adj}[B(\mu)]A. \tag{9.164}$$

From equation (9.156), $B(\mu)$ at $\mu = 0$ is given by

$$B(0) = -I. \tag{9.165}$$

The initial conditions for equations (9.161) and (9.162) are

$$\Delta(0) = \det B(0) = (-1)^n, \tag{9.166}$$

$$C(0) = \operatorname{adj}[B(0)]A = (-1)^{n+1}A, \tag{9.167}$$

where n is the order of the square matrix B.

The initial-value differential equations given by (9.161) and (9.162) with initial conditions (9.166) and (9.167) are of the form of the equations to be used in this section. The roots of the determinantal equation

$$\det B(\mu) = 0 \tag{9.168}$$

give the reciprocals of the eigenvalues λ.

Calculation of Eigenvalues and Eigenvectors

From the theory of complex variables, the number of roots N within a closed contour C_1 in the complex plane is given by

$$N = \frac{1}{2\pi i} \oint_{C_1} \frac{\Delta'(\mu)}{\Delta(\mu)}\, d\mu, \tag{9.169}$$

where

$$\Delta'(\mu) = \frac{d\Delta}{d\mu}. \tag{9.170}$$

Equation (9.169) can be evaluated numerically by integrating differential equations (9.161) and (9.162) around a circle of radius r to obtain $\Delta'(\mu)$ and $\Delta(\mu)$. Then for $N = 1$, the largest eigenvalue, $\lambda_{\max}$, corresponding to the smallest root of equation (9.168) can be obtained by evaluating

$$\mu_1 = \frac{1}{\lambda_{\max}} = \frac{1}{2\pi i} \oint_{C_1} \mu \frac{\Delta'(\mu)}{\Delta(\mu)}\, d\mu. \tag{9.171}$$

Equations (9.169) and (9.171) are evaluated by integrating differential equations

$$\frac{dz}{d\mu} = \frac{1}{2\pi i} \frac{\Delta'(\mu)}{\Delta(\mu)} \tag{9.172}$$

and

$$\frac{dw}{d\mu} = \frac{1}{2\pi i} \frac{\mu\Delta'(\mu)}{\Delta(\mu)} \tag{9.173}$$

around the closed contour C_1 with initial conditions at the start of the closed contour, where $\mu = \mu_s$, given, respectively, by

$$z(\mu_s) = 0, \tag{9.174}$$

$$w(\mu_s) = 0. \tag{9.175}$$

Equations (9.161) and (9.162) have initial conditions at $\mu = 0$ and are integrated first from $\mu = 0$ to $\mu = \mu_s$ and then around the closed contour C_1 to obtain $\Delta'(\mu)$ and $\Delta(\mu)$.

A matrix is called consistent if its elements satisfy the condition $a_{jk} = a_{ik}/a_{ij}$. For a nonconsistent matrix A, the largest eigenvalue is always greater than or equal to the order, n, of the matrix:

$$\lambda_{\max} \geq n. \tag{9.176}$$

The equality holds only if A is consistent. Thus the selection of the radius r of the circular contour C_1 should be such that

$$\frac{1}{\lambda_{\max}} \leq \frac{1}{n} \leq r. \tag{9.177}$$

If r is too large, the number of roots N within the contour C_1 will be greater than unity. The radius of the circle at which this occurs can be used to estimate the magnitude of the second largest eigenvalue.

Any nonzero column of matrix C evaluated at $\mu = \mu_1$ is an eigenvector of matrix A. This may be shown as follows. From the definition of the inverse matrix we have

$$B(\mu) \operatorname{adj} B(\mu) = I \det B(\mu). \tag{9.178}$$

At $\mu = \mu_1$ we have

$$\det B(\mu_1) = 0, \tag{9.179}$$

$$B(\mu_1) \operatorname{adj} B(\mu_1) = 0. \tag{9.180}$$

Using equation (9.156), equation (9.180) can be expressed in the form

$$\mu_1 A \text{ adj } B(\mu_1) = \text{adj } B(\mu_1). \tag{9.181}$$

Multiplying both sides of equation (9.181) by A and dividing by μ_1 gives

$$A \text{ adj}[B(\mu_1)]A = \frac{1}{\mu_1} \text{ adj}[B(\mu_1)]A. \tag{9.182}$$

Then using the definition of the C matrix given by equation (9.160), equation (9.182) becomes

$$AC = \frac{1}{\mu_1} C. \tag{9.183}$$

From the eigenvector equation (Reference 38) we have

$$Av_i = \lambda_i v_i, \tag{9.184}$$

where v_i is the eigenvector corresponding to the eigenvalue λ_i. Thus any nonzero column of matrix C evaluated at $\mu = \mu_1$ is an eigenvector of matrix A. The eigenvector can be obtained by evaluating the integrals

$$C_{ij}(\mu_1) = \frac{1}{2\pi i} \oint_{C_1} \frac{C_{ij}(\mu)}{\mu - \mu_1} \, d\mu \tag{9.185}$$

for a given column $j = k$ of the matrix C and rows $i = 1, 2, \ldots, n$.

Another method for evaluating the largest eigenvalue and the corresponding eigenvector of the matrix A is to integrate the initial-value equations (9.161) and (9.162) along the real axis until the determinant changes sign. The value of μ at crossover is $\mu_1 = 1/\lambda_{\max}$ and the eigenvector is any column of matrix C.

The above methods are appropriate for locating other eigenvalues which may be small in absolute value. This is important in the Saaty approach in judging the internal consistency of the estimated relative weight matrix. The other eigenvalues can be obtained by integrating initial-value equations (9.161) and (9.162) around closed contours in the complex plane such that each contour contains only one root. The eigenvalues, λ_k, $k = 1, 2, \ldots, n$ are obtained by evaluating

$$\mu_k = \frac{1}{\lambda_k} = \frac{1}{2\pi i} \oint_{C_k} \mu \frac{\Delta'(\mu)}{\Delta(\mu)} \, d\mu. \tag{9.186}$$

If the contour contains more than one root, then equation (9.186) yields the sum of these roots.

The numerical results for the largest eigenvalue and associated eigenvector via imbedding are compared with the results obtained using the power method (Reference 30). The iteration for the power method is

$$y_{n+1} = Ax_n, \tag{9.187}$$

$$k_{n+1} = \text{largest modulus in } y_{n+1}, \tag{9.188}$$

$$x_{n+1} = \frac{1}{k_{n+1}} x_n, \tag{9.189}$$

where the initial vector, x_0, is any set of values such as $(1, 1, \ldots, 1)^T$.

Numerical Results

Numerical results were obtained for Saaty's Wealth of Nations matrix given in Reference 27 and repeated in Table 9.4. Saaty had people estimate the relative wealth of nations and showed that the eigenvector corresponding to the matrix agreed closely to the GNP. Using the imbedding method, the initial-value equations were integrated in the complex plane using a fourth-order Runge–Kutta method with FORTRAN single precision complex data type. Equations (9.161) and (9.162) were integrated using 15 grid intervals along the imaginary axis from $\mu = 0$ to $\mu = 0.3i$. Equations (9.172) and (9.173) were adjoined at $\mu = 0.3i$ and all the equations were integrated using another 60 intervals around a circle of radius 0.3. The total number of equations integrated was $n^2 + 3$ or 52. The numerical results were

$$N = 1.0000024 - 0.000027i,$$

$$\mu_1 = 0.1314523 - 0.0000034i.$$

Thus the maximum eigenvalue is

$$\lambda_{\max} = 1/\mu_1 = 7.607322.$$

Once the value of μ_1 has been obtained, the above integration can be repeated with equations (9.185) adjoined to obtain the eigenvector. This was done using the first column of matrix C with $\mu_1 = 0.131445$. The eigenvector was normalized such that the sum of the components of the vector is unity. The numerical results are shown in Table 9.5, where the eigenvector is compared with Saaty's results and with the power method results. The initial-value equation results agree with the power method

Table 9.4. Wealth of Nations Matrix

Country	U.S.	USSR	China	France	U.K.	Japan	West Germany
U.S.	1	4	9	6	6	5	5
USSR	1/4	1	7	5	5	3	4
China	1/9	1/7	1	1/5	1/5	1/7	1/5
France	1/6	1/5	5	1	1	1/3	1/3
U.K.	1/6	1/5	5	1	1	1/3	1/3
Japan	1/5	1/3	7	3	3	1	2
West Germany	1/5	1/4	5	3	3	1/2	1

results to approximately five decimal places. The results agree with Saaty's eigenvector to approximately 2 or 3 decimal places.

As an alternate procedure, initial-value equations (9.161) and (9.162) were integrated along the real axis in steps of 0.01 until the determinant changed sign. The value of μ at crossover is $\mu_1 = 1/\lambda_{\max}$ and the eigenvector is any column of matrix C. The numerical results were

$$\mu = 0.13, \qquad \det = -0.0128296,$$

$$\mu = 0.14, \qquad \det = 0.0769734.$$

The eigenvector obtained by normalizing column 1 of matrix C at $\mu = 0.13$

Table 9.5. Comparison of Numerical Results for Eigenvector

Country	Saaty's eigenvector, $\lambda_{\max} = 7.61$	Power method, $\lambda_{\max} = 7.60772$	Complex plane integration of initial-value equations, $\lambda_{\max} = 7.607322$
U.S.	0.429	0.427115	0.4271147
USSR	0.231	0.230293	0.2302914
China	0.021	0.0208384	0.0208387
France	0.053	0.0523856	0.0523865
U.K.	0.053	0.0523856	0.0523865
Japan	0.119	0.122719	0.1227193
West Germany	0.095	0.0942627	0.094263

Table 9.6. Eigenvector for Real Axis Integration of Initial-value Equations

Country	Saaty's eigenvector	Initial-value equations (column 1 of matrix C)
U.S.	0.429	0.4271408
USSR	0.231	0.2293710
China	0.021	0.0211053
France	0.053	0.0526910
U.K.	0.053	0.0526910
Japan	0.119	0.1226097
West Germany	0.095	0.0943913

is given in Table 9.6. Greater accuracy could easily have been obtained by using interpolation. Even though interpolation was not used, the results agree with Saaty's eigenvector to approximately two or three decimal places also.

To obtain an estimate of the magnitude of the second largest eigenvalue, equations (9.161), (9.162), (9.172), and (9.173) were integrated around a closed circular contour of radius 0.6 using 60 grid intervals after integrating equations (9.161) and (9.162) along the imaginary axis from $\mu = 0$ to $\mu = 0.6i$ with 30 grid intervals. The numerical results were

$$N = 3.0002755 - 0.0008678i,$$

$$\mu_1 + \mu_2 + \mu_3 = 0.1977310 + 0.0002498i.$$

Since there are three roots inside the contour, equation (9.173) gives the sum of the roots. In this particular case the second and third largest eigenvalues are complex conjugates. The absolute magnitudes of these eigenvalues are greater than $1/0.6 = 1.67$, where 0.6 is the radius of the circular contour. The actual values of the second and third largest eigenvalues could be obtained by choosing contours such that only one of the roots is inside.

9.5.3. Comparison of Methods for Determining the Weights of Belonging to Fuzzy Sets

Saaty has solved a basic problem in fuzzy set theory using an eigenvector method, as described in Section 9.5.2, to determine the weights of belonging of each member to the set. In this section a weighted least squares method is utilized to obtain the weights (Reference 39). This method has the advantage that it involves the solution of a set of simultaneous linear algebraic equations and is thus conceptually easier to understand than the eigenvector method. Examples are given for estimating the relative wealth of nations and the relative amount of foreign trade of nations. Numerical solutions are obtained using both the eigenvector method and the weighted least squares method and the results compared.

Weighted Least-Squares Method

Consider the elements, a_{ij}, of Saaty's matrix A in equation (9.145). It is desired to determine the weights, w_i, such that, given a_{ij}, we have

$$a_{ij} \approx w_i/w_j. \tag{9.190}$$

The weights can be obtained by solving the constrained optimization problem

$$S = \sum_{i=1}^{n} \sum_{j=1}^{n} (a_{ij}w_j - w_i)^2, \tag{9.191}$$

$$\sum_{i=1}^{n} w_i = 1, \tag{9.192}$$

$$\text{minimize } S. \tag{9.193}$$

An additional constraint is that $w_i > 0$. However, it is conjectured that the above problem can be solved such that $w_i > 0$ without this constraint. The least-squares solution for

$$S_1 = \sum_{i=1}^{n} \sum_{j=1}^{n} \left(a_{ij} - \frac{w_i}{w_j}\right)^2 \tag{9.194}$$

while more desirable than for the weighted least squares given by equation (9.191), is much more difficult to solve numerically.

In order to minimize S, form the sum

$$S' = \sum_{i=1}^{n} \sum_{j=1}^{n} (a_{ij}w_j - w_i)^2 + 2\lambda \sum_{i=1}^{n} w_i \tag{9.195}$$

where λ is the Lagrange multiplier. Differentiating equation (9.195) with respect to w_m, the following set of equations is obtained:

$$\sum_{i=1}^{n} (a_{im}w_m - w_i)a_{im} - \sum_{j=1}^{n} (a_{mj} - w_m) + \lambda = 0, \quad m = 1, 2, \ldots, n. \tag{9.196}$$

Equations (9.196) and (9.192) form a set of $n + 1$ inhomogeneous linear equations with $n + 1$ unknowns. For example for $n = 2$, the equations are

$$(1 + a_{21}^2)w_1 - (a_{12} + a_{21})w_2 + \lambda = 0, \tag{9.197}$$

$$-(a_{12} + a_{21})w_1 + (1 + a_{12}^2)w_2 + \lambda = 0, \tag{9.198}$$

$$w_1 + w_2 = 1. \tag{9.199}$$

Given the coefficients, a_{ij}, equations (9.197)–(9.199) can be solved for w_1, w_2, and λ using a standard FORTRAN subroutine for solving simultaneous linear equations. In this simple case, however, an analytical solution is possible:

$$w_1 = \frac{(1 + a_{12}^2) + a_{12} + a_{21}}{(1 + a_{12})^2 + (1 + a_{21})^2} \tag{9.200}$$

$$w_2 = \frac{(1 + a_{21}^2) + a_{12} + a_{21}}{(1 + a_{12})^2 + (1 + a_{21})^2}. \tag{9.201}$$

Equations (9.200) and (9.201) show that since the a_{ij}'s > 0, then the w_i's > 0.

In general, equations (9.196) and (9.192) can be expressed in the matrix form

$$B\mathbf{w} = \mathbf{m}, \tag{9.202}$$

where

$$\mathbf{w} = (w_1, w_2, \ldots, w_n, \lambda)^T, \tag{9.203}$$

$$\mathbf{m} = (0, 0, \ldots, 0, 1)^T, \tag{9.204}$$

$$B = n + 1 \times n + 1 \text{ matrix with elements } b_{ij}, \tag{9.205}$$

$$b_{ii} = (n - 1) + \sum_{j \neq i}^{n} a_{ji}^2, \qquad i, j = 1, \ldots, n, \tag{9.206}$$

$$b_{ij} = -a_{ij} - a_{ji}, \qquad i, j = 1, \ldots, n, \tag{9.207}$$

$$b_{k,n+1} = b_{n+1,k} = 1, \qquad k = 1, \ldots, n, \tag{9.208}$$

$$b_{n+1,n+1} = 0. \tag{9.209}$$

Wealth of Nations Matrix

Numerical results were obtained for Saaty's Wealth of Nations matrix given in Table 9.4, Section 9.5.2. Saaty made estimates of the relative wealth of nations and showed that the eigenvector corresponding to the matrix

Table 9.7. Comparison of Numerical Results for Wealth of Nations Matrix

Country	Saaty's eigenvector, $\lambda_{max} = 7.61$	Power method eigenvector, $\lambda_{max} = 7.60772$	Weighted least-squares method
U.S.	0.429	0.427115	0.486711
USSR	0.231	0.230293	0.175001
China	0.021	0.0208384	0.0299184
France	0.053	0.0523856	0.0593444
U.K.	0.053	0.0523856	0.0593444
Japan	0.119	0.122719	0.10434
West Germany	0.095	0.0942627	0.0853411
	$S = \sum_i \sum_j (a_{ij}w_j - w_i)^2$	$S = 0.458232$	$S = 0.288071$
	$S_1 = \sum_i \sum_j (a_{ij} - w_i/w_j)^2$	$S_1 = 187.898$	$S_1 = 124.499$

agreed closely with the GNP. The power method for obtaining the eigenvector was utilized in Section 9.5.2 and compared to Saaty's results. Table 9.7 compares these results with the weighted least-squares results. It is seen that the sums, S and S_1, defined by equations (9.191) and (9.194), are less for the weighted least-squares method than for the power method. The sums, S_1, were computed for comparisons even though the minimizations were made with respect to the sums S.

Taiwan Trade Matrices

Using Saaty's scales Chu (Reference 40) estimated the relative strengths of belonging of the U.S., Japan, South America, and Europe to the fuzzy set of important trading partners with Taiwan. This was done with regard to exports, imports, and total trade. Both methods for determining the relative weights were used and comparisons were made with published trade data for the year 1975. Both methods yielded good agreement with those data. Table 9.8 gives the Taiwan trade matrices and Table 9.9 gives a comparison of the numerical results. The sums, S and S_1, are again lower for the weighted least-squares method than for the power method in all cases except for the imports sum S_1. In all cases, the dominant weight tends to be larger for the weighted least-squares method.

Table 9.8. Taiwan Trade Matrices

	U.S.	Japan	South America	Europe
Exports:				
U.S.	1	3	9	5
Japan	1/3	1	9	1/2
South America	1/9	1/5	1	1/2
Europe	1/5	3	3	1
Imports:				
U.S.	1	1	9	3
Japan	1	1	7	3
South America	1/9	1/9	1	1/7
Europe	1/3	1/2	7	1
Total trade:				
U.S.	1	3	9	3
Japan	1/4	1	7	3
South America	1/9	1/7	1	1/5
Europe	1/5	1/2	5	1

Discussion

The numerical results tend to show that either the eigenvector or the weighted least-squares method can be used to obtain the weights. For the examples used in this paper, the eigenvector method appeared to give answers closer to the expected values. However, the sums, S and S_1, were generally smaller for the weighted least-squares method. Also, the weighted least-squares method, which involves the solution of a set of simultaneous linear algebraic equations, is conceptually easier to understand than the eigenvector method. Using the eigenvector method, it can be proved that the weights, w_i, are all greater than zero (Reference 27). While it is not known whether such a theorem exists for the weighted least-squares method, the numerical results given here indicate that the w_i's obtained by this method are also greater than zero and are comparable to those obtainable by the eigenvector method.

Table 9.9. Comparison of Numerical Results for Taiwan Trade Matrices

Trading partner	Fraction	Power method eigenvector	Weighted least-squares method
		Exports, $\lambda_{max} = 4.93$	
U.S.	0.525	0.540	0.641
Japan	0.204	0.193	0.160
South America	0.038	0.052	0.047
Europe	0.233	0.215	0.152
		$S = 0.526$	$S = 0.277$
		$S_1 = 41.23$	$S_1 = 59.56$
		Imports, $\lambda_{max} = 4.14$	
U.S.	0.394	0.399	0.414
Japan	0.418	0.382	0.396
South America	0.014	0.036	0.042
Europe	0.174	0.183	0.148
		$S = 0.081$	$S = 0.040$
		$S_1 = 22.75$	$S_1 = 18.71$
		Total trade, $\lambda_{max} = 4.09$	
U.S.	0.452	0.533	0.575
Japan	0.323	0.279	0.232
South America	0.024	0.041	0.049
Europe	0.200	0.147	0.144
		$S = 0.179$	$S = 0.126$
		$S_1 = 19.94$	$S_1 = 19.92$

9.5.4. Variational Equations for the Eigenvalues and Eigenvectors of Nonsymmetric Matrices

The tracking of eigenvalues and eigenvectors for parametrized matrices is of major importance in optimization and stability problems. In this section a one-parameter family of matrices with distinct eigenvalues is considered (Reference 41). A complete system of differential equations is developed for both the eigenvalues and the right and left eigenvectors.[†] The computational feasibility of the differential system is demonstrated by means of a numerical example.

Introduction: The Basic Problem

Let $M(\alpha)$ be an $n \times n$ complex matrix-valued differentiable function of a parameter α varying over some simply connected domain $\mathbb{C}^0$ of the complex plane $\mathbb{C}$. It will be assumed that $M(\alpha)$ has n distinct eigenvalues $\lambda_1(\alpha), \ldots, \lambda_n(\alpha)$ in $\mathbb{C}$ for each $\alpha \in \mathbb{C}^0$. Letting superscript T denote transpose, it then follows (Reference 42) that there exist two sets of linearly independent vectors $\{x_1(\alpha), \ldots, x_n(\alpha)\}$ and $\{w_1(\alpha), \ldots, w_n(\alpha)\}$ in $\mathbb{C}^n$ for each $\alpha \in \mathbb{C}^0$ satisfying

$$M(\alpha)x_i(\alpha) = \lambda_i(\alpha)x_i(\alpha), \qquad i = 1, \ldots, n, \tag{9.210a}$$

$$M(\alpha)^T w_i(\alpha) = \lambda_i(\alpha)w_i(\alpha), \qquad i = 1, \ldots, n, \tag{9.210b}$$

$$x_i(\alpha)^T w_j(\alpha) \begin{cases} = 0 & \text{if } i \neq j \\ \neq 0 & \text{if } i = j. \end{cases} \tag{9.210c}$$

The vectors $\{x_1(\alpha), \ldots, x_n(\alpha)\}$ and $\{w_1(\alpha), \ldots, w_n(\alpha)\}$ are generally referred to as the right and left eigenvalues of $M(\alpha)$, respectively.

The differentiability of $M(\alpha)$ over $\mathbb{C}^0$, coupled with the ruling out of exceptional points α in $\mathbb{C}^0$ where eigenvalues coalesce, guarantees (Reference 43, Sections II.1 and II.4-6) that the eigenvalues and right and left eigenvectors of $M(\alpha)$ have differentiable representations over $\mathbb{C}^0$. In this case it is known (Reference 43, p. 81) that

$$\dot{\lambda}_i(\alpha) = \frac{w_i(\alpha)^T \dot{M}(\alpha) x_i(\alpha)}{w_i^T(\alpha) x_i(\alpha)}, \qquad \alpha \in \mathbb{C}^0, \qquad i = 1, \ldots, n, \tag{9.211}$$

where a dot denotes differentiation with respect to α. However, corresponding analytical expressions for the derivatives $\dot{x}_i(\alpha)$ and $\dot{w}_i(\alpha)$ of the right and left eigenvectors of $M(\alpha)$ do not appear to be available in the literature. Without such additional equations, the system of differential equations (9.211) is analytically incomplete in the sense that solutions $\lambda_i(\alpha)$ for (9.211) cannot be obtained by integration from initial conditions.

[†] The right and left eigenvectors of a given $n \times n$ matrix M corresponding to an eigenvalue λ are defined to be the nontrivial solutions x and w^T to $Mx = \lambda x$ and $w^T M = \lambda w^T$, respectively, where superscript T denotes transpose.

In the following paragraphs differential system (9.211) will be completed by providing differential equations for the right and left eigenvectors of $M(\alpha)$. As will be seen, the resulting complete differential system provides a practical tool for numerical work.

Complete Variational Equations for Nonsymmetric Matrices

The exact form of the complete differential system for the eigenvalues and eigenvectors of $M(\alpha)$ depends on the normalization selected for the eigenvectors. We will start by imposing the general normalizations

$$x_i^T(\alpha)x_i(\alpha) = \varphi_i(\alpha), \qquad \alpha \in \mathbb{C}^0, \quad i = 1, \ldots, n, \tag{9.212a}$$

$$w_i^T(\alpha)w_i(\alpha) = \Psi_i(\alpha), \qquad \alpha \in \mathbb{C}^0, \quad i = 1, \ldots, n, \tag{9.212b}$$

for arbitrary differentiable functions $\varphi_i : \mathbb{C}^0 \to \mathbb{C}$ and $\Psi_i : \mathbb{C}^0 \to \mathbb{C}$. Subsequently, (9.212) will be specialized to $\varphi_i(\alpha) \equiv \Psi_i(\alpha) \equiv 1$, and also to another convenient normalization.

Let $i \in \{1, \ldots, n\}$ and $\alpha \in \mathbb{C}^0$ be given. Differentiating (9.210a) with respect to α, and suppressing reference to α for ease of notation, one obtains

$$\dot{M}x_i + M\dot{x}_i = \dot{\lambda}_i x_i + \lambda_i \dot{x}_i. \tag{9.213}$$

Multiplying through (9.213) by w_i^T, we obtain

$$w_i^T\dot{M}x_i + w_i^T M\dot{x}_i = \dot{\lambda}_i w_i^T x_i + \lambda_i w_i^T \dot{x}_i. \tag{9.214}$$

Since $w_i^T M = \lambda_i w_i^T$ by (9.210b), and $w_i^T x_i \neq 0$ by (9.210c), (9.214) reduces to

$$w_i^T \dot{M} x_i / w_i^T x_i = \dot{\lambda}_i, \tag{9.215}$$

the familiar differential equation for $\dot{\lambda}_i$.

Since the set of right eigenvectors $\{x_1, \ldots, x_n\}$ for $M(\alpha)$ spans $\mathbb{C}^n$, there exist coefficients $\beta_{ij}, j = 1, \ldots, n$, such that

$$\dot{x}_i = \sum_{j=1}^{n} \beta_{ij} x_j. \tag{9.216}$$

For $k \neq i$, it follows from (9.210c) and (9.216) that

$$w_k^T \dot{x}_i = \sum_{j=1}^{n} \beta_{ij} w_k^T x_j = \beta_{ik} w_k^T x_k. \tag{9.217}$$

Hence, combining (9.216) and (9.217), we have

$$\dot{x}_i = \sum_{j \neq i} \left(\frac{w_j^T \dot{x}_i}{w_j^T x_j} \right) x_j + \beta_{ii} x_i. \tag{9.218}$$

Multiplying through (9.218) by x_i^T, and solving for β_{ii}, yields

$$\beta_{ii} = \left[\frac{x_i^T \dot{x}_i - \sum_{j \neq i} (w_j^T \dot{x}_i / w_j^T x_j) x_i^T x_j}{x_i^T x_i} \right]. \tag{9.219}$$

It is now necessary to replace the expressions $w_j^T \dot{x}_i$ and $x_i^T \dot{x}_i$ in (9.218) and (9.219) by a suitable expression independent of $\dot{x}_i$. Multiplying through (9.213) by w_j^T, $j \neq i$, we obtain

$$w_j^T \dot{M} x_i + w_j^T M \dot{x}_i = \dot{\lambda}_i w_j^T x_i + \lambda_i w_j^T \dot{x}_i. \tag{9.220}$$

By (9.210b) and (9.210c), (9.220) reduces to

$$w_j^T \dot{M} x_i + \lambda_j w_j^T \dot{x}_i = \lambda_i w_j^T \dot{x}_i. \tag{9.221}$$

Since the roots λ_j and λ_i are distinct by assumption, (9.221) can be restated as

$$\frac{w_j^T \dot{M} x_i}{\lambda_i - \lambda_j} = w_j^T \dot{x}_i. \tag{9.222}$$

Finally, the normalization (9.212a) implies

$$\frac{\dot{\varphi}_i}{2} = x_i^T \dot{x}_i. \tag{9.223}$$

Substituting (9.212a), (9.219), (9.222), and (9.223) into (9.218) yields the desired differential equation for the right eigenvector x_i,

$$\dot{x}_i = \sum_{j \neq i} \left[\frac{w_j^T \dot{M} x_i}{(\lambda_i - \lambda_j) w_j^T x_j} \right] x_j + \left\{ \frac{\frac{1}{2}\dot{\varphi}_i - \sum_{j \neq i} [w_j^T \dot{M} x_i / (\lambda_i - \lambda_j) w_j^T x_j] x_i^T x_j}{\varphi_i} \right\} x_i. \tag{9.224}$$

Equations analogous to (9.224) are similarly obtained for the left eigenvectors w_i. The complete differential system for the eigenvalues and right and left eigenvectors of $M(\alpha)$ thus has the form

$$\dot{\lambda}_i = \frac{w_i^T \dot{M} x_i}{w_i^T x_i}, \qquad i = 1, \ldots, n, \tag{9.225a}$$

$$\dot{x}_i = \sum_{j \neq i} \left[\frac{w_j^T \dot{M} x_i}{(\lambda_i - \lambda_j) w_j^T x_j} \right] x_j + \left\{ \frac{\frac{1}{2}\dot{\varphi}_i - \sum_{j \neq i} [w_j^T \dot{M} x_i / (\lambda_i - \lambda_j) w_j^T x_j] x_i^T x_j}{\varphi_i} \right\} x_i, \qquad i = 1, \ldots, n, \tag{9.225b}$$

$$\dot{w}_i = \sum_{j \neq i} \left[\frac{x_j^T \dot{M} w_i}{(\lambda_i - \lambda_j) x_j^T w_j} \right] w_j + \left\{ \frac{\frac{1}{2}\dot{\Psi}_i - \sum_{j \neq i} [x_j^T \dot{M} w_i / (\lambda_i - \lambda_j) x_j^T w_j] w_i^T w_j}{\Psi_i} \right\} w_i, \qquad i = 1, \ldots, n. \tag{9.225c}$$

Though our primary purpose is to use equations (9.225) for computational purposes, some analytical consequences are immediate. For example, for the Perron root of a positive matrix, equation (9.225a) implies that adding a positive matrix to a given positive matrix cannot decrease the Perron root.

System (9.225) is considerably simplified if the selected normalization is $\varphi_i(\alpha) \equiv \Psi_i(\alpha) \equiv 1$, $i = 1, \ldots, n$. In this case (9.225) reduces to

$$\dot{\lambda}_i = \frac{w_i^T \dot{M} x_i}{w_i^T x_i}, \qquad i = 1, \ldots, n, \tag{9.226a}$$

$$\dot{x}_i = (I - x_i x_i^T) \sum_{j \neq i} \left[\frac{w_j^T \dot{M} x_i}{(\lambda_i - \lambda_j) w_j^T x_j} \right] x_j, \qquad i = 1, \ldots, n, \tag{9.226b}$$

$$\dot{w}_i = (I - w_i w_i^T) \sum_{j \neq i} \left[\frac{x_j^T \dot{M} w_i}{(\lambda_i - \lambda_j) x_j^T w_j} \right] w_j, \qquad i = 1, \ldots, n. \tag{9.226c}$$

An even greater simplification occurs if the $\varphi_i(\cdot)$ and $\Psi_i(\cdot)$ functions are selected so that the final terms in parentheses in (9.225b) and (9.225c) vanish identically. System (9.225) then reduces to

$$\dot{\lambda}_i = \frac{w_i^T \dot{M} x_i}{w_i^T x_i}, \qquad i = 1, \ldots, n, \tag{9.227a}$$

$$\dot{x}_i = \sum_{j \neq i} \left[\frac{w_j^T M x_i}{(\lambda_i - \lambda_j) w_j^T x_j} \right] x_j, \qquad i = 1, \ldots, n, \tag{9.227b}$$

$$\dot{w}_i = \sum_{j \neq i} \left[\frac{x_j^T \dot{M} w_i}{(\lambda_i - \lambda_j) x_j^T w_j} \right] w_j, \qquad i = 1, \ldots, n. \tag{9.227c}$$

At various stages in the integration of (9.227), the right and left eigenvectors can be normalized to unit length to prevent their magnitudes from becoming inconveniently large or small.

A numerical example will be given to illustrate the use of both (9.226) and (9.227). First however, we shall consider symmetric matrices.

Complete Variational Equations for Symmetric Matrices

Suppose $M(\alpha)$ satisfies $M(\alpha) = M(\alpha)^T$ for each $\alpha \in C^0$. In this case the right and left eigenvectors of $M(\alpha)$ coincide over C^0, hence the normalizations (9.210c) and (9.212) together yield

$$x_i^T(\alpha)x_j(\alpha) = \begin{cases} 0 & \text{if } i \neq j \\ \varphi_i(\alpha) & \text{if } i = j. \end{cases} \tag{9.228}$$

The differential system (9.225) thus reduces to

$$\dot{\lambda}_i = \frac{x_i^T \dot{M} x_i}{\varphi_i}, \qquad i = 1, \ldots, n, \tag{9.229a}$$

$$\dot{x}_i = \sum_{j \neq i} \left[\frac{x_j^T \dot{M} x_i}{(\lambda_i - \lambda_j)\varphi_i} \right] x_j + \left(\frac{\frac{1}{2}\dot{\varphi}_i}{\varphi_i} \right) x_i, \qquad i = 1, \ldots, n. \tag{9.229b}$$

If $\varphi_i(\alpha) \equiv 1$, $i = 1, \ldots, n$, (9.229) is further simplified to

$$\dot{\lambda}_i = x_i^T \dot{M} x_i, \qquad i = 1, \ldots, n, \tag{9.230a}$$

$$\dot{x}_i = \sum_{j \neq i} \left(\frac{x_j^T \dot{M} x_i}{\lambda_i - \lambda_j} \right) x_j, \qquad i = 1, \ldots, n. \tag{9.230b}$$

As noted in Reference 43, page 81, equations analogous to (9.230) for partial differential operators are familiar formulas in quantum mechanics. See, e.g., Reference 44, Chapter XI, pages 383–384. The differential form of (9.230a) is

$$d\lambda_i = x_i^T \, dM x_i. \tag{9.231}$$

It is known, by the Courant–Fischer minimax theorem (References 36, 45), that the addition of a positive definite matrix to a given positive definite matrix M increases all of the eigenvalues of M. This result is immediately obtainable from (9.231).

An Illustrative Numerical Example. Consider a matrix-valued function $M(\alpha)$ defined over $\alpha \in R$ by

$$M(\alpha) \equiv \begin{bmatrix} 1 & \alpha \\ \alpha^2 & 3 \end{bmatrix}. \tag{9.232}$$

For this simple example, analytical expressions are easily obtainable for the eigenvalues $\{\lambda_1(\alpha), \lambda_2(\alpha)\}$ and the right and left eigenvectors $\{x_1(\alpha), x_2(\alpha), w_1(\alpha), w_2(\alpha)\}$ of $M(\alpha)$, where the latter are normalized to have unit length. Specifically,

$$\lambda_1(\alpha) = 2 + \gamma(\alpha), \tag{9.233a}$$

$$\lambda_2(\alpha) = 2 - \gamma(\alpha), \tag{9.233b}$$

$$x_1(\alpha) = \left(\frac{\alpha}{1 + \gamma(\alpha)} k_1, k_1 \right)^T, \tag{9.233c}$$

$$x_2(\alpha) = \left(\frac{\alpha}{1 - \gamma(\alpha)} k_2, k_2 \right)^T, \tag{9.233d}$$

$$w_1(\alpha) = \left(\frac{\gamma(\alpha) - 1}{\alpha} k_3, k_3 \right)^T, \tag{9.233e}$$

$$w_2(\alpha) = \left(\frac{-\gamma(\alpha) - 1}{\alpha} k_4, k_4 \right)^T, \tag{9.233f}$$

where $\gamma(\alpha) \equiv (1 + \alpha^3)^{1/2}$, and the constants $k_1, \ldots, k_4$ are given by

$$k_1 \equiv \left\{ \frac{\alpha^2}{[1 + \gamma(\alpha)]^2} + 1 \right\}^{-1/2}, \tag{9.234a}$$

$$k_2 \equiv \left\{ \frac{\alpha^2}{[1 - \gamma(\alpha)]^2} + 1 \right\}^{-1/2}, \tag{9.234b}$$

$$k_3 \equiv \left\{ \frac{[\gamma(\alpha) - 1]^2}{\alpha^2} + 1 \right\}^{-1/2}, \tag{9.234c}$$

$$k_4 \equiv \left\{ \frac{[-\gamma(\alpha) - 1]^2}{\alpha^2} + 1 \right\}^{-1/2}. \tag{9.234d}$$

Note that the eigenvalues of $M(\alpha)$ are real if and only if $\alpha \geqq -1.0$.

A numerical solution was first obtained for the eigenvalues and right and left eigenvectors of $M(\alpha)$ over the α interval [0.5, 2.0] by integrating the unit normalized differential equations (9.226). A fourth-order Runge–Kutta method was used for the integration with the α grid intervals set

Table 9.10. Eigenvalues and Eigenvectors for $\alpha = 2.0$

	Numerical solution				Unit normalized analytical solution	
	Normalized equations (9.227)		Unit normalized equations (9.226)			
	Component		Component		Component	
	1	2	1	2	1	2
λ_1	5.0	—	5.0	—	5.0	—
λ_2	−1.0	—	−1.0	—	−1.0	—
x_1	0.449584	0.899168	0.447214	0.894427	0.447214	0.894427
x_2	−0.73633	0.73633	−0.707107	0.707107	−0.707107	0.707107
w_1	0.73633	0.73633	0.707107	0.707107	0.707107	0.707107
w_2	−0.899168	0.449584	−0.894427	0.447214	−0.894427	0.447214

equal to 0.01. The integration was initialized by solving (9.233) for the eigenvalues and right and left eigenvectors of $M(\alpha)$ at $\alpha = 0.5$. As indicated in Table 9.10, the numerical solution obtained using the unit normalized differential equations (9.226) agreed with the analytical unit normalized solution (9.233) to at least six digits.

A numerical solution was also obtained for the eigenvalues and eigenvectors of $M(\alpha)$ over $[0.5, 2.0]$ by integrating the differential equations (9.227). By initializing the system as before, we were guaranteed that the magnitudes of the right and left eigenvectors $x_1, \ldots, x_n$ and $w_1, \ldots, w_n$ would be positive in some neighborhood of $\alpha = 0.5$, and in fact the magnitudes remained positive over the entire interval $[0.5, 2.0]$. As indicated in Table 9.10, the eigenvalues were obtained with the same accuracy as before. In addition, a subsequent unit normalization of the eigenvectors obtained via (9.227) yielded the same eigenvectors as obtained via the unit normalized differential equations (9.226).

As seen from (9.233a) and (9.233b), the eigenvalues of $M(\alpha)$ coalesce at $\alpha = -1.0$ and are complex for $\alpha < -1.0$. The unit normalized differential equations (9.226) were integrated from 0.5 to -1.0, using the integration step size of -0.01 for α. Six-digit accuracy for the eigenvalues and eigenvectors was obtained over $[-0.97, 0.5]$, degenerating to approximately two-digit accuracy at $\alpha \equiv -1.0$. Similar results were obtained by integrating the differential equations (9.227) from 0.5 to -1.0.

Discussion

This section presents a first step towards the development of a computationally feasible procedure for tracking the eigenvalues and right and left eigenvectors of a parametrized matrix. Our main motivation has been the capability of modern-day computers to integrate, with great speed and accuracy, large-scale systems of ordinary differential equations subject to initial conditions. The computational feasibility of the initial-value differential system developed in this section was illustrated by numerical example.

In the next section it will be shown that initial-value systems can also be developed for tracking a single eigenvalue together with one of its corresponding right or left eigenvectors.

9.5.5. Individual Tracking of an Eigenvalue and Eigenvector of a Parametrized Matrix

Let $M(\alpha)$ be an $n \times n$ complex matrix-valued differentiable function of a parameter α varying over some simply connected region $\mathbb{C}^0$ of the complex plane $\mathbb{C}$. In the previous section a complete initial-value system of differential equations was developed for both the eigenvalues and right and left eigenvectors of $M(\alpha)$, assuming $M(\alpha)$ has n distinct eigenvalues for each α in $\mathbb{C}^0$. In this section it is shown that an initial-value system can also be developed for tracking a single eigenvalue of $M(\alpha)$ together with one of its corresponding right or left eigenvectors (Reference 46). The computational feasibility of the initial-value system is illustrated by numerical example.

Variational Equations for Individual Tracking

Consider the system of equations

$$M(\alpha)x = \lambda x, \tag{9.235a}$$

$$x^T x = 1, \tag{9.235b}$$

where superscript T denotes transpose. Any solution $(\lambda(\alpha), x(\alpha))^T$ for system (9.235) in $\mathbb{C} \times \mathbb{C}^n$ yields an eigenvalue and corresponding unit normalized right eigenvector for $M(\alpha)$, respectively.

If the Jacobian matrix for system (9.235),

$$J(\alpha) \equiv \begin{bmatrix} \lambda(\alpha)I - M(\alpha) & x(\alpha) \\ x(\alpha)^T & 0 \end{bmatrix} \tag{9.236}$$

is nonsingular at α^0, then, totally differentiating (9.235) with respect to α, one obtains

$$\begin{bmatrix} \dot{x}(\alpha) \\ \dot{\lambda}(\alpha) \end{bmatrix} = J(\alpha)^{-1} \begin{bmatrix} \dot{M}(\alpha)x(\alpha) \\ 0 \end{bmatrix}, \tag{9.237}$$

where a dot denotes differentiation with respect to α. Letting $A(\alpha)$ denote the adjoint $\mathrm{adj}(J(\alpha))$ of $J(\alpha)$ and $\delta(\alpha)$ denote the determinant $\mathrm{Det}\,(J(\alpha))$ of $J(\alpha)$, system (9.237) can be expanded (Reference 47) into a complete initial-value differential system of the form

$$\begin{bmatrix} \dot{x}(\alpha) \\ \dot{\lambda}(\alpha) \end{bmatrix} = \frac{A(\alpha)}{\delta(\alpha)} \begin{bmatrix} \dot{M}(\alpha)x(\alpha) \\ 0 \end{bmatrix}, \tag{9.238a}$$

$$\dot{A}(\alpha) = [A(\alpha)\,\mathrm{Tr}(A(\alpha)B(\alpha)) - A(\alpha)B(\alpha)A(\alpha)]/\delta(\alpha), \tag{9.238b}$$

$$\dot{\delta}(\alpha) = \mathrm{Tr}(A(\alpha)B(\alpha)), \tag{9.238c}$$

with initial conditions

$$x(\alpha^0) = x^0, \tag{9.238d}$$

$$\lambda(\alpha^0) = \lambda^0, \tag{9.238e}$$

$$A(\alpha^0) = \mathrm{adj}(J(\alpha^0)), \tag{9.238f}$$

$$\delta(\alpha^0) = \mathrm{Det}\,(J(\alpha^0)), \tag{9.238g}$$

where $B(\alpha) \equiv dJ(\alpha)/d\alpha$, and λ^0 and x^0 denote an arbitrarily selected eigenvalue and corresponding unit normalized right eigenvector for $M(\alpha^0)$, respectively.

The Jacobian $J(\alpha)$ defined by (9.236) is an interesting bordered matrix whose properties do not appear to have been previously explored. It will be shown in Section 9.5.6 that $J(\alpha)$ is nonsingular if $M(\alpha)$ is a positive matrix, $\lambda(\alpha)$ is the Perron root of $M(\alpha)$, and $x(\alpha)$ is a corresponding right eigenvector with elements taken to be positive. Alternatively, $J(\alpha)$ is nonsingular if $M(\alpha)$ is real and symmetric, with complete orthonormal system of real eigenvectors given by $x_1(\alpha), \ldots, x_n(\alpha)$, and $\lambda(\alpha) \equiv \lambda_1(\alpha)$ is a simple root of $M(\alpha)$ corresponding to the eigenvector $x(\alpha) \equiv x_1(\alpha)$. Specifically, it is easily verified that the bordered matrix

$$\begin{bmatrix} \lambda_1(\alpha)I - M(\alpha) & x_1(\alpha) \\ x_1(\alpha)^T & 0 \end{bmatrix}_{(n+1)\times(n+1)} \tag{9.239}$$

then has $n + 1$ linearly independent eigenvectors

$$\begin{bmatrix} x_1(\alpha) \\ 1 \end{bmatrix}, \begin{bmatrix} x_1(\alpha) \\ -1 \end{bmatrix}, \quad \begin{bmatrix} x_2(\alpha) \\ 0 \end{bmatrix}, \ldots, \begin{bmatrix} x_n(\alpha) \\ 0 \end{bmatrix}, \tag{9.240}$$

with corresponding nonzero eigenvalues

$$1, -1, (\lambda_1(\alpha) - \lambda_2(\alpha)), \ldots, (\lambda_1(\alpha) - \lambda_n(\alpha)), \tag{9.241}$$

hence the determinant $\delta(\alpha)$ of $J(\alpha)$ is not zero.

In the following paragraphs the potential usefulness of the initial-value system (9.238) as a practical tool for numerical work will be illustrated by example.

Example. Consider a matrix-valued function $M(\alpha)$ defined over α in R by

$$M(\alpha) = \begin{bmatrix} 1 & \alpha \\ \alpha^2 & 3 \end{bmatrix}. \tag{9.242}$$

For this simple example, analytical expressions are easily obtained for the eigenvalues $\{\lambda_1(\alpha), \lambda_2(\alpha)\}$ and the corresponding unit normalized right eigenvectors $\{x_1(\alpha), x_2(\alpha)\}$ of $M(\alpha)$. Specifically,

$$\lambda_1(\alpha) = 2 + \gamma(\alpha), \tag{9.243a}$$

$$\lambda_2(\alpha) = 2 - \gamma(\alpha), \tag{9.243b}$$

$$x_1(\alpha) = \left(\frac{\alpha}{1 + \gamma(\alpha)} k_1, k_1\right)^T, \tag{9.243c}$$

$$x_2(\alpha) = \left(\frac{\alpha}{1 - \gamma(\alpha)} k_2, k_2\right)^T, \tag{9.243d}$$

where

$$\gamma(\alpha) \equiv (1 + \alpha^3)^{1/2}, \tag{9.243e}$$

$$k_1 \equiv \left\{\frac{\alpha^2}{[1 + \gamma(\alpha)]^2} + 1\right\}^{-1/2}, \tag{9.243f}$$

$$k_2 \equiv \left\{\frac{\alpha^2}{[1 - \gamma(\alpha)]^2} + 1\right\}^{-1/2}, \tag{9.243g}$$

Consider the individual tracking of the first eigenvalue $\lambda_1(\alpha) \equiv \lambda(\alpha)$ and its corresponding right eigenvector $x_1(\alpha)^T \equiv (u(\alpha), v(\alpha))^T$. For this special case the differential system (9.238) reduces to

$$\begin{bmatrix} \dot{u} \\ \dot{v} \\ \dot{\lambda} \end{bmatrix} = \frac{A}{\delta} \begin{bmatrix} v \\ 2\alpha u \end{bmatrix} \tag{9.244a}$$

$$\dot{A} = [A \operatorname{Tr}(AB) - ABA]/\delta, \tag{9.244b}$$

$$\dot{\delta} = \operatorname{Tr}(AB), \tag{9.244c}$$

with initial conditions

$$u(\alpha^0) = \alpha^0 k_1/[1 + \gamma(\alpha^0)], \tag{9.244d}$$

$$v(\alpha^0) = k_1, \tag{9.244e}$$

$$\lambda(\alpha^0) = 2 + \gamma(\alpha^0), \tag{9.244f}$$

$$A(\alpha^0) = \text{adj}(J(\alpha^0)), \tag{9.244g}$$

$$\delta(\alpha^0) = \text{Det}\,(J(\alpha^0)), \tag{9.244h}$$

where

$$J(\alpha) = \begin{bmatrix} \lambda(\alpha) - 1 & -\alpha & u(\alpha) \\ -\alpha^2 & \lambda(\alpha) - 3 & v(\alpha) \\ u(\alpha) & v(\alpha) & 0 \end{bmatrix} \tag{9.244i}$$

and

$$B(\alpha) = \frac{dJ(\alpha)}{d\alpha} = \begin{bmatrix} \dot{\lambda}(\alpha) & -1 & \dot{u}(\alpha) \\ -2\alpha & \dot{\lambda}(\alpha) & \dot{v}(\alpha) \\ \dot{u}(\alpha) & \dot{v}(\alpha) & 0 \end{bmatrix}. \tag{9.244j}$$

A numerical solution was obtained for the first eigenvalue $\lambda_1(\alpha)$ and corresponding right eigenvector $x_1(\alpha)$ of $M(\alpha)$ by integrating the initial-value system (9.244) from $\alpha^0 = 0.5$ to 2.0. A fourth-order Runge–Kutta method was used for the integration with the α grid intervals set equal to 0.01. As indicated in Table 9.11, the numerical results agree with the analytical solution to at least six digits. Similar results were obtained for the second eigenvalue $\lambda_2(\alpha)$.

It is clear from (9.243a) and (9.243b) that the two eigenvalues $\lambda_1(\alpha)$ and $\lambda_2(\alpha)$ coalesce at $\alpha = -1.0$ and become complex for $\alpha < -1.0$. The eigenvector $x_1(\alpha)^T$ becomes (0, 1) at $\alpha = 0.0$. The initial-value system (9.244) was integrated from $\alpha^0 = 0.5$ to -1.0 through 0.0 using an integration step size of -0.01 for α. Six-digit accuracy was obtained integrating

Table 9.11. Eigenvalue $\lambda_1(\alpha) \equiv \lambda(\alpha)$ and Corresponding Eigenvector $x_1(\alpha)^T = (u(\alpha),\ v(\alpha))$ at $\alpha = 2.0$

Quantity	Numerical solution	Analytical solution
$\lambda(2)$	5.0	5.0
$u(2)$	0.447214	0.447214
$v(2)$	0.894427	0.894427

from 0.5 to -0.98. Approximately two-digit accuracy was obtained at the point $\alpha = -1.0$ where the roots coalesce.

9.5.6. A New Differential Equation Method for Finding the Perron Root of a Positive Matrix

A basic problem in linear algebra, such as for Saaty's matrices in Section 9.5.2, is the determination of the largest eigenvalue (Perron root) of a positive matrix. In this section a new differential equation method for finding the Perron root is given (Reference 48). The method utilizes the initial-value differential system developed in the previous section for individually tracking the eigenvalue and corresponding right eigenvector of a parametrized matrix.

Introduction

A basic problem in linear algebra (References 30, 37, 49) is the determination of the largest eigenvalue of an $n \times n$ matrix Q. When the entries of Q are nonnegative, the largest eigenvalue is referred to as the Perron or Perron–Frobenius root.

In Sections 9.5.4 and 9.5.5 it was shown that the eigenvalues and the right and left eigenvectors of a parametrized matrix $M(\alpha)$ can be tracked as functions of a scalar parameter α by integrating a system of ordinary differential equations from initial conditions. In this section it is shown how the initial-value differential system developed in the previous section can be modified to obtain an initial-value differential system for tracking the Perron root and a corresponding unit normalized right eigenvector for a positive matrix Q, i.e., a matrix Q with all positive elements.

The initial-value system developed in Section 9.5.5 is outlined followed by the development of the modified initial-value system for tracking the Perron root. A numerical example is then given.

Individual Tracking of an Eigenvalue and Eigenvector

Let $M(\alpha)$ be an $n \times n$ complex matrix-valued continuously differentiable function of a parameter α varying over a simply connected region of the complex plane. Consider the initial-value system for tracking a single eigenvalue $\lambda(\alpha)$ and corresponding unit normalized right eigenvector $x(\alpha)$ of $M(\alpha)$ if, at the initial point α^0, a certain Jacobian matrix $J(\alpha^0)$ is nonsingular.

Specifically, the differential system takes the form

$$\begin{bmatrix} \dot{x}(\alpha) \\ \dot{\lambda}(\alpha) \end{bmatrix} = \frac{A(\alpha)}{\delta(\alpha)} \begin{bmatrix} M(\alpha)x(\alpha) \\ 0 \end{bmatrix}, \tag{9.245a}$$

$$\dot{A}(\alpha) = [A(\alpha)\,\mathrm{Tr}(A(\alpha)B(\alpha)) - A(\alpha)B(\alpha)A(\alpha)]/\delta(\alpha), \tag{9.245b}$$

$$\dot{\delta}(\alpha) = \mathrm{Tr}(A(\alpha)B(\alpha)), \tag{9.245c}$$

with initial conditions

$$x(\alpha^0) = x^0, \tag{9.245d}$$

$$\lambda(\alpha^0) = \lambda^0, \tag{9.245e}$$

$$A(\alpha^0) = \mathrm{adj}(J(\alpha^0)), \tag{9.245f}$$

$$\delta(\alpha^0) = \mathrm{Det}(J(\alpha^0)), \tag{9.245g}$$

where

$$J(\alpha) \equiv \begin{bmatrix} \lambda(\alpha)I - M(\alpha) & x(\alpha) \\ x(\alpha)^T & 0 \end{bmatrix}, \tag{9.245h}$$

$$B(\alpha) \equiv \dot{J}(\alpha) = \begin{bmatrix} \dot{\lambda}(\alpha)I - \dot{M}(\alpha) & \dot{x}(\alpha) \\ \dot{x}(\alpha)^T & 0 \end{bmatrix}, \tag{9.245i}$$

and a dot denotes differentiation with respect to α.

The initial conditions (9.245d)–(9.245g) are obtained by solving

$$M(\alpha^0)x = \lambda x, \tag{9.246a}$$

$$x^T x = 1, \tag{9.246b}$$

for x^0 and λ^0, and then determining the adjoint $A(\alpha^0)$ and determinant $\delta(\alpha^0)$ of the Jacobian matrix

$$J(\alpha) = \begin{bmatrix} \lambda^0 I - M(\alpha^0) & x^0 \\ x^{0T} & 0 \end{bmatrix} \tag{9.247}$$

for system (9.246).

Variational Equations for the Perron Root

The presently proposed procedure for determining the Perron root λ_p of an arbitrary $n \times n$ positive matrix Q is as follows. Define

$$M(\alpha) \equiv (1 - \alpha)C + \alpha Q, \qquad 0 \leqq \alpha \leqq 1, \tag{9.248}$$

where

$$C \equiv \begin{bmatrix} 1 & \cdots & 1 \\ \vdots & & \vdots \\ 1 & \cdots & 1 \end{bmatrix}_{n \times n} . \tag{9.249}$$

The Perron root $\lambda(0)$ of $M(0) = C$ is easily determined to be $\lambda(0) = n$, and the unit length normalized right eigenvector corresponding to $\lambda(0)$ may be taken to be $x(0)^T = (1/n^{1/2}, \ldots, 1/n^{1/2})$. Clearly the Perron root $\lambda(1)$ of $M(1)$ is the desired Perron root λ_p of Q. Thus, in principle, the Perron root λ_p of Q may be found by integrating system (9.245) from $\alpha^0 = 0.0$ to $\alpha = 1.0$ using the initial conditions

$$x(0)^T = (1/n^{1/2}, \ldots, 1/n^{1/2}), \tag{9.250}$$

$$\lambda(0) = n, \tag{9.251}$$

$$A(0) = \operatorname{adj}(J(0)), \tag{9.252}$$

$$\delta(0) = \operatorname{Det}(J(0)), \tag{9.253}$$

where

$$\begin{aligned} J(0) &= \begin{bmatrix} \lambda(0)I - M(0) & x(0) \\ x(0)^T & 0 \end{bmatrix} \\ &= \begin{bmatrix} nI - C & x(0) \\ x(0)^T & 0 \end{bmatrix}. \end{aligned} \tag{9.254}$$

Owing to the nonlinearity of the differential equation system (9.245), the integration may be stopped prior to reaching $\alpha = 1.0$. However, as the following theorem demonstrates, the Jacobian matrix

$$J(\alpha) \equiv \begin{bmatrix} \lambda(\alpha)I - M(\alpha) & x(\alpha) \\ x(\alpha)^T & 0 \end{bmatrix} \tag{9.255}$$

is nonsingular for $0 \leqq \alpha \leqq 1$, where $\lambda(\alpha)$ is the Perron root of $M(\alpha)$ and $x(\alpha)$ is a corresponding unit normalized right eigenvector with all components taken to be positive. Thus, $\delta(\alpha) = \operatorname{Det}(J(\alpha))$ is uniformly bounded away from zero over the compact α interval [0, 1].

Theorem. Let λ_p be the Perron root of a positive $n \times n$ matrix M, and let x_p be a corresponding right eigenvector with all elements taken to be positive. Then

$$\operatorname{Det} \begin{bmatrix} \lambda_p I - M & x_p \\ x_p^T & 0 \end{bmatrix} \neq 0. \tag{9.256}$$

Proof. It is well known that the dominant (Perron) root λ_p of a positive $n \times n$ matrix M is positive and simple, and that the elements of the right and left eigenvectors x_p and w_p corresponding to λ_p can be taken to be positive.

Define $N \equiv \lambda_p I - M$. Suppose (9.256) is false, i.e., suppose there exists a nonzero vector $(y, s)^T$ satisfying

$$\begin{bmatrix} N & x_p \\ x_p^T & 0 \end{bmatrix} \begin{bmatrix} y \\ s \end{bmatrix} = \mathbf{0}. \tag{9.257}$$

Then

$$Ny + x_p s = \mathbf{0}, \tag{9.258a}$$

$$x_p^T y = 0. \tag{9.258b}$$

Multiplying through (9.258a) by w_p^T, one obtains $w_p^T N y + w_p^T x_p s = \mathbf{0}$, which in turn implies $s = 0$ since $w_p^T N = \mathbf{0}$ and $w_p^T x_p > 0$. It follows from (9.258) that $Ny = \mathbf{0}$ and $x_p^T y = 0$. However, since λ_p is simple, the right eigenvector x_p is unique up to positive linear transformation. Thus $y = \mathbf{0}$, a contradiction.

It follows from this proof by contradiction that condition (9.256) must be true. □

Example

Consider the 2×2 positive matrix Q defined by

$$Q \equiv \begin{bmatrix} 1 & 2 \\ 4 & 5 \end{bmatrix}. \tag{9.259}$$

It is easily verified that, to six decimal places, the Perron root of Q is

$$\lambda_p = 6.464102, \tag{9.260}$$

with corresponding positive unit normalized right eigenvector given by

$$x_p = \begin{bmatrix} u_p \\ v_p \end{bmatrix} = \begin{bmatrix} 0.343724 \\ 0.939071 \end{bmatrix}. \tag{9.261}$$

To find the Perron root of Q by means of the differential equation method developed above, we first define

$$M(\alpha) \equiv (1 - \alpha)\begin{bmatrix} 1 & 1 \\ 1 & 1 \end{bmatrix} + \alpha Q, \qquad 0 \leqq \alpha \leqq 1. \tag{9.262}$$

The appropriate initial-value system (9.245) for the problem at hand then reduces to

$$\begin{bmatrix} \dot{u}(\alpha) \\ \dot{v}(\alpha) \\ \dot{\lambda}(\alpha) \end{bmatrix} = \frac{A(\alpha)}{\delta(\alpha)} \begin{bmatrix} v(\alpha) \\ 3u(\alpha) + 4v(\alpha) \\ 0 \end{bmatrix}, \tag{9.263a}$$

$$\dot{A}(\alpha) = [A(\alpha)\,\mathrm{Tr}(A(\alpha)B(\alpha)) - A(\alpha)B(\alpha)A(\alpha)]/\delta(\alpha), \tag{9.263b}$$

$$\dot{\delta}(\alpha) = \mathrm{Tr}(A(\alpha)B(\alpha)), \tag{9.263c}$$

with initial conditions

$$u(0) = 1/2^{1/2}, \tag{9.263d}$$

$$v(0) = 1/2^{1/2}, \tag{9.263e}$$

$$\lambda(0) = 2, \tag{9.263f}$$

$$A(0) = \mathrm{adj}(J(0)), \tag{9.263g}$$

$$\delta(0) = \mathrm{Det}(J(0)), \tag{9.263h}$$

where

$$J(0) = \begin{bmatrix} 1 & -1 & 1/2^{1/2} \\ -1 & 1 & 1/2^{1/2} \\ 1/2^{1/2} & 1/2^{1/2} & 0 \end{bmatrix} \tag{9.263i}$$

Table 9.12. Perron Eigenvalue $\lambda(\alpha)$ and Corresponding Right Eigenvector $x(\alpha)^T = (u(\alpha), v(\alpha))$ of $M(\alpha)$, and Determinant $\delta(\alpha)$ of $J(\alpha)$; the Perron Root of Q is Given by $\lambda(\alpha)$ at $\alpha = 1$

α	Eigenvector		Eigenvalue $\lambda(\alpha)$	Determinant $\delta(\alpha)$
	$u(\alpha)$	$v(\alpha)$		
0.0	0.707107	0.707107	2.0	−2.0
0.1	0.614441	0.788963	2.41244	−2.42487
0.2	0.545806	0.837912	2.84222	−2.88444
0.3	0.494948	0.868923	3.28226	−3.36452
0.4	0.456485	0.889731	3.72873	−3.85746
0.5	0.426679	0.904403	4.17945	−4.35890
0.6	0.403041	0.915182	4.63311	−4.86621
0.7	0.383904	0.923373	5.08887	−5.37773
0.8	0.368131	0.929774	5.54618	−5.89237
0.9	0.354927	0.934894	6.00468	−6.40937
1.0	0.343724	0.939071	6.46410	−6.92820

and

$$B(\alpha) = \dot{J}(\alpha) = \begin{bmatrix} \dot{\lambda}(\alpha) & -1 & \dot{u}(\alpha) \\ -3 & \dot{\lambda}(\alpha) - 4 & \dot{v}(\alpha) \\ \dot{u}(\alpha) & \dot{v}(\alpha) & 0 \end{bmatrix}. \tag{9.263j}$$

A fourth-order Runge–Kutta method was used to integrate system (9.263) from $\alpha = 0$ to $\alpha = 1$, with grid intervals equal to 0.01. As indicated in Table 9.12, the Perron root $\lambda(1)$ obtained for $M(1) = Q$ agrees to at least six digits with the analytically derived Perron root λ_p of Q given by (9.260).

Exercises

1. Derive the Euler–Lagrange equations for the improved method for the numerical determination of optimal inputs for the example given by equation (9.8).

2. Derive the quasilinearization equations for the solution of the two-point boundary-value problem in Exercise 1.

3. Derive the two-point boundary-value equations for the improved method for the numerical determination of optimal inputs for the van der Pol equation in Exercise 1, Chapter 7. Use Pontryagin's maximum principle and assume that a is the unknown parameter to be estimated and that only x is observed.

4. Derive the two-point boundary-value equations for the improved method for the numerical determination of optimal inputs for the mass attached to a nonlinear spring in Exercise 4, Chapter 7. Use Pontryagin's maximum principle and assume that a is the parameter to be estimated.

5. Using the method of Section 9.2 derive the two-point boundary-value equations for obtaining the optimal input of the van der Pol equation in Exercise 1, Chapter 7, assuming that both a and b are unknown parameters to be estimated. Use Pontryagin's maximum principle and assume that only x is observed.

6. Derive the two-point boundary-value equations in Exercise 5 assuming that both x and $\dot{x}$ are observed.

7. Using the method of Section 9.2 derive the two-point boundary-value equations for obtaining the optimal input of the mass attached to a nonlinear spring in Exercise 4, Chapter 7, assuming that both a and ω_0 are unknown parameters to be estimated. Use Pontryagin's maximum and assume that only x is observed.

8. Given the $n \times n$ matrix $B(\alpha)$

$$B(\alpha) = \begin{bmatrix} 2\cos\alpha & 1 & 0 & 0 & & \\ 1 & 2\cos\alpha & 1 & 0 & & \\ 0 & 1 & 2\cos\alpha & 1 & & \\ & & & \cdot & & \\ & & & & \cdot & \\ & & & & \cdot & \\ & & & & 1 & 2\cos\alpha \end{bmatrix},$$

(a) show that the determinant is given by

$$\det B(\alpha) = \frac{\sin(n+1)\alpha}{\sin\alpha};$$

(b) use the method of Section 9.5.1 to obtain the differential equations for the determinant and adjoint of $B(\alpha)$, and try to integrate numerically from $\alpha = 0$ to $\alpha = 1$ for $n = 3$. Check the results using the analytical solution above noting that $\sin \pi = 0$.

9. Derive the eigenvalues and eigenvectors given in equations (9.233a)–(9.233f).

10

Applications of Optimal Inputs

Two applications of optimal inputs for parameter estimation are given in this chapter. The first application is for blood glucose regulation parameter estimation. The second application is for aircraft parameter estimation.

10.1. Optimal Inputs for Blood Glucose Regulation Parameter Estimation

In Chapter 7, parameter estimation techniques were applied to the estimation of the blood glucose regulation parameters. In this section optimal inputs are derived to enhance the sensitivity of the observed data to the unknown parameters.

In Section 10.1.1 the absolute sensitivity of the observation to the parameter b is to be maximized. The rate of insulin production is proportional to the parameter b. It is assumed that only one, or at most relatively few, measurements of the blood glucose concentration are to be made. The nominal parameter values estimated by Bolie are used to derive the optimal input. The solution is obtained using linear or dynamic programming.

In Section 10.1.2 it is assumed that several measurements of the blood glucose concentration are to be made. The sensitivity of the observations to the parameter b are maximized using a quadratic performance criterion subject to an input energy constraint. The nominal Bolie parameter values are again used. The solution is obtained using the method of complementary functions or the Riccati equation method described in Section 8.2.

In Section 10.1.3 the improved method for the numerical determination of optimal inputs described in Section 9.1 is used. The criterion is the quadratic performance index subject to an input energy constraint. The optimal inputs are derived for the Bergman parameter values estimated in Chapter 7 and are rederived for the Bolie parameter values and compared.

10.1.1. Formulation Using Bolie Parameters for Solution by Linear or Dynamic Programming

The estimation of the parameters for blood glucose regulation is enhanced by using a glucose infusion input which is optimal in some given mathematical sense. Using a linear two-compartment model, techniques are given for deriving the optimal inputs using linear or dynamic programming (Reference 1). Methods of extending the techniques for different performance criteria and more complex glucose regulation models are discussed. These methods may some day prove useful for developing new clinical tests which can differentiate the normal patient from the diabetic.

Introduction

Despite many years of experience the diagnosis of metabolic disease in humans remains an extremely difficult problem. It was once felt that diabetes mellitus could be simply explained as a total lack of insulin in the blood stream, due to the inability of the pancreas to secrete the necessary anabolic hormone. However, precise methods of measuring insulin which have been introduced in recent years have made clinicians aware that the syndrome referred to as diabetes is actually a host of different diseases, referred to as juvenile diabetes, maturity onset diabetes, prediabetes, brittle diabetes, etc. It is necessary to determine for each patient the nature and severity of the disease, so that the appropriate treatment with insulin, sulfonylurea, or dietary regulation can be determined.

Traditionally, diabetes is diagnosed by performing a glucose tolerance test, wherein glucose is taken orally by a patient and the concentration of glucose in the blood is determined after two hours. Although standards vary, in general, a patient given 100 g glucose/kg orally who shows a blood glucose concentration two hours later of 140 mg/100 ml or more is considered diabetic (Reference 2). If the glucose concentration is somewhat lower, some other diagnosis will apply.

Although the standard glucose tolerance test procedure has been successful in elucidating the severity of diabetes, it has been suggested that alternative patterns of glucose infusion may be more successful in differentiating among the various diabetic states. For example, Cerasi and Luft, in Stockholm, have used an intravenous pulse of glucose followed by a constant infusion, and have claimed that this approach is more effective in detecting prediabetes (chemical diabetes) (Reference 3).

A variety of patterns of perturbation of the metabolic regulating system by glucose have been evaluated as to their diagnostic potential. No attempt has been made, however, using mathematical techniques, to find the stimulation pattern which will allow for optimal differentiation between the various diabetic states.

The primary lesion in most forms of diabetes is still felt to be an insensitivity of β cell insulin secretion to stimulation by glucose. The diminished insulin secretion in response to a glucose tolerance test results in a slower than normal rate of glucose uptake by the liver and peripheral tissue (glucose intolerance). Thus, when a glucose tolerance test is performed, the parameter which the physician is attempting to determine (qualitatively, at least) is the β cell sensitivity to glucose. The question proposed then is the following: how may the techniques of optimality be used to design a pattern of glucose stimulation of the metabolic system which will give the most sensitive indication of the responsivity of the pancreas to glucose?

The estimation of the blood glucose regulation parameters has been previously investigated by Bolie (Reference 4) utilizing a linear two-compartment model. The Bolie parameter values are used in this section to derive optimal inputs. Linear models for estimation have also been used by Ackerman *et al.*, Ceresa *et al.*, and Segre *et al.*, References 2–5 in Chapter 7.

Linear Two-Compartment Model

The concentration of glucose in the blood is maintained by those tissues which allow glucose to flow into or out of the glucose space. Insulin, the hormone secreted by the β cells of the pancreas, is the primary regulator of the utilization of glucose. The rate of secretion of insulin into the blood is a function of the blood glucose concentration. The rate of glucose utilization by the peripheral tissue is a function of not only the glucose concentration, but also the plasma insulin level. Thus, in its simplest formulation, the system regulating blood glucose is a classical negative feedback system.

If it is assumed that

1. insulin secretion is directly proportional to the blood glucose concentration;

2. glucose utilization is a linear function of the glucose concentration and the plasma insulin level;
3. glucose production in the fasting state is equal to the glucose utilization, and the glucose production by the liver does not change during glucose infusion;

then the linear two-compartment Bolie model is obtained:

$$\dot{x}_1 = -ax_1 + bx_2, \qquad x_1(0) = 0, \tag{10.1}$$

$$\dot{x}_2 = -cx_1 - dx_2 + y, \qquad x_2(0) = 0, \tag{10.2}$$

where x_1 is the deviation of the extracellular insulin concentration from the mean, in units/liter; x_2 is the deviation of the extracellular glucose concentration from the mean, in g/liter; y is the rate of glucose intravenous injection, in g/hr/liter; $a = 0.78$ 1/hr, $b = 0.208$ units/hr/g, $c = 4.34$ g/hr/unit, $d = 2.92$ 1/hr.

Optimization Problem

The values of the constant parameters, a, b, c, and d are those given by Bolie for the average normal adult. For the diabetes patient, however, the values of these parameters will be different. The rate of insulin production, for example, will be considerably lower than for the normal patient. Since the rate of insulin production is proportional to the parameter, b, it is desired to estimate the magnitude of this parameter. The estimation of the parameter b can be optimized by designing the glucose infusion input such that the sensitivity of the measurements to the parameter b is maximized. It is assumed that only one measurement is to be made. Thus it is desired to maximize the absolute value of the sensitivity at terminal time, T, such that

$$\phi[x_{2b}(T)] = \max_{y} \left| \frac{\partial x_2(T)}{\partial b} \right| \tag{10.3}$$

subject to the input rate constraint

$$0 \leq y(t) \leq \alpha \quad \text{g/hr/liter}. \tag{10.4}$$

Differentiating equations (10.1) and (10.2) with respect to b and setting

$$x_{1b} = \frac{\partial x_1}{\partial b} = x_3, \tag{10.5}$$

$$x_{2b} = \frac{\partial x_2}{\partial b} = x_4, \tag{10.6}$$

we obtain the following equations:

$$\dot{x}_3 = -ax_3 + bx_4 + x_2, \qquad x_3(0) = 0, \tag{10.7}$$

$$\dot{x}_4 = -cx_3 - dx_4, \qquad x_4(0) = 0. \tag{10.8}$$

Equations (10.1), (10.2), (10.7), and (10.8) can be expressed in the matrix form

$$\begin{bmatrix} \dot{x}_1 \\ \dot{x}_2 \\ \dot{x}_3 \\ \dot{x}_4 \end{bmatrix} = \begin{bmatrix} -a & b & 0 & 0 \\ -c & -d & 0 & 0 \\ 0 & 1 & -a & b \\ 0 & 0 & -c & -d \end{bmatrix} \begin{bmatrix} x_1 \\ x_2 \\ x_3 \\ x_4 \end{bmatrix} + \begin{bmatrix} 0 \\ 1 \\ 0 \\ 0 \end{bmatrix} y \tag{10.9}$$

or

$$\dot{\mathbf{x}} = A\mathbf{x} + B\mathbf{y}, \qquad \mathbf{x}(0) = 0. \tag{10.10}$$

The state transition matrix is given by

$$\Phi(t) = \mathscr{L}^{-1}[sI - A]^{-1}, \tag{10.11}$$

where s is the Laplace transform and $\mathscr{L}^{-1}$ is the inverse Laplace transform. The eigenvalues are obtained by setting the determinant $|sI - A|$ equal to zero and solving for the roots. The eigenvalues are

$$\begin{aligned} s_1 &= \tfrac{1}{2}\{-(a+d) + [(a+d)^2 - 4(ad+bc)]^{1/2}\}, \\ s_2 &= \tfrac{1}{2}\{-(a+d) - [(a+d)^2 - 4(ad+bc)]^{1/2}\}, \\ s_3 &= s_1, \\ s_4 &= s_2. \end{aligned} \tag{10.12}$$

At terminal time T

$$\mathbf{x}(T) = \Phi(T-t)\mathbf{x}(0) + \int_t^T \Phi(T-t')y(t')\,dt'. \tag{10.13}$$

The component of $\mathbf{x}$ of interest is the fourth component. Then using the concept of the reduction of dimensionality (Reference 5) and the initial condition given by equation (10.10) we obtain

$$x_{2b}(T) = \int_t^T \Phi_{42}(T-t')y(t')\,dt', \tag{10.14}$$

where

$$\begin{aligned} \Phi_{42}(t) &= (A_1 t + B_1)e^{-s_1 t} + (A_2 t + B_2)e^{-s_2 t}, \\ A_1 &= -\frac{c(a - s_1)}{(s_2 - s_1)^2}, \\ B_1 &= -\frac{c(s_1 + s_2 - 2a)}{(s_2 - s_1)^3}, \\ A_2 &= -\frac{c(a - s_2)}{(s_2 - s_1)^2}, \\ B_2 &= -B_1. \end{aligned} \tag{10.15}$$

Thus the dimensionality of the problem for a single measurement at time T has been reduced from fourth order to unity. The optimization can be performed utilizing either dynamic programming or linear programming.

Linear Programming

In order to solve the above problem using linear programming (Reference 6), assume that it is desired to maximize the absolute value of the ith component of $\mathbf{x}$:

$$\phi[x_i(T)] = \max_y |x_i(T)|. \tag{10.16}$$

Then for zero initial conditions and a scalar input, equation (10.14) can be expressed in the form

$$x_i(T) = \int_t^T \Phi_{ij}(T - t')y(t')\,dt'. \tag{10.17}$$

Using a rectangular quadrature the objective function becomes

$$x_i(T) = \sum_{k=1}^{N} \Phi_{ij}(T - r_k)y(r_k)w_k. \tag{10.18}$$

Thus the linear programming equations are given by the objective function and inequality constraints

$$x_i(T) = \sum_{k=1}^{N} C_k y(r_k), \tag{10.19}$$

$$0 \leq y(r_k) \leq \alpha, \tag{10.20}$$

where

$$C_k = \Phi_{ij}(T - r_k)w_k,$$
$$w_k = T/N. \tag{10.21}$$

Since the absolute value of $x_i(T)$ is to be maximized, the linear programming problem must be solved twice—once as shown above and again with the coefficients in equation (10.19) replaced by $-C_k$. The absolute maximum of $x_i(T)$ is then utilized as the optimum.

Numerical Results

Numerical results were obtained utilizing linear programming. A rectangular quadrature with $N = 10$ was used with input rate constraints

$$0 \leq y(r_k) \leq 1 \text{ g/hr/liter} \tag{10.22}$$

and terminal time $T = 5$ hr. The results are shown in Figure 10.1. The coefficients, C_k, are shown as a function of time. The optimum input is a step input applied three hours before measurement time.

Dynamic Programming

Dynamic programming can also be used to solve the above problem. As before the dimensionality of the problem is reduced using the concept of reduction of dimensionality (Reference 5). Consider the linear system

$$\dot{\mathbf{x}} = A\mathbf{x} + \mathbf{y}, \qquad \mathbf{x}(0) = \mathbf{c}, \tag{10.23}$$

where $\mathbf{x}$ is of dimension n. Assuming that a function of the ith component of $\mathbf{x}$ is to be maximized at terminal time T, define the functional $f(\mathbf{c}, t)$ as in the usual dynamic programming procedure:

$$\begin{aligned} f(\mathbf{c}, t) &= \max_{\mathbf{y}} \phi[x_i(T)] \\ &= \max_{\mathbf{y}} f(\mathbf{c} + \Delta(A\mathbf{x} + \mathbf{y}), t + \Delta). \end{aligned} \tag{10.24}$$

Equation (10.24) requires the solution of an n-dimensional state vector. However, the dimensionality is reduced by defining

$$f(c, t) = \max_{\mathbf{y}} \phi[x_i(T)], \tag{10.25}$$

where c is the state of the ith component of $\mathbf{x}$ at time T if no control is

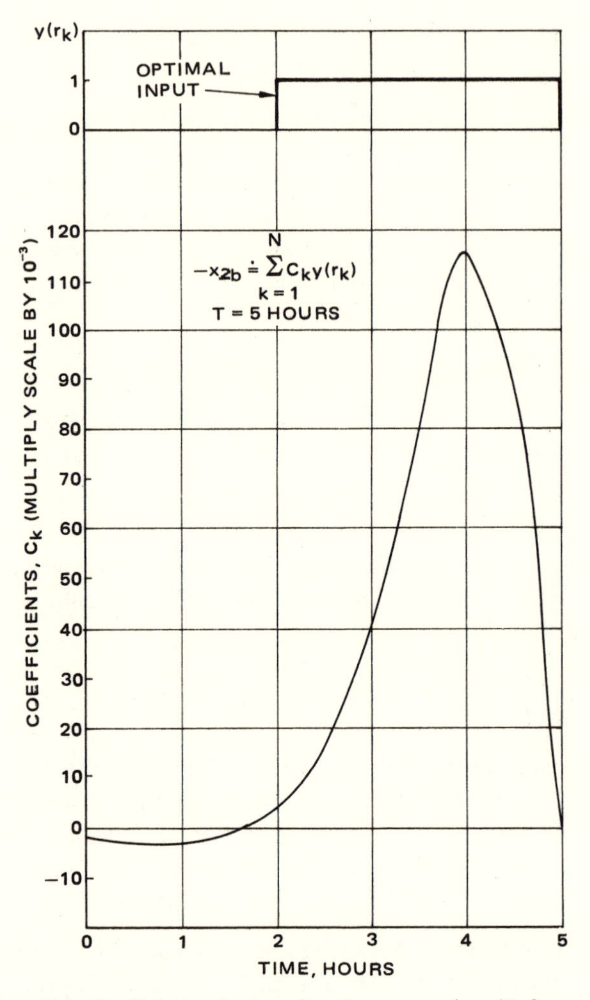

Figure 10.1. Coefficients of return function versus time (Reference 1).

exerted for $t \geq 0$ and

$$c = x_i(T). \tag{10.26}$$

At the terminal time, T, we have

$$\mathbf{x}(T) = \Phi(T - t)\mathbf{c} + \int_t^T \Phi(T - t')\mathbf{y}(t')\,dt'. \tag{10.27}$$

The ith component of $\mathbf{x}(T)$ is

$$x_i(T) = z_i + \int_t^T \sum_{j=1}^n \Phi_{ij}(T - t')y_j(t')\,dt', \tag{10.28}$$

where z_i is the ith component of the vector $\Phi(T-t)\mathbf{c}$. The functional $f(c, t)$ is then given by

$$f(c, t) = \max_{\mathbf{y}} f(c + \Delta \sum_{j=1}^{n} \Phi_{ij}(T-t)y_j, t + \Delta) \tag{10.29}$$

with boundary condition

$$f(c, T) = \phi[x_i(T)]. \tag{10.30}$$

For a scalar input, all the components of y are zero except the jth component. Thus the true dimensionality of the problem is unity with the functional $f(c, t)$ given by

$$f(c, t) = \max_{y} f(c + \Delta\Phi_{ij}(T-t)y, t + \Delta) \tag{10.31}$$

and input constraint

$$0 \leq y \leq \alpha. \tag{10.32}$$

If it is desired to maximize a function of several measurement times, then the dimensionality increases accordingly. Thus we have

$$f(c_1, c_2, \ldots, c_k, t) = \max_{\mathbf{y}} \phi[x_i(T_1), x_i(T_2), \ldots, x_i(T_k)]. \tag{10.33}$$

Extensions Using the Linear Two-Compartment Model

Generally it is desired to measure the blood glucose concentration at more than one time period. However, the optimization rapidly becomes more difficult as the number of measurements increases. Utilizing dynamic programming the dimensionality increases as the number of measurements increases. Utilizing linear programming both positive and negative sensitivities must be investigated at each measurement time.

Different formulations of the problem are also possible. For example, using a quadratic performance criterion, the return function

$$\phi = \max_{y} \int_0^T [x_{2b}^2(t') - qy^2(t')]\, dt' \tag{10.34}$$

could be utilized, where $y^2(t)$ is used to limit the input. This approach will be utilized in the next section.

Discussion

In this section, the possibility has been raised that the techniques of optimality may be useful in generating new clinical tests which may be more effective in diagnosing disease than those presently available. In particular, this section has focused on the diagnosis of diabetes mellitus, which may be related to the sensitivity of pancreatic insulin secretion to the plasma glucose concentration. Using a model of carbohydrate metabolism at the level of the whole animal, it seems possible to determine an optimal pattern of infusion of glucose, such that a determination of the blood glucose at time T will give the most sensitive measure of the insulin secretory function of the pancreas.

Using Bolie's model, it was found that, optimally, glucose should be measured three hours after the beginning of a glucose infusion, when the infusion rate is constrained to some maximum value. However, other possibly more interesting inputs might be suggested if a more complex, but more realistic, model of carbohydrate metabolism is chosen. For example, the Bolie representation of insulin secretory function simply as a constant parameter, b, is not entirely correct. It has been known for some years that a step input of glucose provokes a biphasic insulin secretory response. One model of the pancreas which can simulate the dynamics of insulin secretion was suggested by Srinivasan *et al.*, who assumed the insulin secretion rate to be proportional to the magnitude plus the rate of change of the plasma glucose concentration (References 7–9 in Chapter 7).

Another improvement of the formulation would be to constrain the value of x_2, such that $| x_2(t) | \leq F$, rather than constraining y alone. Optimizing with $x_2(t)$ constrained in a small range would increase the validity of the assumption of linearity and would also diminish the probability that physiological mechanisms not included in the model would have as significant an effect on the state of the system. For example, if x_2 were to be less than 30 mg/100 ml, the sympathetic nervous system would begin to play a significant role in increasing x_2, by increasing the secretion of epinephrine and glucagon. Likewise, if x_2 were greater than 80 mg/100 ml, the renal threshold for glucose would be exceeded, and the kidney would have to be included in the model.

A further improvement in the model would be to replace the linear representation of insulin secretion by one of the mathematical models of the pancreas which are available in the literature (References 10–11 in Chapter 7). The models are nonlinear since the coefficients of the differential equations governing insulin secretion are functions of the plasma glucose concentration.

The design of optimal inputs for determining relevant parameters of the metabolic system has potential for improving the diagnosis of diabetes in humans. Fortunately, a variety of models of the system have been developed during the last decade. It is hoped that by evolving novel testing regimens, a new approach will be generated for differentiating more clearly the normal and the prediabetic states.

10.1.2. Formulation Using Bolie Parameters for Solution by Method of Complementary Functions or Riccati Equation Method

The estimation of the parameters of blood glucose regulation is of interest in developing mathematical models of the regulation process and in possible clinical applications where a patient may be classified as either normal or diabetic on the basis of the magnitude of his parameters. The estimation of parameters can be improved by utilizing optimal inputs to enhance the sensitivity of the observed data to the unknown parameters. Using a quadratic performance criterion, the determination of the optimal input is shown to involve the solution of a two-point boundary-value problem. Methods for the solution of the optimal input are described and numerical results are given (Reference 7).

In the previous section, it was assumed that only one or at most relatively few measurements of the blood glucose concentration were to be made. In this section it is assumed that several measurements are to be made.

The perturbation equations for blood glucose regulation are given by the simplified linear differential equations (Reference 4).

$$\dot{x}_1 = -ax_1 + bx_2, \tag{10.35}$$

$$\dot{x}_2 = -cx_1 - dx_2 + y, \tag{10.36}$$

where x_1 is the deviation of the extracellular insulin concentration from the mean, in units/l; x_2 is the deviation of the extracellular glucose concentration from the mean, in g/l; and y is the rate of glucose intravenous injection, in g/hr/l. It is assumed that the rate of insulin production per gram, b, is the parameter which is to be estimated. The derivations which follow are given only for the single unknown parameter, b, with the values of the other parameters assumed to be known. The equations can easily be modified, however, for the case where any one of the other parameters is unknown or extended to the case where several parameters are unknown (Chapter 9, Section 9.2).

Since b is unknown, an initial estimate of b must be utilized in order to obtain the optimal input. Then, if necessary, utilizing the optimal input with the parameter estimation techniques described in Chapter 7, Section

7.1, a new estimate of b is found and the process repeated until a satisfactory estimate of b is obtained.

Two-Point Boundary-Value Problem

Equations (10.35) and (10.36) can be expressed in vector form by the time-invariant linear system

$$\dot{\mathbf{x}}(t) = F\mathbf{x}(t) + Gy(t), \tag{10.37}$$

where $\mathbf{x}$ is a 2×1 state vector, y is a scalar control or input, and

$$F = \begin{bmatrix} -a & b \\ -c & -d \end{bmatrix}, \tag{10.38}$$

$$G = \begin{bmatrix} 0 \\ 1 \end{bmatrix}. \tag{10.39}$$

Assume that the measurements are contaminated with white noise

$$z(t) = H\mathbf{x}(t) + v(t), \tag{10.40}$$

where z is a scalar measurement,

$$H = [0 \quad 1], \tag{10.41}$$

v is the zero mean Gaussian white noise,

$$E[v(t)] = 0, \tag{10.42}$$

and

$$E[v(t)v^T(\tau)] = R\delta(t - \tau). \tag{10.43}$$

The parameter b is assumed to be unknown in equation (10.35). Then the optimal input is to be determined such that the Fisher information matrix

$$M = \int_0^{T_f} \mathbf{x}_b{}^T H^T R^{-1} H \mathbf{x}_b \, dt \tag{10.44}$$

is maximized subject to the input energy constraint

$$\int_0^{T_f} y^2 \, dt = E, \tag{10.45}$$

where $\mathbf{x}_b$ is the sensitivity vector defined by

$$\mathbf{x}_b = \frac{\partial \mathbf{x}}{\partial b}. \tag{10.46}$$

Equations (10.44) and (10.45) are made up of quadratic forms of the sensitivity and the control, the latter being utilized to limit the input dosage. The inverse of the Fisher information matrix, M^{-1}, is the Cramér–Rao lower bound.

Taking the partial derivative of equation (10.37) with respect to b yields

$$\dot{\mathbf{x}}_b = F\mathbf{x}_b + F_b\mathbf{x}, \tag{10.47}$$

where

$$F_b = \frac{\partial}{\partial b}[F]. \tag{10.48}$$

Let $\mathbf{x}_A$ equal the 4×1 augmented state vector

$$\mathbf{x}_A = \begin{bmatrix} \mathbf{x} \\ \mathbf{x}_b \end{bmatrix}. \tag{10.49}$$

Then

$$\dot{\mathbf{x}}_A = F_A\mathbf{x}_A + G_A y, \tag{10.50}$$

where

$$F_A = \begin{bmatrix} F & 0 \\ F_b & F \end{bmatrix}, \tag{10.51}$$

$$G_A = \begin{bmatrix} G \\ 0 \end{bmatrix}. \tag{10.52}$$

The performance index can then be expressed as the return, J,

$$J = \max_{y} \frac{1}{2} \int_0^{T_f} [\mathbf{x}_A{}^T H_A{}^T R^{-1} H_A \mathbf{x}_A - qy^2]\, dt, \tag{10.53}$$

where H_A is the 1×4 matrix

$$H_A = [0 \quad H]. \tag{10.54}$$

Utilizing Pontryagin's maximum principle, the Hamiltonian function is

$$\mathscr{H} = \tfrac{1}{2}[-\mathbf{x}_A{}^T H_A{}^T R^{-1} H_A \mathbf{x}_A + qy^2] + \boldsymbol{\lambda}^T[F_A\mathbf{x}_A + G_A y]. \tag{10.55}$$

The costate vector $\boldsymbol{\lambda}(t)$ is the solution of the vector differential equation

$$\dot{\boldsymbol{\lambda}} = -\left[\frac{\partial \mathscr{H}}{\partial \mathbf{x}_A}\right]^T$$
$$= H_A{}^T R^{-1} H_A \mathbf{x}_A - F_A{}^T \boldsymbol{\lambda}. \tag{10.56}$$

The input $y(t)$ that maximizes $\mathscr{H}$ is

$$\frac{\partial \mathscr{H}}{\partial y} = qy + G_A{}^T \boldsymbol{\lambda} = 0,$$
$$y = -\frac{1}{q} G_A{}^T \boldsymbol{\lambda}. \tag{10.57}$$

The two-point boundary-value problem is then given by

$$\begin{bmatrix} \dot{\mathbf{x}}_A \\ \dot{\boldsymbol{\lambda}} \end{bmatrix} = \begin{bmatrix} F_A & -(1/q)G_A G_A{}^T \\ H_A{}^T R^{-1} H_A & -F_A{}^T \end{bmatrix} \begin{bmatrix} \mathbf{x}_A \\ \boldsymbol{\lambda} \end{bmatrix} \tag{10.58}$$

with boundary conditions

$$\mathbf{x}_A = \begin{bmatrix} \mathbf{x}(0) \\ 0 \end{bmatrix}, \qquad \boldsymbol{\lambda}(T_f) = 0. \tag{10.59}$$

Equation (10.58) can also be expressed in the form

$$\dot{\mathbf{x}}_B = A\mathbf{x}_B, \tag{10.60}$$

where

$$\mathbf{x}_B = \begin{bmatrix} \mathbf{x}_A \\ \boldsymbol{\lambda} \end{bmatrix}, \tag{10.61}$$

$$A = \begin{bmatrix} F_A & -(1/q)G_A G_A{}^T \\ H_A{}^T R^{-1} H_A & -F_A{}^T \end{bmatrix}. \tag{10.62}$$

The solution to the two-point boundary-value problem can be found by utilizing the method of complementary functions or the Riccati equation method as suggested by Mehra.

Method of Complementary Functions

The method of complementary functions makes use of the 8×8 transition matrix Φ obtained by integrating

$$\dot{\Phi} = A\Phi, \qquad \Phi(0) = I, \tag{10.63}$$

where A is defined by equation (10.62). The solution to equation (10.60) is given by

$$\mathbf{x}_B(t) = \Phi(t)\mathbf{x}_B(0), \tag{10.64}$$

where the vector

$$\mathbf{x}_B(0) = \begin{bmatrix} \mathbf{x}_A(0) \\ \boldsymbol{\lambda}(0) \end{bmatrix} \tag{10.65}$$

is determined from the partitioned transition matrix equation

$$\begin{bmatrix} \mathbf{x}_A(0) \\ \boldsymbol{\lambda}(T_f) \end{bmatrix} = \begin{bmatrix} \Phi_{11}(0) & \Phi_{12}(0) \\ \Phi_{21}(T_f) & \Phi_{22}(T_f) \end{bmatrix} \begin{bmatrix} \mathbf{x}_A(0) \\ \boldsymbol{\lambda}(0) \end{bmatrix}. \tag{10.66}$$

In order to obtain a nontrivial solution, a small initial condition for the glucose (or insulin) concentration must be assumed, i.e., $\mathbf{x}_A(0)$ cannot equal zero. The magnitude of the initial condition determines the energy input, for a given set of conditions, since increasing the initial condition by a factor of m simply increases the input and the state vector by the same factor.

The optimal solution is obtained by integrating equation (10.63) from time $t = 0$ to time $t = T_f$ and storing the values of $\Phi(t)$ at the boundaries. The vector $\mathbf{x}_B(0)$ is then evaluated from equation (10.66) in order to obtain the full set of initial conditions. Equation (10.60) is then integrated from time $t = 0$ to $t = T_f$ and the optimal input is obtained from equation (10.57).

For a given value of q, the optimal return, J, will increase as the terminal time, T_f, approaches a critical length, T_{crit}. The input energy also increases and becomes infinite at the critical length. Thus the input energy can be varied by varying the initial condition or the terminal time. The magnitude of q determines the critical length; the greater the magnitude of q the longer is T_{crit}.

Riccati Equation Method

The Riccati equation method of solution applies for the case where the boundary conditions in equations (10.59) are homogeneous, i.e., $\mathbf{x}_A(0) = 0$. The solution is trivial except for certain values of q which are the eigenvalues of the two-point boundary-value problem. To obtain the optimal input, Mehra (Reference 8) defines the Riccati matrix and the transition matrix as follows (see Chapter 8, Section 8.2.4).

The Riccati matrix, $P(t)$, is defined by the relation

$$\mathbf{x}_A(t) = P(t)\boldsymbol{\lambda}(t). \tag{10.67}$$

Differentiating equation (10.67), substituting from equation (10.58), and rearranging yields

$$\dot{P} = F_A P + P F_A{}^T - P H_A{}^T R^{-1} H_A P - (1/q) G_A G_A{}^T, \tag{10.68}$$

$$P(0) = 0. \tag{10.69}$$

Let $\Phi(t, q)$ denote the transition matrix for a particular value of q. Then

$$\begin{bmatrix} \mathbf{x}_A(T_f) \\ \boldsymbol{\lambda}(T_f) \end{bmatrix} = \begin{bmatrix} \Phi_{11}(T_f, q) & \Phi_{12}(T_f, q) \\ \Phi_{21}(T_f, q) & \Phi_{22}(T_f, q) \end{bmatrix} \begin{bmatrix} \mathbf{x}_A(0) \\ \boldsymbol{\lambda}(0) \end{bmatrix}. \tag{10.70}$$

The second equation in equation (10.70) along with the boundary conditions gives

$$\begin{aligned} \boldsymbol{\lambda}(T_f) &= \Phi_{22}(T_f, q)\boldsymbol{\lambda}(0) \\ &= 0. \end{aligned} \tag{10.71}$$

For a nontrivial solution, the determinant of the matrix must equal zero:

$$| \Phi_{22}(T_f, q) | = 0. \tag{10.72}$$

To obtain the optimal input, matrix Riccati equation (10.68) is integrated forward in time for a particular value of q. When the elements of $P(t)$ become very large, the critical length, $T_f = T_{\text{crit}}$, has been reached. The initial costate vector, $\boldsymbol{\lambda}(0)$, is obtained from equations (10.71) and (10.72) as an eigenvector of $\Phi_{22}(T_f, q)$ corresponding to the zero eigenvalue. A unique value of $\boldsymbol{\lambda}(0)$ is found by using the normalization condition of the input energy constraint, equation (10.45). Equation (10.58) is then integrated forward in time and the optimal input is obtained from equation (10.57).

Numerical Results

To obtain numerical results for the optimal inputs, a fourth-order Runge–Kutta method was utilized for the integration of the differential equations. Grid intervals of 0.02 hr were utilized for the method of complementary functions and 0.01 hr for the Riccati equation method. The optimal return was computed using Simpson's rule.

The two-point boundary-value equation (10.58) is given by

$$\begin{bmatrix} \dot{x}_1 \\ \dot{x}_2 \\ \dot{x}_{1b} \\ \dot{x}_{2b} \\ \dot{\lambda}_1 \\ \dot{\lambda}_2 \\ \dot{\lambda}_3 \\ \dot{\lambda}_4 \end{bmatrix} = \left[\begin{array}{cccc|cccc} -a & b & 0 & 0 & 0 & 0 & 0 & 0 \\ -c & -d & 0 & 0 & 0 & -1/q & 0 & 0 \\ 0 & 1 & -a & b & 0 & 0 & 0 & 0 \\ 0 & 0 & -c & -d & 0 & 0 & 0 & 0 \\ \hline 0 & 0 & 0 & 0 & a & c & 0 & 0 \\ 0 & 0 & 0 & 0 & -b & d & -1 & 0 \\ 0 & 0 & 0 & 0 & 0 & 0 & a & c \\ 0 & 0 & 0 & 1/r & 0 & 0 & -b & d \end{array}\right] \begin{bmatrix} x_1 \\ x_2 \\ x_{1b} \\ x_{2b} \\ \lambda_1 \\ \lambda_2 \\ \lambda_3 \\ \lambda_4 \end{bmatrix}, \tag{10.73}$$

where the parameter values given by Bolie (Reference 4) are

$$a = 0.78 \ \ 1/\text{hr},$$
$$b = 0.208 \ \ \text{units/hr/g},$$
$$c = 4.34 \ \ \text{g/hr/unit},$$
$$d = 2.92 \ \ 1/\text{hr}.$$

The Bolie parameters are utilized because they are given in a form that can readily be utilized directly in equation (10.73). The variance of the zero mean Gaussian white noise process is assumed to be r. Choosing r to be equal to unity is equivalent to maximizing the deterministic sensitivity in equation (10.53). The magnitude of q is selected to give a critical length between 2 and 2.5 hr. For the method of complementary functions, a small initial glucose deviation from the mean, $x_2(0)$, is assumed.

Using the method of complementary functions, the optimal return as a function of the terminal time, T_f, is shown in Figure 10.2. The following parameters were assumed:

$$x_2(0) = 0.1 \ \ \text{g/liter},$$
$$q = 0.06,$$
$$r = 1.$$

The optimal return increases as the terminal time approaches the critical length, T_{crit}. The optimal input for $T_f = 2.32$ hr is shown in Figure 10.3 as a function of time. The optimal glucose, insulin, and sensitivity responses to the optimal input are shown in Figure 10.4.

Using the matrix Riccati equation method, the initial glucose deviation from the mean is assumed to be zero:

$$x_2(0) = 0$$

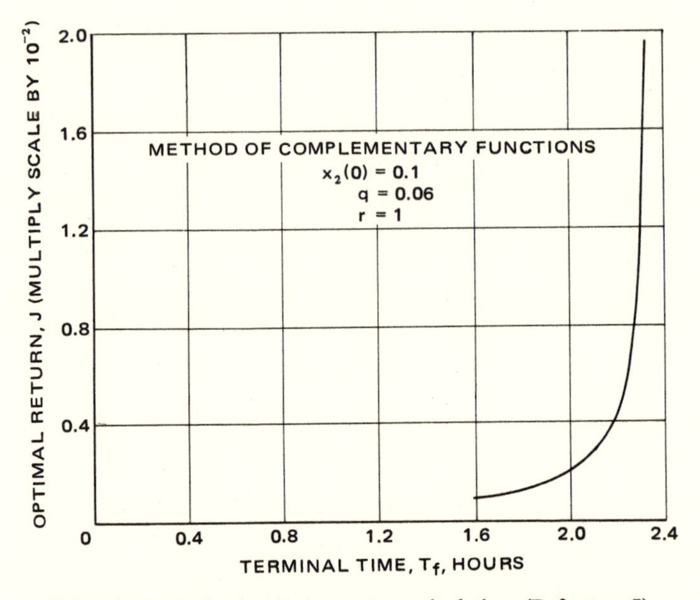

Figure 10.2. Optimal return versus terminal time (Reference 7).

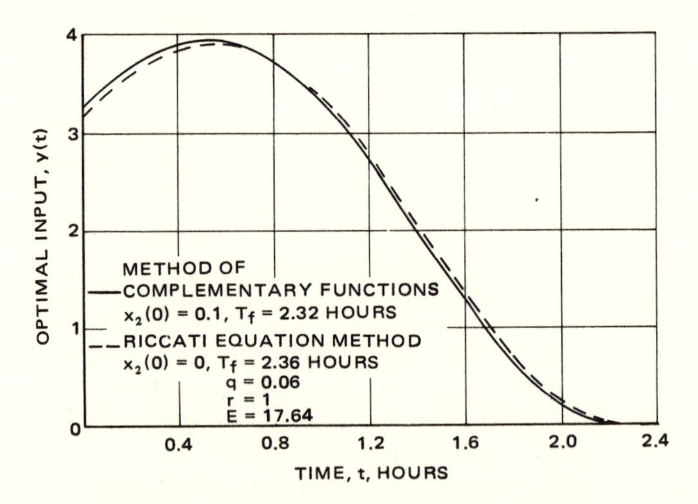

Figure 10.3. Optimal input versus time (Reference 7).

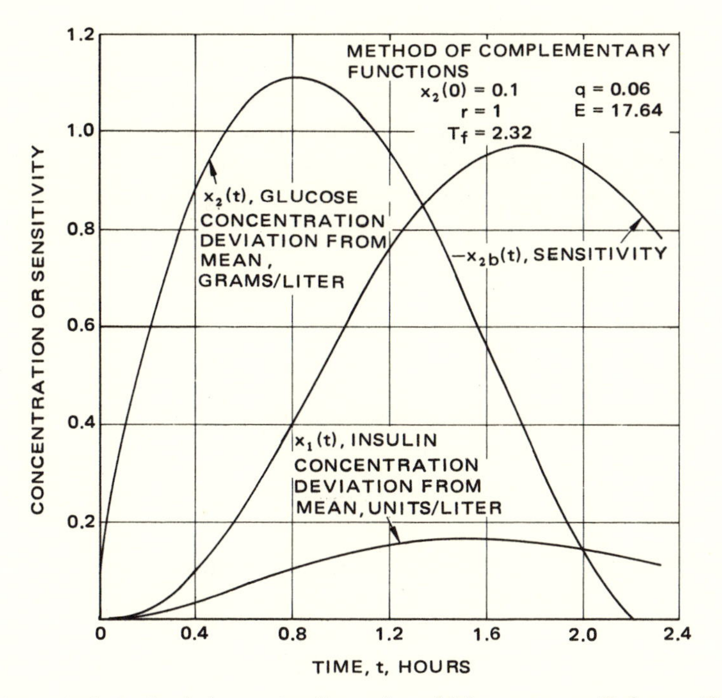

Figure 10.4. Optimal glucose, insulin, and sensitivity responses (Reference 7).

Integrating the matrix Riccati equation yields the critical length

$$T_{\text{crit}} = 2.36 \text{ hr}.$$

The optimal input is found by integrating equation (10.58) using $\boldsymbol{\lambda}(0)$ as an eigenvector of $\Phi_{22}(T_f, q)$ corresponding to the zero eigenvalue. Figure 10.3 shows the optimal input as a function of time with the same energy input as for the method of complementary functions. Comparing the optimal input curves using the Riccati equation method and the method of complementary functions, it is seen that the optimal inputs are almost identical.

10.1.3. Improved Method Using Bolie and Bergman Parameters for Numerical Determination of the Optimal Inputs

The optimal inputs for a linear two-compartment model were derived in the previous section for the Bolie blood glucose regulation parameters. The design of the optimal inputs involves the maximization of a quadratic performance index subject to an input energy constraint. A Lagrange multiplier is introduced whose value is an unknown constant. An improved method for the numerical determination of the optimal inputs was presented in Chapter 9, Section 9.1 in which the Lagrange multiplier is introduced as a state variable and evaluated simultaneously with the optimal input. In this section the equations for the optimal inputs are rederived using the improved method and numerical results are given for both Bolie and Bergman parameters (Reference 9).

Introduction

Consider a quadratic performance criterion for a system described by a first-order differential equation. The optimal input is determined such that the integral

$$M = \int_0^T x_b^2(t)\, dt \tag{10.74}$$

is maximized, subject to the input energy constraint

$$E = \int_0^T y^2(t)\, dt, \tag{10.75}$$

where

$$x_b(t) = \partial x(t)/\partial b. \tag{10.76}$$

Here, $x(t)$ is a scalar state variable, $y(t)$ is the input, and b is an unknown system parameter. The performance index is maximized via the classical method by the maximization of the integral

$$J = \max \frac{1}{2} \int_0^T [x_b^2(t) - q(y^2(t) - E/T)]\, dt, \tag{10.77}$$

where q is the Lagrange multiplier and is equal to a constant. The magnitude of the Lagrange multiplier must be selected such that the input energy constraint is satisfied.

In Section 8.2.4 it was shown that for a linear system with homogeneous boundary conditions, the solution exists only for certain values of q which are the eigenvalues of the two-point boundary-value problem. The eigenvalues q are functions of the interval length T. For a fixed q, the critical length T can be determined by the integration of the Riccati matrix equation. When the elements of the Riccati matrix equation become very large, the critical length has been reached. By integrating for several values of q, a curve relating q to T can be obtained.

In Section 8.2.5 it was shown that for a linear system with nonhomogeneous boundary conditions, the performance index increases as the critical length is approached for a given value of q. The desired input energy is obtained by plotting a curve of input energy versus interval length T and finding the T corresponding to the desired input energy. The optimal input for a system with homogeneous boundary conditions is nearly identical to that of the system with nonhomogeneous boundary conditions when the input energy is the same and the latter has a small initial condition with terminal time near the critical length.

For both homogeneous and nonhomogeneous boundary conditions, the Lagrange multiplier q must be found by trial and error such that the input energy constraint is satisfied. In Section 9.1 a new approach for the numerical determination of optimal inputs was presented. The Lagrange multiplier q is introduced as a state variable. The solution simultaneously yields the optimal input and the value of q for which the input energy constraint is satisfied. The method is applicable to both linear and nonlinear systems.

The optimal inputs for the linear two-compartment model were given in the previous section for the Bolie blood glucose regulation parameters with terminal time $T = 2.32$ hr. The equations for the numerical determination of the optimal inputs were derived using the methods described in Section 8.2. In this section the equations are rederived using the improved method of Section 9.1. Numerical results are given for the Bolie parameters for humans with $T = 1$ hr and for the parameters for dogs derived by Bergman, Kalaba, and Spingarn in Section 7.1.

Two-Point Boundary-Value Problem

The initial development of the equations is the same as in Section 10.1.2 up until the augmented state vector is defined with q as a state variable. These equations are repeated here for convenience. The perturbation equations for blood glucose regulation are

$$\dot{x}_1 = -ax_1 + bx_2, \tag{10.78}$$

$$\dot{x}_2 = -cx_1 - dx_2 + y, \tag{10.79}$$

where as in Section 10.1.2, x_1 is the insulin concentration, x_2 is the glucose concentration, and y is the input glucose rate.

Equations (10.78) and (10.79) can be expressed in vector form by the time-invariant linear system

$$\dot{\mathbf{x}}(t) = F\mathbf{x}(t) + Gy(t), \tag{10.80}$$

where $\mathbf{x}$ is a 2×1 state vector, y is the scalar input, and

$$F = \begin{bmatrix} -a & b \\ -c & -d \end{bmatrix}, \tag{10.81}$$

$$G = \begin{bmatrix} 0 \\ 1 \end{bmatrix}. \tag{10.82}$$

Assume that the measurements are contaminated with white noise

$$z(t) = H\mathbf{x}(t) + v(t), \tag{10.83}$$

where $z(t)$ is a scalar measurement and $v(t)$ is a zero mean Gaussian white noise process,

$$\begin{aligned} E\{v(t)\} &= 0, \\ E\{v(t)v^T(\tau)\} &= R\delta(t-\tau) = \sigma^2\delta(t-\tau) \end{aligned} \tag{10.84}$$

with measurement matrix

$$H = [0 \quad 1]. \tag{10.85}$$

Assume b is an unknown parameter. Then the optimal input is to be determined such that the Fisher information matrix

$$M = \int_0^{T_f} \mathbf{x}_b{}^T H^T R^{-1} H \mathbf{x}_b \, dt \tag{10.86}$$

is maximized, subject to the input energy constraint

$$\int_0^{T_f} y^2 \, dt = E, \tag{10.87}$$

where $\mathbf{x}_b$ is the sensitivity vector defined by

$$\mathbf{x}_b = \frac{\partial \mathbf{x}}{\partial b}. \tag{10.88}$$

Equations (10.86) and (10.87) are made up of quadratic forms of the sensitivity and the control, the latter being utilized to limit the input dosage. The inverse of the Fisher information matrix, M^{-1}, is the Cramér–Rao lower bound.

Two performance indices are used.

Performance index I:

$$J_1 = \max_y \frac{1}{2} \int_0^T (\mathbf{x}_b{}^T H^T R^{-1} H \mathbf{x}_b - q y^2)\, dt + \frac{q(T)E}{2}. \tag{10.89}$$

Performance index II:

$$J_2 = \max_y \frac{1}{2} \int_0^T (\mathbf{x}_b{}^T H^T R^{-1} H \mathbf{x}_b - q y^2)\, dt + \frac{q(T)E}{2} - \frac{q^2(0)}{2}. \tag{10.90}$$

Performance index I is derived by integrating out the term $qE/2T$ in equation (10.77) and placing the result outside of the integral. Performance index II contains the additional term $-q^2(0)/2$ and is found to have the better convergence properties. The equations for performance index I are derived first, with changes for performance index II indicated at the end of the next section.

Taking the partial derivative of equation (10.80) with respect to b yields

$$\dot{\mathbf{x}}_b = F\mathbf{x}_b + F_b\mathbf{x}, \tag{10.91}$$

where

$$F_b = \frac{\partial}{\partial b} F. \tag{10.92}$$

The value of the Lagrange multiplier, q, for which the input energy constraint (10.87) is satisfied can be obtained along with the optimal input by adjoining the differential constraint

$$\dot{q}(t) = 0 \tag{10.93}$$

with unknown initial condition, $q(0)$.

Let $\boldsymbol{\chi}_A$ equal the 5×1 augmented state vector

$$\boldsymbol{\chi}_A = \begin{bmatrix} \mathbf{x} \\ \mathbf{x}_b \\ q \end{bmatrix}. \tag{10.94}$$

Then

$$\dot{\boldsymbol{\chi}}_A = F_A \boldsymbol{\chi}_A + G_A y, \tag{10.95}$$

where

$$F_A = \begin{bmatrix} F & 0 & 0 \\ F_b & F & 0 \\ 0 & 0 & 0 \end{bmatrix}$$

$$= \begin{bmatrix} -a & b & 0 & 0 & 0 \\ -c & -d & 0 & 0 & 0 \\ 0 & 1 & -a & b & 0 \\ 0 & 0 & -c & -d & 0 \\ 0 & 0 & 0 & 0 & 0 \end{bmatrix}, \tag{10.96}$$

$$G_A = [0 \quad 1 \quad 0 \quad 0 \quad 0]^T. \tag{10.97}$$

Performance index I is then expressed in the form

$$J_1 = \max_y \frac{1}{2} \int_0^T (\chi_A{}^T H_A{}^T R^{-1} H_A \chi_A - q y^2)\, dt + \frac{q(T)E}{2}, \tag{10.98}$$

where

$$H_A = [0 \quad 0 \quad 0 \quad 1 \quad 0]. \tag{10.99}$$

Utilizing Pontryagin's maximum principle, the Hamiltonian function is

$$\mathscr{H} = \tfrac{1}{2}\{-\chi_A{}^T H_A{}^T R^{-1} H_A \chi_A + K\chi_A y^2\} + \mathbf{p}^T\{F_A \chi_A + G_A y\}, \tag{10.100}$$

where

$$K = [0 \quad 0 \quad 0 \quad 0 \quad 1]. \tag{10.101}$$

The 5×1 costate vector, $\mathbf{p}(t)$, is the solution of the vector differential equation

$$\dot{\mathbf{p}} = -\left[\frac{\partial \mathscr{H}}{\partial \chi_A}\right]^T$$

$$= H_A{}^T R^{-1} H_A \chi_A - F_A{}^T \mathbf{p} - K^T y^2/2. \tag{10.102}$$

The input $y(t)$ that maximizes $\mathscr{H}$ is

$$\frac{\partial \mathscr{H}}{\partial y} = K\chi_A y + G_A{}^T \mathbf{p}, \tag{10.103}$$

$$y = -\frac{1}{q} G_A{}^T \mathbf{p} = -\frac{1}{q} p_2. \tag{10.104}$$

Upon substituting the value of y into equations (10.95) and (10.102), the two-point boundary-value problem becomes

$$\begin{bmatrix} \dot{\chi}_A \\ \dot{\mathbf{p}} \end{bmatrix} = \begin{bmatrix} F_A & -(1/q) G_A G_A{}^T \\ H_A{}^T R^{-1} H_A & -F_A{}^T \end{bmatrix} \begin{bmatrix} \chi_A \\ \mathbf{p} \end{bmatrix} + \begin{bmatrix} 0 \\ -(1/2q^2) p_2{}^2 K^T \end{bmatrix} \tag{10.105}$$

with boundary conditions

$$\chi_A(0) = \begin{bmatrix} 0 \\ x_2(0) \\ 0 \\ 0 \\ q(0) \end{bmatrix}, \qquad p(T) = \begin{bmatrix} 0 \\ 0 \\ 0 \\ 0 \\ -E/2 \end{bmatrix}, \tag{10.106}$$

where $x_2(0)$ is an initial condition. For convenience, $x_2(0)$ is set equal to 0.1 for most of the numerical calculations; $q(0)$ is the initial estimate of the unknown Lagrange multiplier, q; and

$$p_5(T) = \frac{\partial}{\partial q}\left[-\frac{qE}{2}\right]_{t=T} = -\frac{E}{2}, \tag{10.107}$$

$$p_5(0) = 0. \tag{10.108}$$

The unknown initial conditions are $p_1(0)$, $p_2(0)$, $p_3(0)$, $p_4(0)$, and $q(0)$.

Equation (10.105) can be expressed in the form

$$\begin{bmatrix} \dot{x}_1 \\ \dot{x}_2 \\ \dot{x}_{1b} \\ \dot{x}_{2b} \\ \dot{q} \\ \dot{p}_1 \\ \dot{p}_2 \\ \dot{p}_3 \\ \dot{p}_4 \\ \dot{p}_5 \end{bmatrix} = \left[\begin{array}{ccccc|ccccc} -a & b & 0 & 0 & 0 & 0 & 0 & 0 & 0 & 0 \\ -c & -d & 0 & 0 & 0 & 0 & -1/q & 0 & 0 & 0 \\ 0 & 1 & -a & b & 0 & 0 & 0 & 0 & 0 & 0 \\ 0 & 0 & -c & -d & 0 & 0 & 0 & 0 & 0 & 0 \\ 0 & 0 & 0 & 0 & 0 & 0 & 0 & 0 & 0 & 0 \\ \hline 0 & 0 & 0 & 0 & 0 & a & c & 0 & 0 & 0 \\ 0 & 0 & 0 & 0 & 0 & -b & d & -1 & 0 & 0 \\ 0 & 0 & 0 & 0 & 0 & 0 & 0 & a & c & 0 \\ 0 & 0 & 0 & 1/\sigma^2 & 0 & 0 & 0 & -b & d & 0 \\ 0 & 0 & 0 & 0 & 0 & 0 & 0 & 0 & 0 & 0 \end{array}\right] \begin{bmatrix} x_1 \\ x_2 \\ x_{1b} \\ x_{2b} \\ q \\ p_1 \\ p_2 \\ p_3 \\ p_4 \\ p_5 \end{bmatrix} + \begin{bmatrix} 0 \\ \vdots \\ \\ \\ \\ \\ \\ \\ \\ -(1/2q^2)p_2^2 \end{bmatrix} \tag{10.109}$$

or

$$\dot{\chi} = A\chi + \phi. \tag{10.110}$$

Solution via Newton–Raphson Method

Using the Newton–Raphson method, the unknown initial conditions are assumed to be given by

$$\begin{gathered} p_1(0) = C_1, \qquad p_2(0) = C_2, \qquad p_3(0) = C_3, \\ p_4(0) = C_4, \qquad q(0) = C_5 \end{gathered} \tag{10.111}$$

Expanding the boundary conditions

$$\mathbf{p}(T, \mathbf{C}) = \begin{bmatrix} p_1(T, \mathbf{C}) \\ p_2(T, \mathbf{C}) \\ p_3(T, \mathbf{C}) \\ p_4(T, \mathbf{C}) \\ p_5(T, \mathbf{C}) \end{bmatrix} = \begin{bmatrix} 0 \\ 0 \\ 0 \\ 0 \\ -E/2 \end{bmatrix} \tag{10.112}$$

in a Taylor's series around the kth approximation, retaining only the linear terms, and rearranging, we have

$$\begin{bmatrix} C_1^{k+1} \\ C_2^{k+1} \\ C_3^{k+1} \\ C_4^{k+1} \\ C_5^{k+1} \end{bmatrix} = \begin{bmatrix} C_1^{k} \\ C_2^{k} \\ C_3^{k} \\ C_4^{k} \\ C_5^{k} \end{bmatrix} - \begin{bmatrix} p_{1c1} & p_{1c2} & \cdots & p_{1c5} \\ p_{2c1} & p_{2c2} & \cdots & p_{2c5} \\ \vdots & & & \\ p_{5c1} & p_{5c2} & \cdots & p_{5c5} \end{bmatrix}^{-1} \begin{bmatrix} p_1(T, \mathbf{C}^k) \\ p_2(T, \mathbf{C}^k) \\ p_3(T, \mathbf{C}^k) \\ p_4(T, \mathbf{C}^k) \\ p_5(T, \mathbf{C}^k) + E/2 \end{bmatrix}, \tag{10.113}$$

where

$$p_{icj} = \frac{\partial p_i}{\partial C_j}, \tag{10.114}$$

$$\mathbf{C} = [C_1, C_2, C_3, C_4, C_5]^T. \tag{10.115}$$

The equations for the p_{icj}'s are obtained by differentiating equation (10.110) with respect to the C_i's:

$$\dot{\boldsymbol{\chi}}_{ci} = A\boldsymbol{\chi}_{ci} + \boldsymbol{\phi}_{ci}, \qquad i = 1, 2, \ldots, 5, \tag{10.116}$$

where

$$\boldsymbol{\chi}_{ci} = \frac{\partial \boldsymbol{\chi}}{\partial C_i}, \tag{10.117}$$

$$\begin{aligned} \boldsymbol{\phi}_{ci} &= \frac{\partial A}{\partial C_i}\boldsymbol{\chi} + \frac{\partial \boldsymbol{\phi}}{\partial C_i} \\ &= [0, \phi_{2ci}, 0, 0, 0, 0, 0, 0, 0, \phi_{10ci}]^T, \end{aligned} \tag{10.118}$$

$$\phi_{2ci} = (1/q^2)q_{ci}p_2, \tag{10.119}$$

$$\phi_{10ci} = -\frac{1}{q^2}p_{2ci}p_2 + \frac{1}{q^3}q_{ci}p_2^2. \tag{10.120}$$

The initial conditions are obtained by differentiating equations (10.95) and (10.102):

$$\boldsymbol{\chi}_{Aci}(0) = \begin{bmatrix} 0 \\ 0 \\ 0 \\ 0 \\ \delta_{5i} \end{bmatrix}, \qquad \mathbf{p}_{ci}(0) = \begin{bmatrix} \delta_{1ci} \\ \delta_{2ci} \\ \delta_{3ci} \\ \delta_{4ci} \\ 0 \end{bmatrix}, \tag{10.121}$$

where

$$\begin{aligned} \delta_{jci} &= 0, \qquad j \neq i \\ &= 1, \qquad j = i. \end{aligned} \tag{10.122}$$

Equations (10.105) and (10.116) can be expressed in the matrix form

$$\dot{Z} = AZ + \Phi, \tag{10.123}$$

where

$$Z = \begin{bmatrix} x_1 & x_{1c1} & x_{1c2} & x_{1c3} & x_{1c4} & x_{1c5} \\ x_2 & x_{2c1} & x_{2c2} & x_{2c3} & x_{2c4} & x_{2c5} \\ x_{1b} & x_{1bc1} & x_{1bc2} & x_{1bc3} & x_{1bc4} & x_{1bc5} \\ x_{2b} & x_{2bc1} & x_{2bc2} & x_{2bc3} & x_{2bc4} & x_{2bc5} \\ q & q_{c1} & q_{c2} & q_{c3} & q_{c4} & q_{c5} \\ p_1 & p_{1c1} & p_{1c2} & p_{1c3} & p_{1c4} & p_{1c5} \\ p_2 & p_{2c1} & p_{2c2} & p_{2c3} & p_{2c4} & p_{2c5} \\ p_3 & p_{3c1} & p_{3c2} & p_{3c3} & p_{3c4} & p_{3c5} \\ p_4 & p_{4c1} & p_{4c2} & p_{4c3} & p_{4c4} & p_{4c5} \\ p_5 & p_{5c1} & p_{5c2} & p_{5c3} & p_{5c4} & p_{5c5} \end{bmatrix}, \tag{10.124}$$

$$\Phi = \begin{bmatrix} 0 & 0 & 0 & 0 & 0 & 0 \\ 0 & \phi_{2c1} & \phi_{2c2} & \phi_{2c3} & \phi_{2c4} & \phi_{2c5} \\ 0 & 0 & 0 & 0 & 0 & 0 \\ 0 & 0 & 0 & 0 & 0 & 0 \\ 0 & 0 & 0 & 0 & 0 & 0 \\ 0 & 0 & 0 & 0 & 0 & 0 \\ 0 & 0 & 0 & 0 & 0 & 0 \\ 0 & 0 & 0 & 0 & 0 & 0 \\ 0 & 0 & 0 & 0 & 0 & 0 \\ \phi_{10} & \phi_{10c1} & \phi_{10c2} & \phi_{10c3} & \phi_{10c4} & \phi_{10c5} \end{bmatrix}. \tag{10.125}$$

The initial conditions are

$$Z(0) = \begin{bmatrix} 0 & 0 & 0 & 0 & 0 & 0 \\ x_2(0) & 0 & 0 & 0 & 0 & 0 \\ 0 & 0 & 0 & 0 & 0 & 0 \\ 0 & 0 & 0 & 0 & 0 & 0 \\ C_5 & 0 & 0 & 0 & 0 & 1 \\ C_1 & 1 & 0 & 0 & 0 & 0 \\ C_2 & 0 & 1 & 0 & 0 & 0 \\ C_3 & 0 & 0 & 1 & 0 & 0 \\ C_4 & 0 & 0 & 0 & 1 & 0 \\ 0 & 0 & 0 & 0 & 0 & 0 \end{bmatrix}. \quad (10.126)$$

To obtain the numerical solution, an initial guess is made for the set of initial conditions, C_i, $i = 1, 2, \ldots, 5$ in equation (10.111). The initial-value equations (10.123) are then integrated from time $t = 0$ to $t = T$. A new set of values for the C_i's is calculated from equation (10.113), and the above sequence is repeated to obtain a second approximation, etc.

For performance index II, the boundary conditions in equations (10.107), (10.108), and the last component of equation (10.121) are

$$p_5(T) = -(E/2) + q(0), \quad (10.127)$$

$$p_5(0) = q(0), \quad (10.128)$$

$$\frac{\partial p_5(0)}{\partial C_5} = 1. \quad (10.129)$$

Optimal Input Solution for Bolie Parameters

Numerical results were obtained using the Newton–Raphson method. A fourth-order Runge–Kutta method was utilized with grid intervals of 1/50 hr when the terminal time, T, is expressed in hours, and grid intervals of 1 min when the terminal time is expressed in minutes. The Bolie parameter values are given in Table 10.1 along with other assumed parameters and the final numerical results for q and M. The units of the parameters are given in Table 10.2. The terminal time and the input energy constraint were assumed to be as follows:

$$\begin{aligned} T &= 1 \text{ hr}, \\ E &= 0.0489402. \end{aligned} \quad (10.130)$$

The final converged value of q for these conditions is $q = 0.015$. The al-

Table 10.1. Parameter Values

	Parameters	Bolie parameter values	Bolie parameter values, Bergman units	Bergman parameter values
Differential equation parameters	a	0.78	0.013	0.185
	b	0.208	0.034667	0.342
	c	4.34	0.007233	0.0349
	d	2.92	0.048667	0.0263
Other assumed parameters	(1) $x_2(0)$	0.1	10	10
	(2) T	1	60	60
	(3) σ^2	1	10^4	10^4
	(4) E	0.04894	8.153	8.153
Numerical results for above parameters	q	0.015	0.1944	0.05607
	M	0.00136	2.9367	0.766

Table 10.2. Parameter Units and Conversions

Parameters, state variables, and input	Bolie units	Bergman units	Multiplication factors for conversion of Bolie units to Bergman units
a	1/hr	1/min	1/60
b	(units/g)/hr	(100 μunits/mg)/min	1/6
c	(g/unit)/hr	(mg/100 μunits)/min	1/600
d	1/hr	1/min	1/60
σ^2	$(\text{g/liter})^2$	$(\text{mg/100 ml})^2$	10^4
E	$(\text{g/liter})^2/\text{hr}$	$(\text{mg/100 ml})^2/\text{min}$	166.67
q	$(\text{hr})^4/(\text{units/liter})^2$	$(\text{min})^4/(\mu\text{units/ml})^2$	12.96
$x_1(t)$	units/liter	μunits/ml	10^3
$x_2(t)$	g/liter	mg/100 ml	10^2
$x_{1b}(t)$	g hr/liter	mg min/100 ml	6000
$x_{2b}(t)$	g^2 hr/liter-units	mg^2 min/100 ml-100 μunits	600
$y(t)$	(g/liter)/hr	(mg/100 ml)/min	1.667

gorithm is initialized as follows:

$$p_1(0) = 0, \qquad p_2(0) = 0, \qquad p_3(0) = 0, \qquad p_4(0) = 0,$$
$$q(0) = 0.02. \tag{10.131}$$

The input energy constraint, E, is introduced via the boundary condition

$$p_5(T) = -E/2. \tag{10.132}$$

At the end of the first iteration, q is reset to 0.02, so that initial estimates of $p_1(0)$, $p_2(0)$, $p_3(0)$, and $p_4(0)$ can be obtained for a given q. Table 10.3 shows the convergence of q and $E(q)$ for performance index I. The solution converges within nine iterations.

Table 10.4 compares the regions of convergence for performance indices I and II. For performance index I, convergence is obtained for an initial estimate of q at least 33% greater than the final value of q. For performance index II however, convergence is obtained for an initial estimate of q at least 10 times greater than the final value. If the input energy constraint is increased to $E = 0.713753$ with the other parameters kept the same, then the final value of q is 0.01. Table 10.4 shows that for this

Table 10.3. Convergence of q and $E(q)$ for Bolie Parameters and Performance Index I; $q(0) = 0.02$, $E = 0.0489402$

Iteration number	q	$E(q)$
0	0.02	0.096991
1	0.02	0.0145198
2	0.00670042	0.661891
3	0.0136762	1.17911
4	0.0118949	0.296771
5	0.0124931	0.10936
6	0.0139345	0.0619369
7	0.0148426	0.0503973
8	0.0149968	0.0489675
9	0.015	0.0489402

Table 10.4. Regions of Convergence

$q(0)$	Convergence	
	Performance index I	Performance index II
	$q(\text{final}) = 0.015$, $E = 0.0489402$	
0.15	No	Yes
0.025	No	Yes
0.02	Yes	Yes
	$q(\text{final}) = 0.01$, $E = 0.713753$	
0.012	No	No
0.011	Yes	Yes

case convergence for both performance indices I and II is obtained for an initial estimate of q of only about 10% greater than the final value of q.

Figure 10.5 shows the input energy, $E(q)$ versus q, obtained by the method of complementary functions. Using this one-sweep method, E is obtained for a given input value of q (final). The curve shows that a possible cause of the small region of convergence for the larger input energy constraint is the steep slope of the energy versus q curve in the vicinity of $q = 0.01$. A much larger region of convergence is obtained for $q = 0.015$, as shown in Table 10.4, where the slope of the curve is considerably less steep. These results seem to indicate that when the Newton–Raphson method is used, the input energy constraint should be small in order to obtain a large region of convergence. Once a good estimate of q is obtained for a given value of E, the latter can be increased to determine another estimate of q. This situation appears to be similar to the case given in Section 10.1.2, where the terminal time is $T = 2.32$ hr. Using the Newton–Raphson method, convergence is extremely difficult to obtain for the large value of E used in that section.

The optimal input for a linear system need not be determined for a large value of E, however. If the optimal input is obtained for a small value of E, it needs only to be scaled upwards to obtain the optimal input for

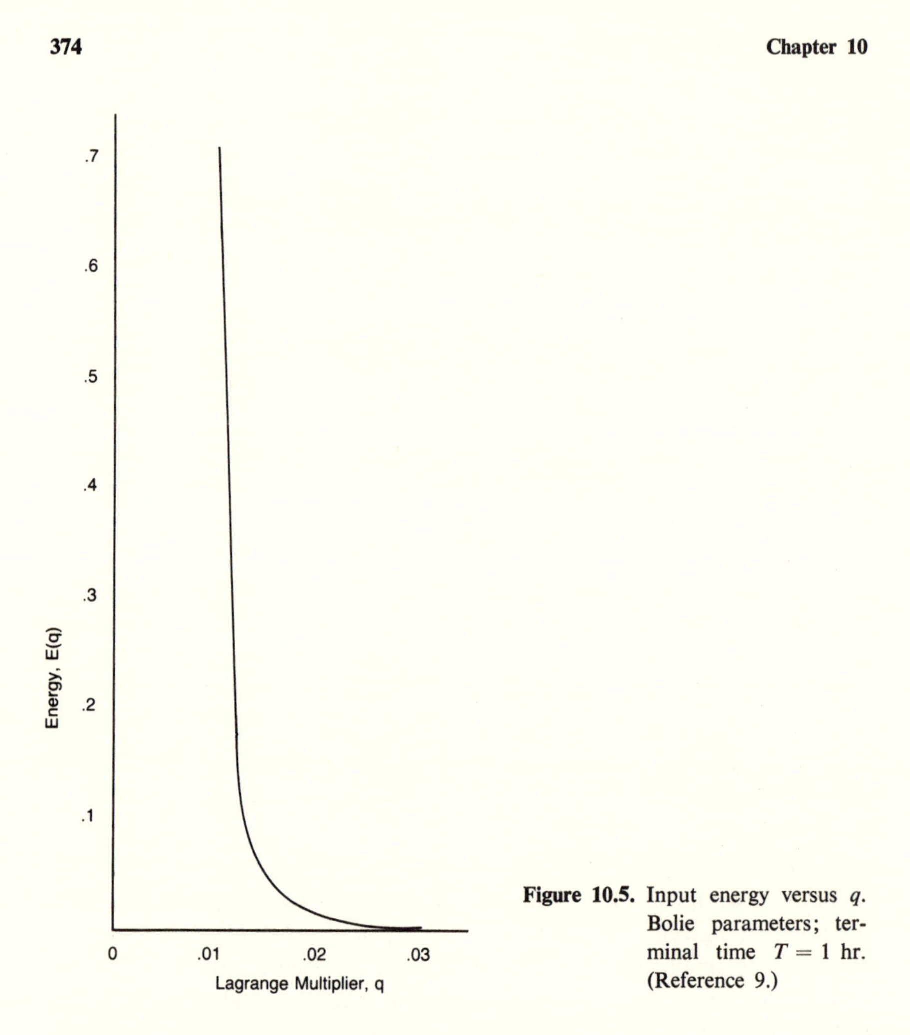

Figure 10.5. Input energy versus q. Bolie parameters; terminal time $T = 1$ hr. (Reference 9.)

any desired value of E. For example, Figure 10.6 shows the optimal input curves, $y(t)$, for

$$E = 0.0489402, \qquad q = 0.015 \tag{10.133}$$

and

$$E = 0.713753, \qquad q = 0.01. \tag{10.134}$$

In the first case, $y(0) = 0.417284$ and in the second case, $y(0) = 1.5603$. Multiplying the curve for $q = 0.015$ by $1.5603/0.417284 = 3.739$, the dashed curve in Figure 10.6 is obtained. The dashed curve is extremely close to the optimal input curve for $q = 0.01$.

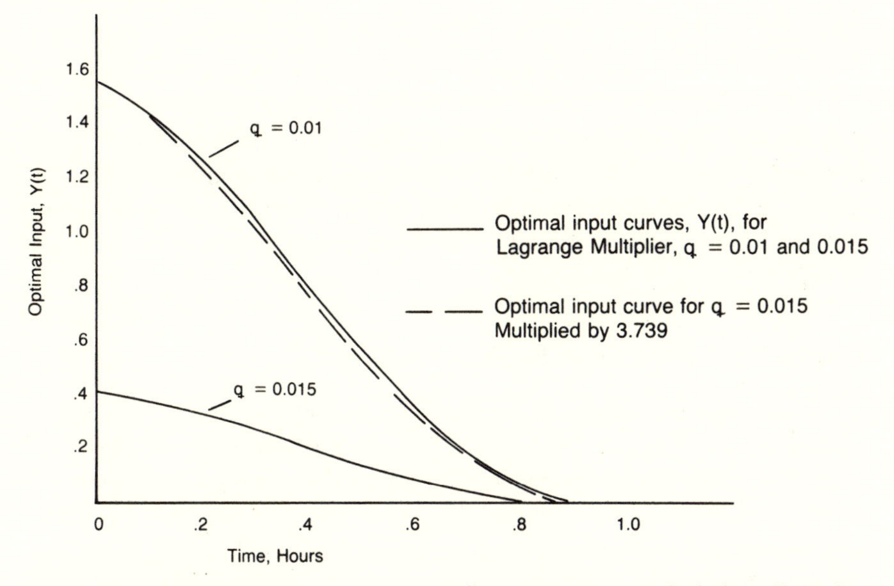

Figure 10.6. Optimal input versus time. Bolie parameters; terminal time $T = 1$ hr. (Reference 9.)

Conversion of Units

The blood glucose regulation parameters for dogs were estimated in Section 7.1. The units of the parameters in that section, referred to for simplicity as the Bergman units, were different from those used by Bolie for estimating the human parameters. Table 10.2 gives the conversion factors for converting the Bolie units to Bergman units. The Bolie parameter values in Bergman units are given in Table 10.1. Note the large increase in the parameter values for the initial condition, $x_2(0)$, the variance, σ^2, and the input energy constraint, E, in the Bergman units. For the conditions given,

$$\begin{aligned} T &= 60 \text{ min}, \\ E &= 8.15302, \\ x_2(0) &= 10, \end{aligned} \tag{10.135}$$

the final converged value of q is 0.1944. For an initial estimate of $q(0)$ equal to 5 times the final value of q, the Newton–Raphson method diverged using performance index II. The problem in this case appeared again to be that the input energy constraint, E, was too large. Decreasing $x_2(0)$ from 10 to 2 decreases $y(t)$ by a factor of 5, and E by a factor of 25. The final value of q remains the same. Thus for the conditions

$$E = 0.326121, \qquad x_2(0) = 2, \tag{10.136}$$

convergence was obtained in 12 iterations for an initial estimate of $q(0)$ equal to 5 times the final value of q, using performance index II. Convergence was also obtained for $x_2(0) = 1$; however, in this case convergence was much slower. The initial rate of decrease of q was approximately equal to $E/2$.

Optimal Input Solution for Bergman Parameters

The Bergman parameter estimates for the blood glucose regulation parameters are given in Table 10.1. The optimal input for the Bergman parameter values was obtained for the same input energy constraint, E, as for the Bolie parameter values in Bergman units. Figure 10.7 shows the optimal inputs for the Bolie and Bergman parameters. The optimal inputs are not considerably different.

The variance, $\sigma^2 = 10^4$, is unrealistically large in the Bergman units because of the assumption that $\sigma^2 = 1$ in Bolie units. Decreasing σ^2 by 100 has the following effects: $y(t)$, $x_1(t)$, $x_2(t)$, $x_{1b}(t)$, $x_{2b}(t)$, and E all remain the same; q and M are increased by 100. The effect is obviously to decrease the Cramér–Rao lower bound, M^{-1}.

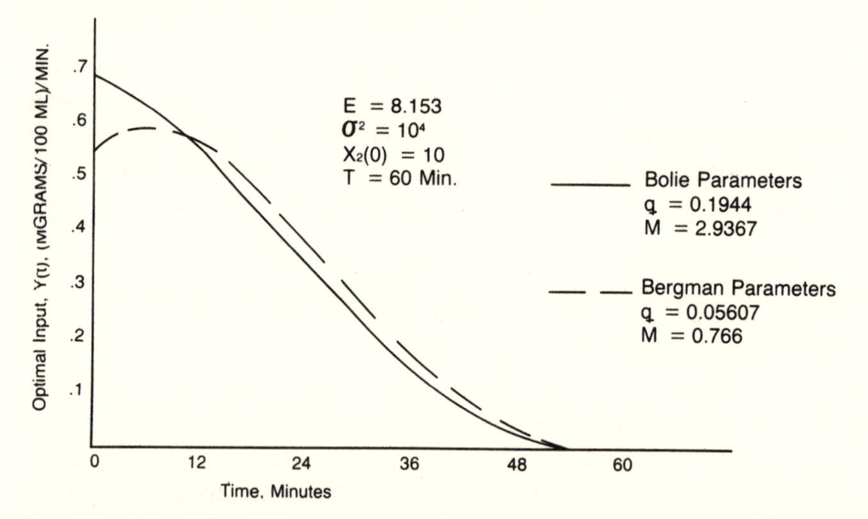

Figure 10.7. Optimal inputs for Bolie and Bergman parameters with the same input energy E (Reference 9).

10.2. Optimal Inputs for Aircraft Parameter Estimation

Optimal input design criteria have been used to design inputs to estimate aircraft stability and control derivatives from flight test data (References 8 and 10). Some important limitations in the design of aircraft inputs must be considered, however, (Reference 10). For example, aircraft maneuvers cannot be excessively large. Long flight test times may be undesirable. The simultaneous estimation of all aircraft parameters from a single maneuver is not feasible because of unidentifiability. The measured aircraft response should be sensitive to the parameters being estimated, etc.

Consider the optimal elevator deflection inputs for estimating the parameters in the longitudinal short period equations of the C-8 aircraft (Reference 8). The equations for the short period dynamics are

$$\begin{bmatrix} \dot{q} \\ \dot{\alpha} \end{bmatrix} = \begin{bmatrix} a & b \\ c & d \end{bmatrix} \begin{bmatrix} q \\ \alpha \end{bmatrix} + \begin{bmatrix} g_1 \\ g_2 \end{bmatrix} \delta_e,$$

where q is the pitch rate, α is the angle of attack, δ_e is the elevator command, and a, b, c, d, g_1, and g_2 are the unknown aircraft parameters. *A priori* estimates of these parameters for use in computing the optimal elevator deflection inputs are assumed to be available from wind tunnel tests.

The measurement equations are

$$\begin{bmatrix} y_1 \\ y_2 \end{bmatrix} = \begin{bmatrix} 1 & 0 \\ 0 & 1 \end{bmatrix} \begin{bmatrix} q \\ \alpha \end{bmatrix} + \begin{bmatrix} n_q \\ n_\alpha \end{bmatrix},$$

where the measurement noise sources, n_q and n_α, are assumed to have the noise spectral density matrix

$$R = \begin{bmatrix} \sigma_q^2 & 0 \\ 0 & \sigma_\alpha^2 \end{bmatrix}.$$

The experiment data length is assumed to be fixed at T_f seconds. The optimal input is determined such that the trace of the Fisher information matrix

$$\mathrm{Tr}(M) = \int_0^{T_f} \mathrm{Tr}[X_p^T H^T R^{-1} H X_p]\, dt$$

is maximized subject to the input energy constraint

$$\int_0^{T_f} \delta_e^2\, dt = E,$$

where X_p is the parameter influence coefficient matrix defined in Chapter 9, Section 9.2.1. Using Mehra's method, the matrix Riccati equation is integrated several times, each time for a different value of q, to obtain a curve of the Lagrange multiplier q versus the critical time T for estimating the five parameters, a, b, d, g_1, and g_2. The value of q corresponding to T_f is found from the curve. The eigenvector of $\Phi_{\lambda\lambda}(T_f; q)$ corresponding to the zero eigenvalue is obtained and the two-point boundary-value equations are integrated forward in time to obtain the optimal elevator deflection.

In Reference 8 numerical results for the optimal elevator deflection are obtained and compared with a doublet input of the same energy and time duration.

Exercises

1. Verify that equations (10.15) are correct.
2. Derive the two-point boundary-value equations for the improved method for the numerical determination of optimal inputs for the blood glucose regulation equations (10.78) and (10.79), assuming that both a and c are unknown parameters to be estimated. Use Pontryagin's maximum principle and assume that only x_2 is observed.
3. Derive the two-point boundary-value equations in Exercise 2 assuming that both x_1 and x_2 are observed.
4. Derive the two-point boundary-value equations for the improved method for the numerical determination of optimal inputs for the van der Pol equation in Exercise 1, Chapter 7, assuming that both a and b are unknown parameters to be estimated. Use Pontryagin's maximum principle and assume that both x and $\dot{x}$ are observed.
5. Derive the two-point boundary-value equations for the improved method for the numerical determination of optimal inputs for the mass attached to a nonlinear spring in Exercise 4, Chapter 7. Use Pontryagin's maximum principle and assume that only x is observed.

PART V

Computer Programs

11

Computer Programs for the Solution of Boundary-Value and Identification Problems

The computer program listings are given in this section. The listings are divided into two sections: the first for the solution of two-point boundary-value problems and the second for the solution of system identification problems. The two-point boundary-value problems include examples from the linear first-order system state-regulator problem, optimal inputs for estimating one of the parameters of a nonlinear first-order system, and optimal inputs for estimating one of the parameters of the blood glucose regulation two-compartment model. The system identification problems include examples for estimating a parameter of a first-order linear system and for estimating the four parameters of the blood glucose regulation two-compartment model.

Since the listings contain very few comment statements, a brief description of the listing is given with each program. References to the text are given for a more complete description of the method of solution for each problem.

The programs are written in the BASIC programming language (Reference 1). An advantage of BASIC is the availability of commands for matrix computations. These commands are identified by the word MAT. For example MAT C = A*B indicates multiply matrix A by matrix B, MAT C = (A)*B indicates multiply the matrix B by the scalar A, etc.

11.1. Two-Point Boundary-Value Problems

The program PONTBV is used to obtain the numerical solution to the linear two-point boundary-value problem for the first-order system state-regulator problem. The solution is obtained via the method of complementary functions.

The programs NLCO44 and NEWT3 are used to obtain the numerical solution of the optimal input for estimating the parameter a in the equation

$$\dot{x} = -ax - bx^2 + y$$

using the quasilinearization and Newton–Raphson methods, respectively. The solutions in both programs are obtained for a given value of the Lagrange multiplier q. The improved method for the determination of the optimal input for this problem is used in the program NEWTQ to simultaneously determine the Lagrange multiplier q.

The program NEGLUQ is used to determine the optimal input for estimating the parameter a of the blood glucose regulation two-compartment model. The improved method for the determination of the optimal input is used to simultaneously determine the Lagrange multiplier q. The solution is obtained via the Newton–Raphson method.

Program name: PONTBV

Problem: Linear two-point boundary-value problem.

Method of solution: Method of complementary functions.

Example: State-regulator problem for the first-order system

$$\dot{x} = a_1 x + y.$$

The two-point boundary-value equations are derived using Pontryagin's maximum principle.

Reference in text: Chapter 3.2.2.

Definition of variables:

Z = Vector of particular and homogeneous variables (the particular solution is zero),

$$\mathbf{Z} = [p_1, p_2, h_{11}, h_{12}, h_{21}, h_{22}].$$

F = Vector of derivatives, $F = \dot{\mathbf{Z}} = A^*\mathbf{Z}$.

Listing:

70–150: Components of the matrix A, equal to the coefficients of the particular and homogeneous equations.

220–240: Initial conditions of the vector $\mathbf{Z}$.

250–590: Integration of the homogeneous equations, $\dot{\mathbf{Z}}$.

510–560: Storage of the solutions of the homogeneous equations at some arbitrary desired time periods.

600–620: Solution of the linear algebraic equations at the boundaries.

630–690: Solution of the two-point boundary-value problem at the desired time periods.

Output: Optimal state, $x(t)$, as a function of time.

PONTBV

```
5 PRINT " T","   X "
10 READ L
15 LET T=L
20 DIM P(1,10),L(3,10),D(5),E(5),G(5),F(5)
30 DIM Z(5),A(5,5),B(5),C(5),Y(5)
40 DIM H(5),I(5),J(5),K(5)
50 LET H=1/100
60 LET T1=0
70 MAT A=ZER
80 LET A(2,2)=-1
90 LET A(2,3)=1
100 LET A(3,2)=1
110 LET A(3,3)=1
120 LET A(4,4)=-1
130 LET A(4,5)=1
140 LET A(5,4)=1
150 LET A(5,5)=1
200 LET C=0
210 LET F=0
220 MAT Z=ZER
230 LET Z(2)=1
240 LET Z(5)=1
249 LET L9=INT((L+.1*H)/H)
250 FOR N=0 TO L9
260 MAT Y=(1)*Z
270 IF N=0 THEN 500
280 GOSUB 800
290 MAT H=(H/2)*F
300 MAT Z=Y+H
310 GOSUB 800
320 MAT I=(H/2)*F
330 MAT Z=Y+I
```

```
340 GOSUB 800
350 MAT J=(H)*F
360 MAT Z=Y+J
370 GOSUB 800
380 MAT K=(H)*F
390 MAT H=(2)*H
400 MAT I=(4)*I
410 MAT J=(2)*J
420 MAT I=H+I
430 MAT J=I+J
440 MAT K=J+K
450 MAT K=(1/6)*K
460 MAT Z=Y+K
470 LET M1=INT((T1+.1*H)/H)
480 IF N=M1 THEN 510
490 GO TO 590
500 LET M=0
510 LET P(0,M)=Z(0)
520 LET P(1,M)=Z(1)
530 LET L(0,M)=Z(2)
540 LET L(1,M)=Z(3)
550 LET L(2,M)=Z(4)
560 LET L(3,M)=Z(5)
570 LET M=M+1
580 LET T1=T1+L/10
590 NEXT N
600 LET D=L(0,0)*L(3,10) - L(1,10)*L(2,0)
610 LET A=(L(3,10)*(1-P(0,0))-L(2,0)*(-P(1,10)))/D
620 LET B=(L(0,0)*(-P(1,10))-L(1,10)*(1-P(0,0)))/D
630 FOR M=10 TO 0 STEP -1
640 LET X=P(0,M)+A*L(0,M)+B*L(2,M)
670 PRINT T,X
680 LET T=T-L/10
682 IF T<0 THEN 686
684 GO TO 690
686 LET T=0
690 NEXT M
695 GO TO 860
700 DATA 10
800 MAT F=A*Z
850 RETURN
860 END
```

Program name: NLCO44

Problem: Nonlinear two-point boundary-value problem.

Method of solution: Quasilinearization.

Example: Optimal input for estimating the parameter a in the equation

$$\dot{x} = -ax - bx^2 + y$$

for a given value of the Lagrange multiplier q.

Reference in text: Chapter 8.3.1.

Definition of variables:

$R =$ Vector of state and costate variables,

$$\mathbf{x} = (x, x_a, \lambda_1, \lambda_2)^T.$$

$X =$ Matrix for storage of kth approximation of the solution, $\mathbf{x}_k(t)$.

Listing:

115–140: Initial approximation and storage of $\mathbf{x}(t)$,

$$\mathbf{x}_0(t) = \mathbf{c}.$$

150–180: Initial conditions of the matrix Z, where Z is composed of the particular solution, 4×1 vector $\mathbf{p}$, and the homogeneous solution, 4×4 matrix H

$$Z = [\mathbf{p} \;\; H].$$

200–266: Integration of equation

$$\dot{\mathbf{x}}_{k+1} = A_k(t)\mathbf{x}_{k+1} + \mathbf{u}_k(t)$$

for the $k + 1$ approximation, $\mathbf{x}_{k+1}(t)$, using the initial conditions $\mathbf{x}_{k+1}(0) = \mathbf{c}_A$. This computation is not done on the first iteration.

270–470: Integration of the quasilinearized differential equations $\dot{\mathbf{p}}$ and $\dot{H}$.

471–510: Evaluation of the initial conditions from the boundary conditions, $\mathbf{c}_A$.

706 GO TO 160: Repeat the above process to obtain $\mathbf{x}_{k+1}(t)$ for the new initial conditions.

Output: Optimal input $y(t)$ and responses $x(t)$, $x_a(t)$, as a function of time.

NLC044

```
1 LET K4 = 1
2 LET A=.1
3 LET G=.075
4 LET E=3
5 LET B=0
10 READ L
15 DIM U(3,4),V(3),G(3),D(3),R(3)
17 DIM T(3,3),C(3),B(3)
20 DIM X(3,100)
25 DIM W(3),L(3),M(3),N(3),O(3),S(3,3)
30 DIM Z(3,4),A(3,3),Y(3,4),F(3,4)
40 DIM H(3,4),I(3,4),J(3,4),K(3,4)
45 LET A9=.1
50 LET H=1/100
55 LET M9=INT(L/H+.1*H)
60 LET C=.1
65 MAT A=ZER
66 LET A(0,2)=-1/G
67 LET A(3,1)=1
75 MAT V=ZER
80 MAT U=ZER
84 MAT D=ZER
85 LET D(0)=C
90 LET N2=0
95 MAT C=ZER
100 LET C(0)=C
105 MAT T=IDN
115 REM INITIAL APPROXIMATION
117 MAT X=ZER(E,M9)
120 FOR N=0 TO M9
125 LET X(0,N)=C
140 NEXT N
150 REM INITIAL CONDITIONS MAT Z
160 MAT Z=ZER
170 FOR I=0 TO E
175 LET Z(I,I+1)=1
180 NEXT I
200 FOR N=1 TO M9
205 IF N2=0 THEN 270
210 MAT W=(1)*R
220 LET N1=0
222 GOSUB 800
224 MAT L=(H/2)*G
226 MAT R=W+L
228 GOSUB 980
230 MAT M=(H/2)*G
232 MAT R=W+M
234 GOSUB 980
```

```
236 MAT N=(H)*G
238 MAT R=W+N
240 GOSUB 980
242 MAT O=(H)*G
244 MAT L=(2)*L
246 MAT M=(4)*M
248 MAT N=(2)*N
250 MAT M=L+M
252 MAT N=M+N
254 MAT O=N+O
256 MAT O=(1/6)*O
258 MAT R=W+O
260 FOR I=0 TO E
262 LET X(I,N-1)=D(I)
264 NEXT I
266 MAT D=(1)*R
270 MAT Y=(1)*Z
272 LET N1=1
280 GOSUB 800
290 MAT H=(H/2)*F
300 MAT Z=Y+H
310 GOSUB 990
320 MAT I=(H/2)*F
330 MAT Z=Y+I
340 GOSUB 990
350 MAT J=(H)*F
360 MAT Z=Y+J
370 GOSUB 990
380 MAT K=(H)*F
390 MAT H=(2)*H
400 MAT I=(4)*I
410 MAT J=(2)*J
420 MAT I=H+I
430 MAT J=I+J
440 MAT K=J+K
450 MAT K=(1/6)*K
460 MAT Z=Y+K
470 NEXT N
471 LET N=M9
472 FOR I=0 TO E
473 LET X(I,N)=D(I)
474 NEXT I
475 FOR I=2 TO E
476 FOR J=0 TO E
477 LET T(I,J)=Z(I,J+1)
478 NEXT J
480 LET C(I)=-Z(I,0)
485 NEXT I
490 MAT S=INV(T)
495 MAT B=S*C
```

```
500 MAT T=IDN
510 MAT R=T*B
515 MAT D=(1)*R
520 LET N2=1
555 LET T1=0
557 LET K4=1
560 PRINT
565 PRINT " T","    X","    XA","    Y"
615 FOR N=0 TO M9
620 IF N=M9 THEN 645
630 LET M1=INT((T1+.1*H)/H)
635 IF N=M1 THEN 645
640 GO TO 660
645 LET T=N*H
650 PRINT T,X(0,N),X(1,N),-X(2,N)/G
655 LET T1=T1+A9
660 LET W2=X(1,N)*X(1,N)
665 LET Y2=(X(2,N)/G)**2
666 IF N<>0 THEN 670
667 LET K1=W2
668 LET K2=Y2
669 GO TO 694
670 IF N=INT(L/H+.1*H) THEN 690
672 IF K4=1 THEN 682
674 LET K1=K1+2*W2
676 LET K2=K2+2*Y2
678 LET K4=1
680 GO TO 694
682 LET K1=K1+4*W2
684 LET K2=K2+4*Y2
686 LET K4=0
688 GO TO 694
690 LET K1=K1+W2
692 LET K2=K2+Y2
694 NEXT N
696 LET K1=H/3*K1
698 LET K2=H/3*K2
700 PRINT
702 PRINT "RETURN=" K1-G*K2
704 PRINT "N="N,"IY2="K2,"IX2="K1
706 GO TO 160
750 DATA 1
795 REM CALCULATE A(T) AND U(T) AT TIME N-1
800 LET X2=A+2*B*X(0,N-1)
805 LET X3=1+2*B*X(1,N-1)
810 LET A(0,0)=-X2
815 LET A(1,0)=-X3
820 LET A(1,1)=-X2
825 LET A(2,0)=2*B*X(2,N-1)
827 LET A(2,1)=2*B*X(3,N-1)
```

```
830 LET A(2,2)=X2
835 LET A(2,3)=X3
840 LET A(3,0)=A(2,1)
845 LET A(3,3)=X2
900 LET U(0,0)=B*X(0,N-1)*X(0,N-1)
910 LET U(1,0)=2*B*X(0,N-1)*X(1,N-1)
920 LET U(2,0)=-2*B*(X(0,N-1)*X(2,N-1)+X(1,N-1)*X(3,N-1))
925 LET U(3,0)=-2*B*X(0,N-1)*X(3,N-1)
950 IF N1=1 THEN 990
960 LET V(0)=U(0,0)
962 LET V(1)=U(1,0)
964 LET V(2)=U(2,0)
966 LET V(3)=U(3,0)
980 MAT G=A*R
985 MAT G=G+V
988 GO TO 998
990 MAT F=A*Z
995 MAT F=F+U
998 RETURN
999 END
```

Program name: NEWT3

Problem: Nonlinear two-point boundary-value problem.

Method of solution: Newton–Raphson method.

Example: Optimal input for estimating the parameter a in the equation

$$\dot{x} = -ax - bx^2 + y$$

for a given value of the Lagrange multiplier q.

Reference in the text: Chapter 8.3.1.

Definition of variables:

Z = Vector of state, costate, and auxiliary equation variables, $\mathbf{Z}$.

F = Vector of derivatives, $\dot{\mathbf{Z}} = \mathbf{f}$.

Listing:

150–180: Initial conditions of the differential equations, $\mathbf{Z}(0)$.

250–660: Integration of the differential equations.

700–740: New estimate of the unknown initial conditions

$$\mathbf{c}^{k+1} = \mathbf{c}^k - \Lambda_c^{-1}\boldsymbol{\lambda}(T).$$

780 GO

TO 155: Repeat the above process to obtain the solution of the two-point boundary-value problem for the new initial conditions.

Output: Optimal input $y(t)$ and responses $x(t)$, $x_a(t)$ as a function of time.

NEWT3

```
2 LET S2=.25
5 LET B=0
8 LET R=2
9 LET R1=0
10 READ L
20 DIM P(2),L(2,2),M(2),N(2,2)C(2)
30 DIM Z(12),Y(12),F(12),H(12),I(12),J(12),K(12)
50 LET H=1/100
60 LET C=.1
70 LET G=.075
75 LET A=.1
80 LET M9=INT(L/H+.1*H)
90 LET A9=.1
115 REM INITIAL APPROXIMATION
120 MAT C=ZER
150 REM INITIAL CONDITIONS MAT Z
155 MAT Z=ZER
160 LET Z(1)=C
165 LET Z(3)=C(1)
170 LET Z(4)=C(2)
175 LET Z(7)=1
180 LET Z(12)=1
195 LET K4=1
200 LET T1=0
205 PRINT
210 PRINT "C="
215 MAT PRINT C
220 PRINT
240 PRINT " T","     X","     XA","     Y"
250 FOR N=0 TO M9
260 MAT Y=(1)*Z
270 IF N=0 THEN 470
280 GOSUB 800
290 MAT H=(H/2)*F
300 MAT Z=Y+H
310 GOSUB 800
320 MAT I=(H/2)*F
330 MAT Z=Y+I
340 GOSUB 800
350 MAT J=(H)*F
```

```
360 MAT Z=Y+J
370 GOSUB 800
380 MAT K=(H)*F
390 MAT H=(2)*H
400 MAT I=(4)*I
410 MAT J=(2)*J
420 MAT I=H+I
430 MAT J=I+J
440 MAT K=J+K
450 MAT K=(1/6)*K
460 MAT Z=Y+K
470 LET W2=Z(2)*Z(2)
472 LET Y2=(Z(3)/G)**2
474 IF N<>0 THEN 482
476 LET K1=W2
478 LET K2=Y2
480 GO TO 630
482 IF N=INT(L/H+.1*H) THEN 505
484 IF K4=1 THEN 494
486 LET K1=K1+2*W2
488 LET K2=K2+2*Y2
490 LET K4=1
492 GO TO 630
494 LET K1=K1+4*W2
496 LET K2=K2+4*Y2
498 LET K4=0
500 GO TO 630
505 LET K1=K1+W2
510 LET K2=K2+Y2
630 LET M1=INT((T1+.1*H)/H)
635 IF N=M1 THEN 645
640 GO TO 660
645 LET T=N*H
650 PRINT T,Z(1),Z(2),-1/G*Z(3)
655 LET T1=T1+A9
660 NEXT N
662 LET K=2*K
664 LET K=1
665 LET K1=K1*H/3
670 LET K2=K2*H/3
675 PRINT "RETURN="K1/S2-G*K2
680 PRINT "N="N,"IY2="K2,"IX2="K1
685 PRINT "J="1/2*(K1-G*K2)
695 REM COMPUTE NEW INITIAL CONDITIONS
700 LET L(1,1)=Z(7)
705 LET L(2,1)=Z(8)
710 LET L(1,2)=Z(11)
715 LET L(2,2)=Z(12)
720 LET M(1)=Z(3)
725 LET M(2)=Z(4)
```

```
730 MAT N=INV(L)
735 MAT P=N*M
740 MAT C=C-P
750 LET R1=R1+1
752 IF R1<R THEN 780
755 PRINT
760 PRINT "R1="R1,"DESIRED R="
765 INPUT R
770 IF R1=R THEN 999
780 GO TO 155
790 DATA 1
795 REM CALCULATE A(T) AND U(T) AT TIME N-1
800 LET X1=2*B*Z(1)
802 LET X?=2*B*X(2)
805 LET X3=A+X1
806 LET X4=2*B*Z(5)
807 LET X5=2*B*Z(9)
810 LET F(1)=-A*Z(1)-B*Z(1)*Z(1)-1/G*Z(3)
815 LET F(2)=-Z(1)-X3*Z(2)
820 LET F(3)=X3*Z(3)+(1+X2)*Z(4)
825 LET F(4)=Z(2)/S2+X3*Z(4)
830 LET F(5)=-X3*Z(5)-1/G*Z(7)
835 LET F(6)=-Z(5)-A*Z(6)-2*B*(Z(1)*Z(6)+Z(2)*Z(5))
840 LET F(7)=X3*Z(7)+X4*Z(3)+(1+X2)*Z(8)+2*B*Z(6)*Z(4)
845 LET F(8)=Z(6)/S2+X3*Z(8)+X4*Z(4)
850 LET F(9)=-X3*Z(9)-1/G*Z(11)
855 LET F(10)=-Z(9)-A*Z(10)-2*B*(Z(1)*Z(10)+Z(2)*Z(9))
860 LET F(11)=X3*Z(11)+X5*Z(3)+(1+X2)*Z(12)+2*B*Z(10)*Z(4)
870 LET F(12)=Z(10)/S2+X3*Z(12)+X5*Z(4)
875 RETURN
999 END
```

Program name: NEWTQ

Problem: Nonlinear two-point boundary-value problem.

Method of solution: Newton–Raphson method.

Example: Optimal input for estimating the parameter a in the equation

$$\dot{x} = -ax - bx^2 + y$$

and for the simultaneous determination of the Lagrange multiplier q.

Reference in text: Chapter 9.1.

Definition of variables:

Z = Vector of state, costate, and auxiliary equation variables, $\mathbf{Z}$.

F = Vector of derivatives, $\dot{\mathbf{Z}} = \mathbf{f}$.

Listing:

150–188: Initial conditions of the differential equations, $\mathbf{Z}(0)$.

250–660: Integration of the differential equations.

700–740: New estimate of the unknown initial conditions

$$\mathbf{c}^{k+1} = \mathbf{c}^k - P_c^{-1}\mathbf{p}(T).$$

780 GO TO 155: Repeat the above process to obtain the solution of the two-point boundary-value problem for the new initial conditions.

Output: Optimal input $y(t)$ and responses $x(t)$, $x_a(t)$ as a function of time.

```
NEWTQ

1 LET I5=1
2 LET S2=1
5 LET B=0
8 LET R=2
9 LET R1=0
10 READ L
20 DIM P(3),L(3,3),M(3),N(3,3),C(3)
30 DIM Z(24),Y(24),F(24),H(24),I(24),J(24),K(24)
50 LET H=1/100
60 LET C=.1
70 LET G=.075
75 LET A=.1
80 LET M9=INT(L/H+.1*H)
90 LET A9=.1
95 LET E=17.2742
115 REM INITIAL APPROXIMATION
120 MAT C=ZER
125 LET C(3)=G
150 REM INITIAL CONDITIONS MAT Z
155 MAT Z=ZER
160 LET Z(1)=C
165 LET Z(3)=C(1)
170 LET Z(4)=C(2)
175 LET Z(7)=1
180 LET Z(12)=1
182 LET Z(17)=C(3)
184 LET Z(20)=1
186 LET Z(21)=C(3)
188 LET Z(24)=1
190 LET G=C(3)
195 LET K4=1
200 LET T1=0
```

```
205 PRINT
210 PRINT "C="
215 MAT PRINT C
220 PRINT
240 PRINT " T","     X","     XA","     Y"
250 FOR N=0 TO M9
260 MAT Y=(1)*Z
270 IF N=0 THEN 470
280 GOSUB 800
290 MAT H=(H/2)*F
300 MAT Z=Y+H
310 GOSUB 800
320 MAT I=(H/2)*F
330 MAT Z=Y+I
340 GOSUB 800
350 MAT J=(H)*F
360 MAT Z=Y+J
370 GOSUB 800
380 MAT K=(H)*F
390 MAT H=(2)*H
400 MAT I=(4)*I
410 MAT J=(2)*J
420 MAT I=H+I
430 MAT J=I+J
440 MAT K=J+K
450 MAT K=(1/6)*K
460 MAT Z=Y+K
470 LET W2=Z(2)*Z(2)
472 LET Y2=(Z(3)/G)**2
474 IF N<>0 THEN 482
476 LET K1=W2
478 LET K2=Y2
480 GO TO 630
482 IF N=INT(L/H+.1*H) THEN 505
484 IF K4=1 THEN 494
486 LET K1=K1+2*W2
488 LET K2=K2+2*Y2
490 LET K4=1
492 GO TO 630
494 LET K1=K1+4*W2
496 LET K2=K2+4*Y2
498 LET K4=0
500 GO TO 630
505 LET K1=K1+W2
510 LET K2=K2+Y2
630 LET M1=INT((T1+.1*H)/H)
635 IF N=M1 THEN 645
640 GO TO 660
645 LET T=N*H
650 PRINT T,Z(1),Z(2),-1/G*Z(3)
```

```
655 LET T1=T1+A9
660 NEXT N
662 LET K=2*K
664 LET K=1
665 LET K1=K1*H/3
670 LET K2=K2*H/3
675 PRINT "RETURN="1/2*(K1-G*K2-G*G),"G="G
680 PRINT "N="N,"IY2="K2,"IX2="K1
685 PRINT "J="1/2*(K1-G*K2)
695 REM COMPUTE NEW INITIAL CONDITIONS
700 LET L(1,1)=Z(7)
701 LET L(1,3)=Z(15)
702 LET L(2,3)=Z(16)
703 LET L(3,1)=Z(22)
704 LET L(3,2)=Z(23)
705 LET L(2,1)=Z(8)
706 LET L(3,3)=Z(24)
710 LET L(1,2)=Z(11)
715 LET L(2,2)=Z(12)
720 LET M(1)=Z(3)
725 LET M(2)=Z(4)
726 LET M(3)=Z(21)+E/2-C(3)
730 MAT N=INV(L)
735 MAT P=N*M
740 MAT C=C-P
741 IF I5=2 THEN 745
742 LET C(3)=G
743 LET I5=2
744 GO TO 750
745 LET I5=2
746 MAT PRINT L;N;M;P;
750 LET R1=R1+1
752 IF R1<R THEN 780
755 PRINT
760 PRINT "R1="R1,"DESIRED R="
765 INPUT R
770 IF R1=R THEN 999
780 GO TO 155
790 DATA 1
795 REM CALCULATE A(T) AND U(T) AT TIME N-1
800 LET X1=2*B*Z(1)
802 LET X2=2*B*Z(2)
805 LET X3=A+X1
806 LET X4=2*B*Z(5)
807 LET X5=2*B*Z(9)
808 LET X6=2*B*Z(13)
810 LET F(1)=-A*Z(1)-B*Z(1)*Z(1)-1/G*Z(3)
815 LET F(2)=-Z(1)-X3*Z(2)
820 LET F(3)=X3*Z(3)+(1+X2)*Z(4)
825 LET F(4)=Z(2)/S2+X3*Z(4)
```

```
830 LET F(5)=-X3*Z(5)-1/G*Z(7)
835 LET F(6)=-Z(5)-A*Z(6)-2*B*(Z(1)*Z(6)+Z(2)*Z(5))
840 LET F(7)=X3*Z(7)+X4*Z(3)+(1+X2)*Z(8)+2*B*Z(6)*Z(4)
845 LET F(8)=Z(6)/S2+X3*Z(8)+X4*Z(4)
850 LET F(9)=-X3*Z(9)-1/G*Z(11)
855 LET F(10)=-Z(9)-A*Z(10)-2*B*(Z(1)*Z(10)+Z(2)*Z(9))
860 LET F(11)=X3*Z(11)+X5*Z(3)+(1+X2)*Z(12)+2*B*Z(10)*Z(4)
870 LET F(12)=Z(10)/S2+X3*Z(12)+X5*Z(4)
875 LET F(13)=-X3*Z(13)-1/G*Z(15)+1/(G**2)*Z(20)*Z(3)
880 LET F(14)=-Z(13)-A*Z(14)-2*B*(Z(1)*Z(14)+Z(2)*Z(13))
885 LET F(15)=X3*Z(15)+X6*Z(3)+(1+X2)*Z(16)+2*B*Z(14)*Z(4)
890 LET F(16)=Z(14)+X3*Z(16)+X6*Z(4)
895 LET F(17)=0
900 LET F(18)=0
905 LET F(19)=0
910 LET F(20)=0
915 LET Q2=1/(G**2)
920 LET Q3=1/(G**3)
925 LET Q4=Z(3)*Z(3)
926 LET Q2=-Q2
927 LET Q3=-Q3
930 LET F(21)=Q2*Q4/2
935 LET F(22)=Q2*Z(7)*Z(3)-Q3*Z(18)*Q4
940 LET F(23)=Q2*Z(11)*Z(3)-Q3*Z(19)*Q4
950 RETURN
999 END
```

Program name: NEGLUQ

Problem: Nonlinear two-point boundary-value problem.

Method of solution: Newton–Raphson method.

Example: Optimal input for estimating the parameter b in the blood glucose regulation model

$$\dot{x}_1 = -ax_1 + bx_2$$
$$\dot{x}_2 = -cx_1 - dx_2 + y$$

and for the simultaneous determination of the Lagrange multiplier q. Performance index II with Bergman parameters.

Reference in text: Chapter 10.1.3.

Definition of variables:

$Z = 10\times 6$ matrix of state, costate, and auxiliary equation variables.
$F = 10\times 6$ matrix of derivatives, $\dot{Z} = F$.

Listing:

91–111: 10×10 matrix A of the two-point boundary-value equation, $\dot{\mathbf{x}} = A\mathbf{x} + \phi$.

115–184: Initial conditions of the differential equations, $Z(0)$.

250–660: Integration of the differential equations.

700–740: New estimate of the unknown initial conditions

$$\mathbf{c}^{k+1} = \mathbf{c}^{k} - P_c^{-1}\mathbf{p}(T).$$

780 GO TO 155: Repeat the above process to obtain the solution of the two-point boundary-value problem for the new initial conditions.

Output: Optimal input $y(t)$ and responses $x_1(t)$, $x_2(t)$, and $x_{2b}(t)$ as a function of time.

```
NEGLUQ

1 LET I5=1
2 LET S2=1
5 LET B=0
8 LET R=2
9 LET R1=0
10 READ L
20 DIM P(5),L(5,5),M(5),N(5,5),C(5),O(10,6)
30 DIM Z(10,6),Y(10,6),F(10,6),H(10,6),I(10,6)
35 DIM A(10,10),J(10,6),K(10,6)
36 READ A5,B5,C5,D5
50 LET H=1
60 LET C=10
70 LET G=60
75 LET A=.1
80 LET M9=INT(L/H+.1*H)
85 LET E=8.15302
90 LET A9=6
91 MAT A=ZER
92 LET A(1,1)=-A5
93 LET A(1,2)=B5
94 LET A(2,1)=-C5
95 LET A(2,2)=-D5
96 LET A(3,2)=1
97 LET A(3,3)=-A5
98 LET A(3,4)=B5
99 LET A(4,3)=-C5
100 LET A(4,4)=-D5
101 LET A(9,4)=.1
102 LET A(2,7)=-1/G
103 LET A(6,6)=A5
```

```
104 LET A(6,7)=C5
105 LET A(7,6)=-B5
106 LET A(7,7)=D5
107 LET A(8,8)=A5
108 LET A(8,9)=C5
109 LET A(9,8)=-B5
110 LET A(9,9)=D5
111 LET A(7,8)=-1
113 PRINT "A="
114 MAT PRINT A;
115 REM INITIAL APPROXIMATION
120 MAT C=ZER
125 LET C(5)=G
150 REM INITIAL CONDITIONS MAT Z
155 MAT Z=ZER
160 LET Z(2,1)=C
165 LET Z(5,1)=C(5)
170 LET Z(6,1)=C(1)
175 LET Z(7,1)=C(2)
176 LET Z(8,1)=C(3)
177 LET Z(9,1)=C(4)
178 LET Z(6,2)=1
179 LET Z(7,3)=1
180 LET Z(8,4)=1
181 LET Z(9,5)=1
182 LET Z(5,6)=1
183 LET Z(10,1)=C(5)
184 LET Z(10,6)=1
185 IF R1>0 THEN 190
186 PRINT "INITIAL CONDITIONS,Z"
187 MAT PRINT Z;
190 LET G=C(5)
195 LET K4=1
200 LET T1=0
205 PRINT
210 PRINT "C="
215 MAT PRINT C
220 PRINT
240 PRINT "T", "X1","X2","X2B","Y"
250 FOR N=0 TO M9
260 MAT Y=(1)*Z
270 IF N=0 THEN 470
280 GOSUB 800
290 MAT H=(H/2)*F
300 MAT Z=Y+H
310 GOSUB 800
320 MAT I=(H/2)*F
330 MAT Z=Y+I
340 GOSUB 800
350 MAT J=(H)*F
```

```
360 MAT Z=Y+J
370 GOSUB 800
380 MAT K=(H)*F
390 MAT H=(2)*H
400 MAT I=(4)*I
410 MAT J=(2)*J
420 MAT I=H+I
430 MAT J=I+J
440 MAT K=J+K
450 MAT K=(1/6)*K
460 MAT Z=Y+K
470 LET W2=Z(4,1)*Z(4,1)
472 LET Y2=(Z(7,1)/G)**2
474 IF N<>0 THEN 482
476 LET K1=W2
478 LET K2=Y2
480 GO TO 630
482 IF N=INT(L/H+.1*H) THEN 505
484 IF K4=1 THEN 494
486 LET K1=K1+2*W2
488 LET K2=K2+2*Y2
490 LET K4=1
492 GO TO 630
494 LET K1=K1+4*W2
496 LET K2=K2+4*Y2
498 LET K4=0
500 GO TO 630
505 LET K1=K1+W2
510 LET K2=K2+Y2
630 LET M1=INT((T1+.1*H)/H)
632 IF N=M9 THEN 645
635 IF N=M1 THEN 645
640 GO TO 660
645 LET T=N*H
650 PRINT T,Z(1,1),Z(2,1),Z(4,1),-1/G*Z(7,1)
655 LET T1=T1+A9
660 NEXT N
664 LET K=1
665 LET K1=K1*H/3
670 LET K2=K2*H/3
675 PRINT "RETURN="1/2*(K1-G*K2-G*G),"G="G
680 PRINT "N="N,"IY2="K2,"IX2="K1
685 PRINT "J="1/2*(K1-G*K2)
695 REM COMPUTE NEW INITIAL CONDITIONS
700 FOR I=1 TO 5
701 FOR J=1 TO 5
702 LET L(I,J)=Z(I+5,J+1)
703 NEXT J
704 LET M(I)=Z(I+5,1)
705 NEXT I
```

```
706 LET M(5)=M(5)+E/2-C(5)
730 MAT N=INV(L)
732 PRINT "R1="R1,"DET="DET
735 MAT P=N*M
736 IF R1>2 THEN 740
737 PRINT "M,P="
738 MAT PRINT M;P;
740 MAT C=C-P
741 IF I5=2 THEN 745
742 LET C(5)=G
743 LET I5=2
744 GO TO 750
745 REM REPEAT TWICE
750 LET R1=R1+1
752 IF R1<R THEN 780
755 PRINT
760 PRINT "R1="R1,"DESIRED R="
765 INPUT R
770 IF R1=R THEN 999
780 GO TO 155
790 DATA 60
792 DATA .185,.342,.0349,.0263
795 REM CALCULATE A(T) AND U(T) AT TIME N-1
800 LET Q2=1/(G**2)
802 LET Q3=1/(G**3)
805 LET Q4=Z(7,1)*Z(7,1)
806 FOR J=1 TO 5
807 LET O(10,J+1)=-Q2*Z(7,J+1)*Z(7,1)+Q3*Z(5,J+1)*Q4
808 LET O(2,J+1)=Q2*Z(5,J+1)*Z(7,1)
809 NEXT J
810 LET O(10,1)=-.5*Q2*Q4
812 LET A(2,7)=-1/G
813 IF R1>0 THEN 815
815 MAT F=A*Z
820 MAT F=F+O
825 RETURN
999 END
```

11.2. System Identification Problems

The program QUAC is used to estimate the unknown parameter a in the equation

$$\dot{x} = -ax + y,$$

where $y(t)$ is a sine wave input. The observations are

$$b_i = x(t_i) + v(t_i),$$

where $v(t)$ is a zero mean Gaussian white noise process. The solution is obtained via the method of quasilinearization.

The program WLSIDD is used to estimate the four unknown parameters a, b, c, and d of the blood glucose regulation model

$$\dot{x}_1 = -ax_1 + bx_2, \qquad x_1(0) = c_1,$$

$$\dot{x}_2 = -cx_1 - dx_2, \qquad x_2(0) = c_2.$$

The solution is obtained via the Gauss–Newton method.

Program name: QUAC.

Problem: System identification.

Method of solution: Quasilinearization.

Example: Estimation of the parameter a in the equation

$$\dot{x} = -ax + y$$

from the observations

$$b_i = x(t_i) + v(t_i)$$

Also calculation of the parameter estimate/observation sensitivity, $\partial\hat{a}/\partial b_j$.

Reference in text: Chapter 8, Sections 8.2.1 and 8.2.2

Definition of variables:

P, Q, W, and U correspond to the variables p, h, x_a, and x_{aa}, respectively.

Listing:

200–250: Generation of observations, b_i, for sine wave input $y(t)$.

320–510: Integration of $\dot{x} = -ax + y$ to obtain initial approximation and storage of $x(t)$.

600–750: Integration of the quasilinearized differential equations $\dot{p}$ and $\dot{h}$ and the differential equations $\dot{x}_a$ and $\dot{x}_{aa}$.

760: New estimate of the parameter a

$$a = \frac{\sum_{i=1}^{N} [b_i - p(t_i)]h(t_i)}{\sum_{i=1}^{N} h^2(t_i)}.$$

820 GO TO 550: Repeat the integration of the differential equations using the new estimate of the parameter a.

Output: Estimate of the parameter a. Also $\partial\hat{a}/\partial b_j$, x_a, x_{aa}, and x as a function of time for the last estimate of the parameter a.

```
QUAC

10 DIM X(100),P(100),Q(100)
15 DIM Y(100,3)
20 DIM M(10),B(10)
22 DIM V(10),U(10),W(10)
25 READ E1,E2
30 LET D=10
40 LET C=1
43 READ L
45 READ Y
46 LET L6=L6+1
47 PRINT "RUN NO. ="L6
48 IF L6=11 THEN 999
50 PRINT "INPUT DATA"
60 PRINT "T","X","X+N","N"
80 LET A=.1
90 LET H=1/100
100 LET T2=0
110 LET T1=0
120 LET X(0)=C
130 LET V4=0
140 LET V5=0
145 LET Z8=0
150 LET B(0)=C
160 PRINT T1,B(0)
200 FOR I=1 TO D
210 LET M(I)=INT((L/D*I+.1*H)/H)
220 LET T1=M(I)*H
226 LET E3=A**2+E1**2
228 LET E4=EXP(-A*T1)
230 LET B(I)=(E1*E2*E4)/E3+E2/SQR(E3)*SIN(E1*T1-ATN(E1/A))+C*E4
232 LET V9=B(I)
236 LET V1=RND(Z1)
237 LET V2=RND(Z2)
238 LET V3=(SQR(-2*LOG(V1)))*COS(6.283154*V2)
241 LET V4=V4+V3
```

```
242 LET V5=V5+V3**2
245 LET B(I)=B(I)+.1*V3
246 PRINT T1,V9,B(I),.1*V3
250 NEXT I
255 PRINT
260 PRINT "MEAN="V4/D,"SD="SQR(V5/D)
270 PRINT
300 LET A0=.15
310 LET X1=C
320 FOR N=1 TO L/H
330 LET T1=(N-1)*H
335 LET J=0
340 LET T=T1
350 LET X=X1
360 GOSUB 830
370 LET K1=H*F
380 LET T=T1+H/2
390 LET X=X1+K1/2
400 GOSUB 830
410 LET K2=H*F
420 LET X=X1+K2/2
430 GOSUB 830
440 LET K3=H*F
450 LET T=T1+H
460 LET X=X1+K3
470 GOSUB 830
480 LET K4=H*F
490 LET X1=X1+(K1+2*(K2+K3)+K4)/6
500 LET X(N)=X1
510 NEXT N
515 PRINT "   A0","B","S"
520 LET I9=0
530 GO TO 765
550 LET P1=C
560 LET Q1=0
562 LET W1=0
563 LET U1=0
570 LET I=1
580 LET B=0
590 LET S=0
592 LET L5=0
593 LET M5=0
595 LET I9=1
600 FOR N=1 TO L/H
605 LET J=0
610 LET P=P1
615 LET Q=Q1
617 LET W=W1
618 LET U=U1
620 GOSUB 850
```

```
625 LET K1=H*F1
630 LET C1=H*F2
632 LET L1=H*F3
633 LET M1=H*F4
635 LET P=P1+K1/2
640 LET Q=Q1+C1/2
642 LET W=W1+L1/2
643 LET U=U1+M1/2
645 GOSUB 850
650 LET K2=H*F1
655 LET C2=H*F2
657 LET L2=H*F3
658 LET M2=H*F4
660 LET P=P1+K2/2
665 LET Q=Q1+C2/2
667 LET W=W1+L2/2
668 LET U=U1+M2/2
670 GOSUB 850
675 LET K3=H*F1
680 LET C3=H*F2
682 LET L3=H*F3
683 LET M3=H*F4
685 LET P=P1+K3
690 LET Q=Q1+C3
692 LET W=W1+L3
693 LET U=U1+M3
695 GOSUB 850
700 LET K4=H*F1
705 LET C4=H*F2
707 LET L4=H*F3
708 LET M4=H*F4
710 LET P1=P1+(K1+2*(K2+K3)+K4)/6
715 LET Q1=Q1+(C1+2*(C2+C3)+C4)/6
717 LET W1=W1+(L1+2*(L2+L3)+L4)/6
718 LET U1=U1+(M1+2*(M2+M3)+M4)/6
720 LET P(N)=P1
725 LET Q(N)=Q1
730 IF N<>M(I) THEN 750
732 LET V(I)=P1+A0*Q1
735 LET B=B+(B(I)-P1)*Q1
740 LET S=S+Q1**2
741 LET U(I)=U1
742 LET L5=L5+W1**2
743 LET M5=M5+(P1+A0*Q1-B(I))*U1
744 LET W(I)=W1
745 LET I=I+1
750 NEXT N
760 LET A0=B/S
765 PRINT A0,B,S
770 LET Z8=Z8+1
```

```
775 IF Z8=7 THEN 46
790 FOR N=1 TO L/H
792 IF I9=0 THEN 820
795 LET X(N)=P(N)+A0*Q(N)
815 NEXT N
820 GO TO 550
825 REM INPUT FUNCTION Y(T)
830 LET Y=E2*SIN(E1*T)
835 LET F=-A0*X+Y
836 LET Y(N-1,J)=Y
837 LET J=J+1
840 RETURN
850 LET F0=-A0*X(N-1)+Y(N-1,J)
870 LET D1=-X(N-1)
880 LET D2=-A0
890 LET F1=F0-A0*D1+(P-X(N-1))*D2
900 LET F2=D1+D2*Q
902 LET F3=-P-A0*Q-A0*W
903 LET F4=-2*W-A0*U
905 LET J=J+1
910 RETURN
980 DATA .1,291.322
990 DATA 1
995 DATA 0
999 PRINT "I","AH/BJ","XA","XAA","X"
1010 FOR I=0 TO D
1020 PRINT I,W(I)/(L5+M5),W(I),U(I),V(I)
1030 NEXT I
1040 PRINT
1050 PRINT "L5="L5,"M5="M5
1060 END
```

Program name: WLSIDD

Problem: System identification.

Method of solution: Gauss–Newton method.

Example: Identification of the parameter a, b, c, and d of the blood glucose regulation model

$$\dot{x}_1 = -ax_1 + bx_2,$$
$$\dot{x}_2 = -cx_1 - dx_2.$$

Reference in text: Chapter 7.1.3.

Definition of variables:

P = Matrix of state and auxiliary equation variables

$$= \begin{bmatrix} x_1 & x_{1a} & x_{1b} & x_{1c} & x_{1d} \\ x_2 & x_{2a} & x_{2a} & x_{2c} & x_{2d} \end{bmatrix}.$$

X = Parameter influence matrix, dimensions = $2 \times$(number of observations)$\times 4$.

C = Least-squares estimate of the parameter vector,

$$\Delta \mathbf{p}_j = (X^T X)^{-1} X^T \mathbf{e}.$$

Listing:

215,

470–602: Read in the insulin and glucose measurements as a function of time.

250–265: Print out measured data.

300–465: Integration of the differential equations, $\dot{\mathbf{Z}} = [\dot{x}_1, \dot{x}_2]$ for internal generation of observations (not used when measured data is read in).

685–850: Integration of the matrix, $\dot{P}$.

795–810: Formation of the parameter influence matrix, X.

851–860: Least-squares estimate of the parameter vector $\Delta \mathbf{p}_j$.

880 GO TO 615: Repeat the above process to obtain the next estimate of the parameters a, b, c, and d.

Output: Least-squares estimates of the unknown parameters a, b, c, and d, and the model responses $x_1(t)$ and $x_2(t)$ as a function of time.

```
WLSIDD

 10 DIM X(31,3),A(3,31),E(31),B(31),U(3,31),R(20)
 20 DIM D(1,1),C(3),S(3),V(3,3),T(3,3)
 30 DIM Z(1),W(1),G(1),L(1),M(1),N(1),O(1)
 40 DIM P(1,4),Y(1,4),F(1,4),H(1,4),I(1,4),J(1,4),K(1,4)
 50 READ A,B,C,D
 52 DATA .1,.2,.04,.001
 55 PRINT "A="A,"B="B,"C="C,"D="D
 60 LET C1=34
 70 LET C2=143
 80 READ L
 85 DATA 25
 90 LET H=.5
 95 LET E9=1
120 LET E1=10
130 LET N1=INT(L/H+.1*H)
140 LET S1=0
170 LET T1=0
180 LET E2=10
```

```
185 LET E2=2*E2-2
190 LET E3=E2+1
200 LET I1=1
210 LET I2=0
215 IF S1=0 THEN 470
220 LET E1=2*E1-1
225 MAT X=ZER(E1,3)
230 MAT A=ZER(3,E1)
235 MAT E=ZER(E1)
236 FOR I=0 TO E1
237 LET E(I)=B(I)
238 NEXT I
240 MAT B=ZER(E1)
242 MAT B=(1)*E
243 MAT E=ZER
245 MAT U=ZER(3,E1)
250 IF E9=0 THEN 280
252 LET E1=E1-1
255 FOR I=0 TO E1 STEP 2
260 PRINT H*R(I/2+1),B(I),B(I+1)
265 NEXT I
266 PRINT
267 PRINT "ABOVE IS ACTUAL DATA"
268 PRINT
270 GO TO 605
280 LET Z(0)=C1
290 LET Z(1)=C2
300 FOR N=1 TO N1
310 MAT W=(1)*Z
320 GOSUB 900
330 MAT L=(H/2)*G
340 MAT Z=W+L
350 GOSUB 900
360 MAT M=(H/2)*G
370 MAT Z=W+M
380 GOSUB 900
390 MAT N=(H)*G
400 MAT Z=W+N
405 GOSUB 900
410 MAT O=(H)*G
415 MAT L=(2)*L
420 MAT M=(4)*M
425 MAT N=(2)*N
430 MAT M=L+M
435 MAT N=M+N
440 MAT O=N+O
445 MAT O=(1/6)*O
450 MAT Z=W+O
455 IF N<> R(I1) THEN 465
457 LET B(I2)=Z(0)
```

```
459 LET B(I2+1)=Z(1)
461 PRINT H*R(I1),B(I2),B(I2+1),I1,N
463 LET I1=I1+1
464 LET I2=I2+2
465 NEXT N
470 IF S1=1 THEN 605
475 FOR I=0 TO E2 STEP 2
480 READ B(I)
485 NEXT I
490 FOR I=1 TO E3 STEP 2
492 READ B(I)
494 NEXT I
495 DATA 166,123,149,89,202
496 DATA 170,83,59,35,26
497 DATA 118,100,75,72,61
498 DATA 46,32,22.5,13,6
500 PRINT "INPUT DATA"
505 PRINT "T","B1","B2","I","N"
510 LET T1=0
515 FOR I=1 TO E1
520 LET R(I)=INT((L/E1*I+.1*H)/H)
525 LET T1=R(I)*H
530 LET N=R(I)
550 NEXT I
595 LET R(0)=0
600 LET S1=1
602 GO TO 220
605 READ A,B,C,D
607 PRINT
608 PRINT "FIRST APPROX"
609 PRINT "A="A,"B="B,"C="C,"D="D
610 PRINT "X1(0)="C1,"X2(0)="C2
615 PRINT
620 PRINT "T","X1","X2"
625 LET I1=1
630 LET I2=0
660 LET D(0,0)=-A
661 LET D(0,1)=B
662 LET D(1,0)=-C
663 LET D(1,1)=-D
665 MAT P=ZER
670 LET P(0,0)=C1
675 LET P(1,0)=C2
685 FOR N=1 TO N1
690 MAT Y=(1)*P
695 GOSUB 950
700 MAT H=(H/2)*F
705 MAT P=Y+H
710 GOSUB 950
715 MAT I=(H/2)*F
```

```
720 MAT P=Y+I
725 GOSUB 950
730 MAT J=(H)*F
735 MAT P=Y+J
740 GOSUB 950
745 MAT K=(H)*F
750 MAT H=(2)*H
755 MAT I=(4)*I
760 MAT J=(2)*J
765 MAT I=H+I
770 MAT J=I+J
775 MAT K=J+K
780 MAT K=(1/6)*K
785 MAT P=Y+K
790 IF N<>R(I1) THEN 850
795 FOR I=1 TO 4
800 LET X(I2,I-1)=P(0,I)
805 LET X(I2+1,I-1)=P(1,I)
810 NEXT I
815 LET E(I2)=B(I2)-P(0,0)
820 LET E(I2+1)=B(I2+1)-P(1,0)
825 PRINT H*R(I1),P(0,0),P(1,0)
830 LET I1=I1+1
835 LET I2=I2+2
850 NEXT N
851 MAT A=TRN(X)
854 MAT U=(1)*A
855 MAT V=U*X
856 MAT T=INV(V)
858 MAT S=U*E
860 MAT C=T*S
865 LET A=A+C(0)
866 LET B=B+C(1)
867 LET C=C+C(2)
868 LET D=D+C(3)
870 PRINT
875 PRINT "A="A,"B="B,"C="C,"D="D
880 GO TO 615
895 DATA .1,.2,.04,.001
900 LET G(0)=-A*Z(0)+B*Z(1)
910 LET G(1)=-C*Z(0)-D*Z(1)
920 RETURN
950 MAT F=D*P
955 LET F(0,1)=F(0,1)-P(0,0)
960 LET F(0,2)=F(0,2)+P(1,0)
965 LET F(1,3)=F(1,3)-P(0,0)
970 LET F(1,4)=F(1,4)-P(1,0)
1000 RETURN
1100 END
```

References

Chapter 1

1. COLLATZ, L., *The Numerical Treatment of Differential Equations*, Springer-Verlag, Berlin, 1960.
2. BOYCE, W. E., and DIPRIMA, R. C., *Elementary Differential Equations and Boundary Value Problems*, John Wiley and Sons, Inc., New York, 1969.
3. BELLMAN, R. E., and KALABA, R. E., *Quasilinearization and Nonlinear Boundary-Value Problems*, American Elsevier Publishing Company, Inc., New York, 1965.
4. CARNAHAN, B., LUTHER, H. A., and WILKES, J. O., *Applied Numerical Methods*, John Wiley and Sons, Inc., New York, 1969.
5. FORTHSYTHE, G., and MOLER, C. B., *Computer Solution of Linear Algebraic Systems*, Prentice-Hall, Inc., Englewood Cliffs, New Jersey, 1967.

Chapter 2

1. GELFAND, I. M., and FOMIN, S. V., *Calculus of Variations*, Prentice-Hall, Inc., Englewood Cliffs, New Jersey, 1963.
2. KALMAN, R. E., The theory of optimal control and the calculus of variations, in *Mathematical Optimization Techniques*, edited by R. Bellman, University of California Press, Berkeley, California, pp. 309–331, 1963.
3. COURANT, R., and HILBERT, D., *Methods of Mathematical Physics*, Vol. 1, Wiley-Interscience, New York, 1953.
4. MIELE, A., Introduction to the calculus of variations in one independent variable, in *Theory of Optimum Aerodynamic Shapes*, edited by A. Miele, Academic Press, New York, pp. 3–19, 1965.
5. GOTTFRIED, B. S., and WEISMAN, J., *Introduction to Optimization Theory*, Prentice-Hall, Inc., Englewood Cliffs, New Jersey, 1973.
6. BELLMAN, R., *Dynamic Programming*, Princeton University Press, Princeton, New Jersey, 1957.

7. DREYFUS, S. E., *Dynamic Programming and the Calculus of Variations*, Academic Press, New York, 1965.
8. BELLMAN, R., and KALABA, R., *Dynamic Programming and Modern Control Theory*, Academic Press, New York, 1965.
9. PONTRYAGIN, L. S., *et al.*, *The Mathematical Theory of Optimal Processes*, Wiley-Interscience, New York, 1962.
10. LANCZOS, C., *The Variational Principles of Mechanics*, University of Toronto Press, Toronto, 1957.
11. KALABA, R. E., *A New Approach to Optimal Control and Filtering*, University of Southern California Report USCEE-316, Los Angeles, California, 1968.
12. BELLMAN, R., and KALABA, R., On the fundamental equations of invariant imbedding—I, *Proceedings of the National Academy of Sciences*, Vol. 47, pp. 336–338, 1961.
13. KAGIWADA, H. and KALABA, R., *Derivation and Validation of an Initial-Value Method for Certain Nonlinear Two-Point Boundary-Value Problems*, RM-5566-PR, The Rand Corporation, January, 1968.
14. KAGIWADA, H., KALABA, R., SCHUMITZKY, A., and SRIDHAR, R., *Invariant Imbedding and Sequential Interpolating Filters for Nonlinear Processes*, RM-5507-PR, The Rand Corporation, November, 1967.
15. KALABA, R., and SRIDHAR, R., *Invariant Imbedding and the Simplest Problem in the Calculus of Variations*, RM-5781-PR, The Rand Corporation, October, 1968.
16. KAGIWADA, H., KALABA, R., SCHUMITZKY, A., and SRIDHAR, R., Cauchy and Fredholm methods for Euler equations, *Journal of Optimization Theory and Applications*, Vol. 2, pp. 157–163, 1968.
17. BELLMAN, R., KAGIWADA, H., and KALABA, R., Invariant imbedding and the numerical integration of boundary-value problems for unstable linear systems of ordinary differential equations, *Communications of the ACM*, Vol. 10, pp. 100–102, 1967.
18. BELLMAN, R., and KALABA, R., On the principle of invariant imbedding and propagation through inhomogeneous media, *Proceedings of the National Academy of Sciences*, Vol. 42, pp. 629–632, 1956.

Suggested Reading

AXELBAND, E. I., An approximation technique for the optimal control of linear distributed parameter systems, *IEEE Transactions on Automatic Control*, Vol. 11, pp. 42–45, January, 1966.

BALAKRISHNAN, A. V., and NEUSTADT, L. W., editors, *Computing Methods in Optimization Problems*, Academic Press, New York, 1964.

LARSON, R. E., *State Increment Dynamic Programming*, American Elsevier Publishing Company, Inc., New York, 1968.

LEONDES, C. T., editor of the series in *Control and Dynamic Systems, Advances in Theory and Applications*, Academic Press, New York, Volumes 1–16, 1965–1980.

THEIL, H., *Principles of Econometrics*, John Wiley and Sons, Inc., New York, 1971.

Chapter 3

1. ATHANS, M., and FALB, P. L., *Optimal Control*, McGraw-Hill Book Company, New York, pp. 757–758, 1966.

2. Sage, A., *Optimum Systems Control*, Prentice-Hall, Inc., Englewood Cliffs, New Jersey, pp. 88–89, 1968.
3. Bellman, R. E., and Kalaba, R. E., *Quasilinearization and Nonlinear Boundary-Value Problems*, American Elsevier Publishing Company, New York, p. 78, 1965.
4. Bellman, R., Kagiwada, H., and Kalaba, R., Invariant imbedding and the numerical integration of boundary-value problems for unstable linear systems of ordinary differential equations, *Communications of the ACM*, Vol. 10, pp. 100–102, 1967.
5. Kagiwada, H., and Kalaba, R., Derivation and validation of an initial value method for certain nonlinear two-point boundary-value problems, *Journal of Optimization Theory and Applications*, Vol. 2, pp. 378–385, 1968.
6. Spingarn, K., Some numerical aspects of optimal control, *J. Franklin Institute*, Vol. 289, pp. 351–359, May, 1970.
7. Courant, R., and Hilbert, D., *Methods of Mathematical Physics*, Interscience Publishers, Inc., New York, p. 183, 1966.
8. Bellman, R., and Kalaba, R., *Dynamic Programming and Modern Control Theory*, Academic Press, New York, p. 35, 1966.
9. Kalaba, R. E., *A New Approach to Optimal Control and Filtering*, University of Southern California Report USCEE-316, Los Angeles, California, November, 1968.
10. Kagiwada, H., and Kalaba, R., *Optimal Trajectories for Quadratic Variational Processes via Invariant Imbedding*, RM-5929-PR, The Rand Corporation, Santa Monica, California, January, 1969.
11. Spingarn, K., Some numerical results using Kalaba's new approach to optimal control and filtering, *Proceedings of the Fourth Hawaii International Conference on System Sciences*, Honolulu, Hawaii, pp. 186–187, January 1971. Also, Technical Note, *IEEE Transactions on Automatic Control*, pp. 713–715, October, 1972.
12. Bryson, A., Jr., and Ho, Y. C., *Applied Optimal Control*, Blaisdell Publishing Co., Waltham, Massachusetts, 1969.
13. Spingarn, K., A comparison of numerical methods for solving optimal control problems, *IEEE Transactions on Aerospace and Electronic Systems*, Vol. 7, pp. 73–78, January, 1971.

Chapter 4

1. Bellman, R. E., and Kalaba, R. E., *Quasilinearization and Nonlinear Boundary-Value Problems*, American Elsevier Publishing Company, Inc., New York, p. 127, 1965.
2. Roberts, S. M., and Shipman, J. S., *Two-Point Boundary-Value Problems: Shooting Methods*, American Elsevier Publishing Company, Inc., New York, 1972.
3. Kalaba, R. E., and Spingarn, K., Optimization of functionals subject to integral constraints, *Journal of Optimization Theory and Applications*, Vol. 24, No. 2, pp. 325–335, 1978.
4. Gottfried, B. S., and Weisman, J., *Introduction to Optimization Theory*, Prentice-Hall, Inc., Englewood Cliffs, New Jersey, 1973.
5. Bellman, R., *Introduction to the Mathematical Theory of Control Processes*, Vol. 1, Academic Press, New York, p. 76, 1967.

6. Kalaba, R., and Ruspini, E., Identification of parameters in nonlinear boundary value problems, *Journal of Optimization Theory and Applications*, Vol. 4, No. 6, pp. 371–377, 1969.
7. Miele, A., and Levy, A. V., Modified quasilinearization and optimal initial choice of the multipliers, Part 1, Mathematical programming problems, *Journal of Optimization Theory and Applications*, Vol. 6, No. 5, pp. 364–380, 1970.
8. Miele, A., Iyer, R. R., and Well, K. H., Modified quasilinearization and optimal initial choice of the multipliers, Part 2, Optimal control problems, *Journal of Optimization Theory and Applications*, Vol. 6, No. 5, pp. 381–409, 1970.
9. Kalaba, R. E., and Spingarn, K., Design of linear regulators with energy constraints, *IEEE Conference on Decision and Control*, Albuquerque, pp. 301–305, December, 1980.
10. Kalman, R. E., Contributions to the theory of optimal control, *Boletin Sociedad Matematica Mexicana*, Vol. 5, No. 1, pp. 102–119, 1960.
11. Athans, M., and Falb, P., *Optimal Control*, McGraw-Hill Book Company, New York, pp. 750–814, 1966.
12. Bryson, A. E., Jr, and Ho, Y., *Applied Optimal Control*, Blaisdell Publishing Co., Waltham, Massachusetts, 1969.
13. Sage, A. P., *Optimum Systems Control*, Prentice-Hall, Inc., Englewood Cliffs, New Jersey, 1968.
14. Harvey, C. A., and Stein, G., Quadratic weights for asymptotic regulator properties, *IEEE Transactions on Automatic Control*, Vol. 23, No. 3, pp. 378–387, 1978.
15. Stein, G., Generalized quadratic weights for asymptotic regulator properties, *IEEE Conference on Decision and Control*, San Diego, California, pp. 831–837, 1978.
16. Kalaba, R. E., Spingarn, K., and Zagustin, E., On the integral equation method for buckling loads, *Applied Mathematics and Computation*, Vol. 1, No. 3, pp. 253–261, 1975.
17. Vol'mir, A. S., *Stability of Elastic Systems*, Fitmatgiz, Moscow, 1965.
18. Popov, E. P., *Mechanics of Materials*, Prentice-Hall, Inc., New York, 1952.
19. Kagiwada, H., and Kalaba, R., *An Initial Value Method for Fredholm Resolvents*, Biomedical Engineering Program, University of Southern California, Los Angeles, California, 1970.
20. Kalaba, R. E., and Spingarn, K., Numerical solution of the integral equations for optimal sequential filtering, *International Journal for Numerical Methods in Engineering*, Vol. 6, No. 3, pp. 451–454, 1973.
21. Kalaba, R. E., Spingarn, K., and Zagustin, E., An imbedding method for buckling loads, *Computers and Mathematics with Applications*, Vol. 1, No. 3/4, pp. 277–284, 1975.
22. Casti, J., and Kalaba, R., *Imbedding Methods in Applied Mathematics*, Addison-Wesley Publishing Company, Reading, Massachusetts, 1973.
23. Alspaugh, D., Kagiwada, H., and Kalaba, R., Applications of invariant imbedding to the buckling of columns, *Journal of Computational Physics*, Vol. 5, No. 1, pp. 56–69, February 1970.
24. Kalaba, R. E., and Spingarn, K., Numerical solution of a nonlinear two-point boundary value problem by an imbedding method, *Nonlinear Analysis, Theory, Methods, & Applications*, Vol. 1, No. 2, pp. 129–133, 1977.
25. Kagiwada, H., and Kalaba, R., *Integral Equations via Imbedding Methods*, Addison-Wesley Publishing Company, Reading, Massachusetts, 1974.

26. KALABA, R., and ZAGUSTIN, E., Reduction of a nonlinear two-point boundary value problem with nonlinear boundary conditions to a Cauchy system, *International Journal of Nonlinear Mechanics*, Vol. 9, pp. 221–228, 1974.
27. HUSS, R., and KALABA, R., Invariant imbedding and the numerical determination of Green's functions, *Journal of Optimization Theory and Applications*, Vol. 6, pp. 415–423, 1970.
28. MIKHLIN, S., and SMOLITSKIY, K., *Approximate Methods for Solution of Differential and Integral Equations*, American Elsevier Publishing Company, Inc., New York, 1967.
29. KALABA, R. E., and SPINGARN, K., Post buckling beam configurations via an imbedding method, *Computers and Mathematics with Applications*, Vol. 4, No. 1, pp. 1–10, 1978.
30. KALABA, R. E., SPINGARN, K., and ZAGUSTIN, E., Imbedding methods for bifurcation problems and post-buckling behavior of nonlinear columns, in *Nonlinear Systems and Applications*, edited by V. Lakshmikantham, Academic Press, New York, pp. 173–187, 1977.
31. KALABA, R., Dynamic programming and the variational principles of classical and statistical mechanics, *Developments in Mechanics*, Vol. 1, Plenum Press, New York, pp. 1–9, 1961.
32. ABRAMOWITZ, M., and STEGUM, I. A., editors, *Handbook of Mathematical Functions*, National Bureau of Standards, Applied Mathematics Series-55, U. S. Government Printing Office, Washington, D.C., 1964.
33. KALABA, R., SPINGARN, K., and TESFATSION, L., A sequential method for nonlinear filtering: Numerical implementation and comparisons, *Journal of Optimization Theory and Applications*, Vol. 34, No. 4, pp. 543–561, 1981.
34. BELLMAN, R., KAGIWADA, H., KALABA, R., and SRIDHAR, R., Invariant imbedding and nonlinear filtering theory, *Journal of the Astronautical Sciences*, Vol. 13, pp. 110–115, 1966.
35. DETCHMENDY, D., and SRIDHAR, R., Sequential estimation of states and parameters in noisy nonlinear dynamical systems, *Transactions of the ASME*, pp. 362–368, 1966.
36. KAGIWADA, H., KALABA, R., SCHUMITSKY, A., and SRIDHAR, R., Invariant imbedding and sequential interpolating filters for nonlinear processes, *Journal of Basic Engineering*, Vol. 91, pp. 195–200, 1969.
37. SUGISAKA, M., and SAGARA, S., A non-linear filter with higher-order weight functions functions via invariant imbedding, *International Journal of Control*, Vol. 21, pp. 801–823, 1975.
38. KALABA, R., and TESFATSION, L., Exact sequential solutions for a class of discrete time nonlinear estimation problems, *IEEE Transactions on Automatic Control*, Vol. 26, No. 5, 1981.

Chapter 5

1. SAGE, A. P., and MELSA, J. L., *Estimation Theory with Applications to Communications and Control*, McGraw-Hill Book Company, New York, 1971.
2. VAN TREES, H. L., *Detection, Estimation, and Modulation Theory*, John Wiley and Sons, Inc., New York, 1968.

3. NAHI, N. E., *Estimation Theory and Applications*, John Wiley and Sons, Inc., New York, 1969.
4. GOODWIN, G. C., and PAYNE, R. L., *Dynamic System Identification: Experiment Design and Data Analysis*, Academic Press, New York, 1977.
5. EYKHOFF, P. *System Identification, Parameter and State Estimation*, John Wiley and Sons, Inc., New York, 1974.
6. GIESE, C., and MCGHEE, R. B., Estimation of nonlinear system states and parameters by regression methods, *Joint Automatic Control Conference*, Troy, New York, pp. 46–55. 1965.
7. GOODWIN, G. C., Optimal input signals for nonlinear-system identification, *Proceedings of the IEE*, Vol. 118, No. 7, pp. 922–926, 1971.
8. SAGE, A., *Optimum Systems Control*, Prentice-Hall, Inc., Englewood Cliffs, New Jersey, 1968.
9. GRAUPE, D., *Identification of Systems*, Van Nostrand Reinhold Company, New York, 1972.
10. BELLMAN, R. E., and KALABA, R. E., *Quasilinearization and Nonlinear Boundary-Value Problems*, American Elsevier Publishing Company, New York, 1965.
11. KAGIWADA, H. H., *System Identification, Methods and Applications*, Addison-Wesley Publishing Company, Reading, Massachusetts, 1974.
12. SAGE, A. P., and MELSA, J. L., *System Identification*, Academic Press, New York, 1971.
13. MENDEL, J. M., *Discrete Techniques of Parameter Estimation*: *The Equation Error Formulation*, Marcel Dekker, Inc., New York, 1973.
14. MEHRA, R. K., and LAINIOTIS, D. G., *System Identification*: *Advances and Case Studies*, Academic Press, New York, 1976.
15. IEEE Special Issue on System Identification and Time-Series Analysis, (Guest editors: Kailath, T., Mayne, D. Q., and Mehra, R. K.) *Transactions on Automatic Control*, Vol. 19, No. 6, 1974.
16. BEKEY, G. A., System identification—An introduction and survey, *Simulation*, pp. 151–166, October, 1970.

Chapter 6

1. BELLMAN, R. E., and KALABA, R. E., *Quasilinearization and Nonlinear Boundary-Value Problems*, American Elsevier Publishing Company, New York, 1965.
2. BUELL, J., and KALABA, R. E., Quasilinearization and the fitting of nonlinear models of drug metabolism to experimental kinetic data, *Mathematical Biosciences*, Vol. 5, pp. 121–132, 1969.
3. KAGIWADA, H. H., *System Identification*, *Methods and Applications*, Addison-Wesley Publishing Company, Reading, Massachusetts, 1974.
4. EYKHOFF, P., *System Identification*, *Parameter and State Estimation*, John Wiley and Sons, Inc., New York, 1974.
5. KALABA, R. E., and SPINGARN, K., Optimal inputs and sensitivities for parameter estimation, *Journal of Optimization Theory and Applications*, Vol. 11, No. 1, pp. 56–67, 1973.
6. BELLMAN, R., JACQUEZ, J., KALABA, R., and SCHWIMMER, S., Quasilinearization and

the estimation of chemical rate constants from raw kinetic data, *Mathematical Biosciences*, Vol. 1, pp. 71–76, 1976.
7. BELLMAN, R., KAGIWADA, H., and KALABA, R., Orbit determination as a multi-point boundary-value problem and quasilinearization, *Proceedings of the National Academy of Sciences*, Vol. 48, No. 8, pp. 1327–1329, 1962.
8. BUELL, J. D., KAGIWADA, H. H., and KALABA, R. E., A proposed computational method for estimation of orbital elements, drag coefficients, and potential fields parameters from satellite measurements, *Annales de Géophysique*, Vol. 23, No. 1, pp. 35–39, 1967.
9. KUMAR, K. S. P., and SRIDHAR, R., On the identification of control systems by the quasilinearization method, *IEEE Transactions on Automatic Control*, Vol. 9, No. 2, pp. 151–154, 1964.

Chapter 7

1. BOLIE, V. W., Coefficients of normal blood glucose regulation, *Journal of Applied Physiology*, Vol. 16, pp. 783–788, 1961.
2. ACKERMAN, E., ROSEVEAR, J. W., and MCGUCKIN, W. F., A mathematical model of the glucose-tolerance test, *Physics in Medicine and Biology*, Vol. 9, No. 2, pp. 203–213, 1964.
3. ACKERMAN, E., GATEWOOD, L. C., ROSEVEAR, J. W., and MOLNAR, G. D., Model studies of blood-glucose regulation, *Bulletin of Mathematical Biophysics, Supplement*, Vol. 27, pp. 21–37, 1965.
4. CERESA, F., GHEMI, F., MARTINI, P. F., MARTINO, P., SEGRE, G., and VITELLI, A., Control of blood glucose in normal and diabetic subjects, *Diabetes*, Vol. 17, No. 9, pp. 570–578, 1968.
5. SEGRE, G., TURCO, G. L., and VERCELLONE, G., Modeling blood glucose and insulin kinetics in normal, diabetic, and obese subjects, *Diabetes*, Vol. 22, No. 2, pp. 94–103, 1973.
6. BERGMAN, R. N., KALABA, R. E., and SPINGARN, K., Optimizing inputs for diagnosis of diabetes, I. Fitting a minimal model to data, *Journal of Optimization Theory and Applications*, Vol. 20, No. 1, pp. 47–62, September 1976.
7. CURRY, D., BENNET, L., and GRODSKY, G., Dynamics of insulin secretion from the isolated pancreas, *Endocrinology*, Vol. 83, p. 572, 1968.
8. BERGMAN, R. N., and URQUHART, J., The pilot gland approach to the study of insulin secretory dynamics, *Recent Progress in Hormone Research*, Vol. 27, pp. 583–605, 1971.
9. SRINIVASAN, R., KADISH, A. H., and SRIDHAR, R., A mathematical model for the control mechanism of free fatty acid—Glucose metabolism in normal humans, *Computers and Biomedical Research*, Vol. 3, No. 2, pp. 146–165, 1970.
10. GRODSKY, G. M., A threshold distribution hypothesis for packet storage of insulin and its mathematical modelling, *Journal of Clinical Investigation*, Vol. 51, No. 8, pp. 2047–2059, 1972.
11. BERGMAN, R. N., and BUCOLO, R. J., Nonlinear metabolic dynamics of the pancreas and liver, *ASME Journal of Dynamic Systems, Measurements, and Control*, Vol. 95, pp. 296–300, 1973.

12. BELLMAN, R. E., and KALABA, R. E., *Quasilinearization and Nonlinear Boundary-Value Problems*, American Elsevier Publishing Company, Inc., New York, 1965.
13. SAGE, A. P., and MELSA, J. L., *System Identification*, Academic Press, New York, p. 157, 1971.
14. BUELL, J., and KALABA, R. E., Quasilinearization and the fitting of nonlinear drug metabolism to experimental kinetic data, *Mathematical Biosciences*, Vol. 5, pp. 121–132, 1969.
15. KRÜGER-THIEMER, E., Formal theory of drug dosage regimens: I, *Journal of Theoretical Biology*, Vol. 13, pp. 212–235, 1966.
16. JELLIFFE, R., A mathematical analysis of digitalis kinetics in patients with normal and reduced renal function, *Mathematical Biosciences*, Vol. 1, pp. 305–325, 1967.
17. KRÜGER-THIEMER, E., and LEVINE, R., The solution of pharmacological problems with computers, Part 8: Non-first-order models of drug metabolism, *Arzneimittel-Forschung*, Vol. 18, pp. 1575–1579, 1968.
18. BUELL, J., KAGIWADA, H., and KALABA, R., Quasilinearization and inverse problems for Lanchester equations of conflict, *Operations Research*, Vol. 16, pp. 437–442, 1968.
19. BUELL, J., KAGIWADA, H., and KALABA, R., A proposed computational method for estimation of orbital elements, drag coefficients, and potential field parameters from satellite measurements, *Annales de Géophysique*, Vol. 23, pp. 35–39, 1967.
20. BELLMAN, R., JACQUEZ, J., KALABA, R., and SCHWIMMER, S., Quasilinearization and the estimation of chemical rate constants from raw kinetic data, *Mathematical Biosciences*, Vol. 1, pp. 71–76, 1967.
21. ROTHENBERGER, F., and LAPIDUS, L., The control of nonlinear systems: Part 4. Quasilinearization as a numerical method, *AICHE Journal*, Vol. 13, No. 5, pp. 973-981.
22. BELLMAN, R., and KALABA, R., *Dynamic Programming and Modern Control Theory*, Academic Press, Inc., New York, 1965.

Chapter 8

1. MEHRA, R. K., Optimal input signals for parameter estimation in dynamic systems—Survey and new results, *IEEE Transactions on Automatic Control*, Vol. 19, No. 6, pp. 753–768, 1974.
2. LEVIN, M. J., Optimal estimation of impulse response in the presence of noise, *IRE Transactions on Circuit Theory*, Vol. 7, pp. 50–56, March, 1960.
3. LEVADI, V. S., Design of input signals for parameter estimation, *IEEE Transactions on Automatic Control*, Vol. 11, No. 2, pp. 205–211, 1966.
4. NAHI, N. E., and WALLIS, D. E., Jr., Optimal inputs for parameter estimation in dynamic systems with white observation noise, Preprints, *Joint Automatic Control Conference*, Boulder, Colorado, pp. 506–513, August 1969.
5. AOKI, M., and STALEY, R. M., On input signal synthesis in parameter identification, *Automatica*, Vol. 6, pp. 431–440, 1970.
6. NAHI, N. E., and NAPJUS, G. A., Design of optimal probing signals for vector parameter estimation, *IEEE Decision and Control Conference*, Miami, Florida, pp. 162–168, 1971.
7. GOODWIN, G. C., Optimal input signals for nonlinear-system identification, *Proceedings of the IEE*, Vol. 118, No. 7, pp. 922–926, 1971.

8. MEHRA, R. K., *et al.*, *Dual Control and Identification Methods for Avionic Systems—Part II, Optimal Inputs for Linear System Identification*, Final Report, SCI Project 5971, AFOSR Contract F44620-71-C-0077, pp. 1–74, May, 1972.
9. MEHRA, R. K., Optimal inputs for linear system identification, *Joint Automatic Control Conference*, Stanford University, Stanford, California, pp. 811–820, August 1972; also in *IEEE Transactions on Automatic Control*, Vol. 19, No. 3, pp. 192–200, June, 1974.
10. MEHRA, R. K., and STEPNER, D. E., Optimal inputs for aircraft parameter identification, *International Conference on Systems and Control*, Coimbatore, India, August, 1973.
11. MEHRA, R. K., *Frequency-Domain Synthesis of Optimal Inputs for Linear System Parameter Estimation*, TR645 Division of Engineering and Applied Physics, Harvard University, July, 1973.
12. GUPTA, N. K., MEHRA, R. K., and HALL, W. E., Jr., Frequency domain input design for aircraft parameter identification, *ASME Winter Annual Meeting*, November, 1974.
13. MEHRA, R. K., *Synthesis of Optimal Inputs for Multiinput–Multioutput (MIMO) Systems with Process Noise; Part I: Frequency Domain Synthesis; Part II: Time Domain Synthesis*, TR649, Division of Engineering and Applied Physics, Harvard University, February, 1974; also in *System Identification, Advances and Case Studies*, edited by R. K. Mehra and D. G. Lainiotis, Academic Press, New York, pp. 211–249, 1976.
14. MEHRA, R. K., Time domain synthesis of optimal inputs for system identification, *IEEE Conference on Decision and Control*, Phoenix, Arizona, pp. 480–487, 1974.
15. MEHRA, R. K., Frequency domain synthesis of optimal inputs for multiinput–multioutput(MIMO) systems with process noise, *IEEE Conference on Decision and Control*, Phoenix, Arizona, pp. 410–418, 1974.
16. MEHRA, R. K., and GUPTA, N. K., Status of input design for aircraft parameter identification, *AGARD Specialists Meeting on Methods for Aircraft State and Parameter Identification*, NASA Langely Research Center, Hampton, Virginia, pp. 12-1 to 12-21, November, 1974.
17. KALABA, R. E., and SPINGARN, K., Optimal inputs and sensitivities for parameter estimation, *Journal of Optimization Theory and Applications*, Vol. 11, pp. 56–67, January, 1973.
18. KALABA, R. E., and SPINGARN, K., Optimal inputs for nonlinear process parameter estimation, *IEEE Transactions on Aerospace and Electronic Systems*, Vol. 10, pp. 339–345, May, 1974.
19. KALABA, R. E., and SPINGARN, K., Optimal input system identification for homogeneous and nonhomogeneous boundary conditions, *Journal of Optimization Theory and Applications*, Vol. 16, Nos. 5/6, pp. 487–496, 1975.
20. KALABA, R. E., and SPINGARN, K., Optimal input system identification for nonlinear dynamic systems, *Journal of Optimization Theory and Applications*, Vol. 21, No. 1, pp. 91–102, January, 1977.
21. KALABA, R. E., and SPINGARN, K., On the numerical determination of optimal inputs, *Journal of Optimization Theory and Applications*, Vol. 25, No. 2, pp. 219–227, 1978.
22. KALABA, R. E., and SPINGARN, K., Sensitivity of parameter estimates to observations, system identification, and optimal inputs, *Applied Mathematics and Computation*, Vol. 7, No. 3, pp. 225–235, 1980.

23. Georganas, N. D., Optimal inputs and sensitivities for nonlinear process parameter estimation using imbedding techniques, *IEEE Transactions on Automatic Control*, (Technical Note), Vol. 21, No. 3, pp. 415–417, 1976.
24. Sugisaka, M., and Sagara, S., Imbedding approach for optimal input design of linear system identification, *5th IFAC Symposium on Identification and System Parameter Estimation*, Darmstadt, Federal Republic of Germany, September, 1979.
25. Zarrop, M. B., A Chebyshev system approach to optimal input design, *IEEE Transactions on Automatic Control*, Vol. 24, No. 5, pp. 687–698, 1979.
26. Grove, T., Bekey, G., and Haywood, J., Analysis of errors in parameter estimation with applications to physiological systems, *American Journal of Physiology*, Vol. 239, pp. R390–R400, November, 1980.

Chapter 9

1. Kalaba, R. E., and Spingarn, K., On the numerical determination of optimal inputs, *Journal of Optimization Theory and Applications*, Vol. 25, No. 2, pp. 219–227, 1978.
2. Kalaba, R. E., and Spingarn, K., Optimization of functionals subject to integral constraints, *Journal of Optimization Theory and Applications*, Vol. 24, No. 2, pp. 325–335, 1978.
3. Mehra, R. K., Optimal inputs for linear system identification, *IEEE Transactions on Automatic Control*, Vol. 19, No. 3, pp. 192–200, 1974.
4. Mehra, R. K., Optimal input signals for parameter estimation in dynamic systems—Survey and new results, *IEEE Transactions on Automatic Control*, Vol. 19, No. 6, pp. 753–768, 1974.
5. Mehra, R. K., *et al.*, *Dual Control and Identification Methods for Avionic Systems—Part II, Optimal Inputs for Linear System Identification*, Final Report, SCI Project 5971, AFOSR Contract F44620-71-C-0077, May, 1972.
6. Zarrop, M. B., and Goodwin, G. C., Comments on optimal inputs for system identification, *IEEE Transactions on Automatic Control* (correspondence), Vol. 20, No. 2, pp. 299–300, 1975.
7. Zarrop, M. B., *Optimal Experiment Design for Dynamic System Identification*, Springer-Verlag, New York, p. 35, 1979.
8. Goodwin, G. C., and Payne, R. L., *Dynamic System Identification: Experiment Design and Data Analysis*, Academic Press, New York, p. 127, 1977.
9. Tse, E., and Anton, J. J., On the identifiability of parameters, *IEEE Transactions on Automatic Control*, Vol. 17, No. 5, pp. 637–646, 1972.
10. Kalman, R. E., On the general theory of control systems, *Proceedings of the 1st IFAC Congress*, Moscow, pp. 481–492, 1960.
11. Eykhoff, P., *System Identification: Parameter and State Estimation*, John Wiley & Sons, Inc., pp. 108–112, 1974.
12. Sage, A. P., *Optimum Systems Control*, Prentice-Hall, Inc., Englewood Cliffs, New Jersey, pp. 292–305, 1968.
13. Graupe, D., *Identification of Systems*, Van Nostrand Reinhold Company, New York, pp. 19–30, 1972.
14. Young, P. C., An instrumental variable method for real-time identification of a noisy process, *Automatica*, Vol. 6, pp. 271–287, 1970.

15. Denery, D. G., An identification algorithm that is insensitive to initial parameter estimates, AIAA Journal, Vol. 9, No. 3, pp. 371–377, 1971.
16. Staley, R. M., and Yue, P. C., On system parameter identifiability, *Information Sciences*, Vol. 2, pp. 127–138, 1970.
17. Tse, E., Information matrix and local identifiability of parameters, *Joint Automatic Control Conference*, Columbus, Ohio, pp. 611–619, 1973.
18. Budin, M., Minimal realization of discrete linear systems from input–output observations, *IEEE Transactions on Automatic Control*, Vol. 16, No. 5, pp. 395–401, 1971.
19. Aström, K. J., Numerical identification of linear dynamic systems from normal operating records, *Second IFAC Symposium on the Theory of Self-Adaptive Control Systems*, Teddington, England; in *Theory of Self-Adaptive Control Systems*, edited by P. H. Hammond, Plenum Press, New York, pp. 96–111, 1966.
20. Brown, R. F., and Godfrey, K. R., Practical difficulties in testing identifiability of linear structural models, *IEEE Transactions on Automatic Control*, Vol. 23, No. 6, pp. 1028–1030, 1978.
21. Beforte, G., and Milanese, M., Comments on practical difficulties in testing identifiability of linear structural models, *IEEE Transactions on Automatic Control* (correspondence), Vol. 25, No. 1, pp. 140–141, 1980.
22. Distefano, J. J., III, and Cobelli, C., On parameter and structural identifiability: Nonunique observability/reconstructibility for identifiable systems, other ambiguities, and new definitions, *IEEE Transactions on Automatic Control* (correspondence), Vol. 25, No. 4, pp. 830–833, 1980.
23. Baptistella, L. F. B., and Ollero, A., Fuzzy methodologies for interactive multicriteria optimization, *IEEE Transactions on Systems, Man, and Cybernetics*, Vol. 10, No. 7, pp. 355–365, 1980.
24. White, C. C., III, and Sage, A. P., A multiple objective optimization-based approach to choicemaking, *IEEE Transactions on Systems, Man, and Cybernetics*, Vol. 10, No. 6, pp. 315–326, 1980.
25. Keeney, R. L., and Raiffa, H., *Decisions with Multiple Objectives: Preferences and Value Trade-Offs*, John Wiley & Sons, Inc., New York, 1976.
26. Kalaba, R., and Spingarn, K., Numerical approaches to the eigenvalues of Saaty's matrices for fuzzy sets, *Computers and Mathematics with Applications*, Vol. 4, No. 4, pp. 369–375, 1978.
27. Saaty, T. L., A scaling method for priorities in hierarchical structures, *Journal of Mathematical Psychology*, Vol. 15, No. 3, pp. 234–281, 1977.
28. Saaty, T. L., Modeling unstructured decision problems, The theory of analytical hierarchies, *Proceedings of the First International Conference on Mathematical Modeling*, Vol. 1, University of Missouri, Rolla, Missouri, pp. 59–77, 1977.
29. Kalaba, R. E., and Spingarn, K., A criterion for the convergence of the Gauss–Seidel method, *Applied Mathematics and Computation*, Vol. 4, No. 4, pp. 359–367, 1978.
30. Goult, R. J., Hoskins, R. F., Milner, J. A., and Pratt, M. J., *Computational Methods in Linear Algebra*, John Wiley and Sons, Inc., New York, 1974.
31. Varga, R. S., *Matrix Iterative Analysis*, Prentice-Hall, Inc., Englewood Cliffs, New Jersey, 1962.
32. Pipes, L. A., and Hovanessian, S. A., *Matrix-Computer Methods in Engineering*, John Wiley and Sons, Inc., New York, 1969.

33. Steinberg, D. I., *Computational Matrix Algebra*, McGraw-Hill Book Company, New York, 1974.
34. Kalaba, R. E., Scott, M., and Zagustin, E., An imbedding method for nonlinear matrix eigenvalue problems of stability and vibration, *Proceedings of the 1974 Army Numerical Analysis Conference*, ARO Report No. 74-2, pp. 145–159, April, 1974.
35. Bellman, R., Kalaba, R., and Zadeh, L., Abstraction and pattern classification, *Journal of Mathematical Analysis and Applications*, Vol. 13, pp. 1–7, 1966.
36. Bellman, R., *Introduction to Matrix Analysis*, McGraw-Hill Book Company, New York, 1960.
37. Bellman, R., An iterative method for finding the Perron root of a matrix, *Proceedings of the American Mathematical Society*, Vol. 6, pp. 719–725, 1955.
38. Hovanessian, S. A., and Pipes, L. A., *Digital Computer Methods in Engineering*, McGraw-Hill Book Company, New York, 1969.
39. Chu, A. T. W., Kalaba, R. E., and Spingarn, K., A comparison of two methods for determining the weights of belonging to fuzzy sets, *Journal of Optimization Theory and Applications*, Vol. 27, No. 4, pp. 531–538, April 1979.
40. Chu, A. T. W., Fuzzy sets in economics, Masters Thesis, Department of Economics, University of Southern California, Los Angeles, California, 1978.
41. Kalaba, R., Spingarn, K., and Tesfatsion, L., Variational equations for the eigenvalues and eigenvectors of nonsymmetric matrices, *Journal of Optimization Theory and Applications*, Vol. 33, No. 1, pp. 1–8, 1981.
42. Gantmacher, F., *The Theory of Matrices*, Vol. I, Chelsea Publishing Company, New York, 1959.
43. Kato, T., *Perturbation Theory for Linear Operators*, Springer-Verlag, New York, 1976.
44. Kemble, E., *The Fundamental Principles of Quantum Mechanics*, Dover Publications, New York, 1958.
45. Courant, R., and Hilbert, D., *Methoden der Mathematischen Physik*, Springer-Verlag, Berlin, 1931.
46. Kalaba, R., Spingarn, K., and Tesfatsion, L., Individual tracking of an eigenvalue and eigenvector of a parametrized matrix, *Nonlinear Analysis*, Vol. 5, No. 4, pp. 337–340, 1981.
47. Kalaba, R., and Tesfatsion, L., *Complete Comparative Static Differential Equations for Economic Analysis*, Modelling Research Group Discussion Paper No. 7931, Department of Economics, University of Southern California, Los Angeles, California, October, 1979.
48. Kalaba, R., Spingarn, K., and Tesfatsion, L., A new differential equation method for finding the Perron root of a positive matrix, *Applied Mathematics and Computation*, Vol. 7, No. 3, pp. 187–193, 1980.
49. Bellman, R., Selective computation V: The largest characteristic root of a matrix, *Nonlinear Analysis*, Vol. 6, pp. 905–908, 1979.

Chapter 10

1. Bergman, R. N., Kalaba, R. E., and Spingarn, K., Optimal inputs for blood glucose regulation parameter estimation, *Eleventh Asilomar Conference*, Pacific Grove, California, pp. 405–409, 1977.

2. American Diabetes Association, *Diabetes Mellitus; Diagnosis and Treatment*, New York, p. 101, 1971.
3. Cerasi, E., and Luft, R., The pathogenesis of diabetes mellitus, *Diabetes*, Vol. 21, Supplement 2, pp. 685–694, 1972.
4. Bolie, V. W., Coefficients of normal blood glucose regulation, *Journal of Applied Physiology*, Vol. 16, pp. 783–788, 1961.
5. Bellman, R., and Kalaba, R., Reduction of dimensionality, dynamic programming, and control processes, *ASME Journal of Basic Engineering*, pp. 82–84, March, 1961.
6. Vajda, S., *Mathematical Programming*, Addison-Wesley Publishing Company, Reading, Massachusetts, 1961.
7. Bergman, R. N., Kalaba, R. E., and Spingarn, K., Optimal inputs for blood glucose regulation parameter estimation—II, *Summer Computer Simulation Conference*, Newport Beach, California, pp. 621–624, 1978.
8. Mehra, R. K., Optimal inputs for linear system identification, *IEEE Transactions on Automatic Control*, Vol. 19, No. 3, pp. 192–200, 1974.
9. Kalaba, R. E., and Spingarn, K., Optimal inputs for blood glucose regulation parameter estimation—III, *Journal of Optimization Theory and Applications*, Vol. 33, No. 2, pp. 267–285, 1981.
10. Mehra, R. K., and Gupta, N. K., Status of input design for aircraft parameter identification, *AGARD Specialists Meeting on Methods for Aircraft State and Parameter Identification*, NASA Langely Research Center, Hampton, Virginia, pp. 12-1 to 12-21, November, 1974.

Chapter 11

1. Digital Equipment Corporation, DEC System 10 basic, Maynard, Massachusetts, 1974.

Author Index

Subject Index